ENVIRONMENTAL DESIGN

ENVIRONMENTAL DESIGN

RICHARD P. DOBER, AIP

ROBERT E. KRIEGER PUBLISHING COMPANY

Huntington, New York/1975

Original edition 1969
Reprint 1975

Printed and Published by
ROBERT E. KRIEGER PUBLISHING COMPANY, INC.
645 New York Avenue
Huntington, New York 11743

© Copyright 1969 by
LITTON EDUCATIONAL PUBLISHING, INC.
Reprinted by arrangement with
VAN NOSTRAND REINHOLD CO.

Printed in the United States of America.

Library of Congress Cataloging in Publication Data

Dober, Richard P
    Environmental design.

    Reprint of the 1969 ed. published by Van Nostrand
Reinhold, New York.
    1. Urban beautification. 2. Cities and
towns—Planning. I. Title.
[NA9052.D6 1975]   711'.4   75-11961
ISBN 0-88275-331-2

TO VERONICA R. DOBER

PREFACE

"Change is the nursery of musicke, joy, life, and eternity," wrote the poet John Donne. True, but today change is also accompanied by obsolescence, anonymity, a general diminishment of social well-being as upper class, nether class, and neither class struggle for political power. A root cause of this general malaise, I believe, is a universal sense of dissatisfaction about an impersonal environment that seems to many to be beyond control and constructive guidance. The poisoned air, polluted water, unworkable and dangerous transport systems, erosion of public amenity and public services, and turbulence and uncertainty in the institutions which traditionally have provided relief and redress are signs of the times.

It is my view that there is nothing inevitable about these conditions. The environment can be designed and does not have to be left to adventitious circumstance. This raises the question of how this might be done, by whom, and whether such designs are inimical to the normal processes of a representative form of government. I see this as a consequential issue but one settled by transferring the rights and resources to plan, to design, and to construct the elements that comprise the environment to the smallest unit of government available to effect this one common objective: to make human habitation as varied, enjoyable, stimulating, healthy, and rewarding as possible. These last words, of course, have different meanings to different people. The definitions of those meanings, or general criteria if you will, would be central to any decisions made. To suggest a single standard applicable to all places and situations is to trap ourselves again in a cycle of inaction and generality in which a thousand minor improvements are postponed while waiting for a perfect plan.

Events have taken us past the point of intervention where a singular aesthetic act can have deep-felt impact. A consort of actions is required, and this demands group work as well as individual ingenuity and invention. But the conditions of life also demand some way to measure the consequences of any design action. Survival itself is at stake — real danger, not Mumfordian speculation. We pump wastes into underground fissures, and earthquakes follow. The incidence of respiratory fatalities parallels the curve of automobile density. We invent a new media, television, and it debases our lives and empties our communities. In this context the ideal contribution of environmental design is to help perfect a process, not a product. Then the identification of needs can rightfully evoke achievement and utility, which the arbitrariness of taste and fashion today so easily blurs.

I define the term "environmental design" as an art larger than architecture, more comprehensive than planning, more sensitive than engineering. An art pragmatic, one that preempts traditional concerns. The practice of this art is intimately connected with man's ability to function, to bring visual order to his surroundings, to enhance and embellish the territory he occupies. The titles of the subsections of this book — "Human Habitat," "Design Structure," "A Sense of Place" — parallel each of these three themes. They are frames of reference which indicate varying degrees of emphasis in environmental design; and as such are convenient, though arbitrary, categories for sorting out problems, solutions, and commentary.

In discussing principles and concepts in each section, the weight of examples and explication is frankly biased towards the urban situation. First, this is where the predominant number of people live now and will live in the future. Secondly, a successful resolution of the urban dilemma is likely to have constructive consequences for the non-urban environment. Thirdly, as various themes are developed, the word "urban" will refer not so much to a specialized geographic area but the territory occupied and used, fully or in part, because of urban activities.

The illustrative material selected to support my themes covers the physical elements that fill the visual world with artifacts, conditions, and forces which man senses and to which he is capable of responding. And since the designed environment is a composite of skyscraper and park, railroad and highway, house and shopping center, and other physical elements, each of these components is examined critically as to the role it has or could have in shaping and forming life and settlement.

There is a truism that authors write what they themselves wish to read. This book is aimed at a wider audience. I hope it can inform the general reader about how the environment in which he lives can be improved, fill the need for a general introductory text in environmental design, and serve as a casebook for encouraging an interprofessional view about environmental design — an art that crosses boundaries, spatial and professional, focusing on man and community.

Richard P. Dober
Cambridge, Massachusetts

CONTENTS

The Environment Within Reach Mixed housing, family life, privacy, communality, openspace. *Rendering: George F. Connolly.*

HUMAN HABITATION

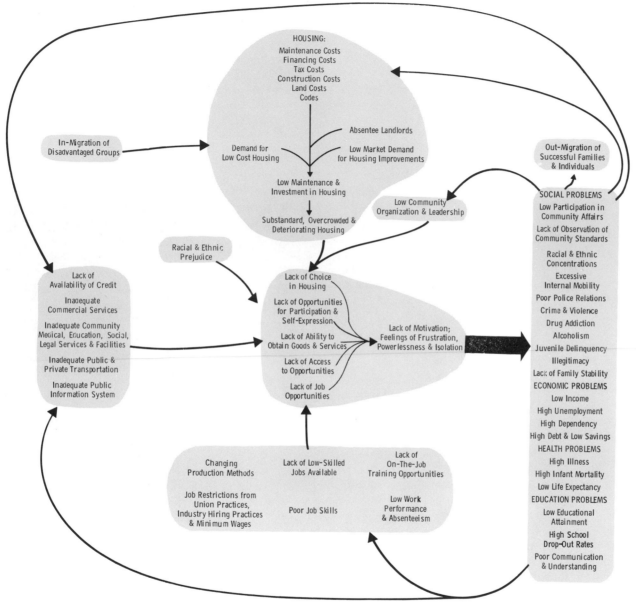

HOUSING:
Maintenance Costs
Financing Costs
Tax Costs
Construction Costs
Land Costs
Codes

Absentee Landlords

Low Market Demand
for Housing Improvements

Demand for
Low Cost Housing

In-Migration of
Disadvantaged Groups

Out-Migration of
Successful Families
& Individuals

Low Maintenance &
Investment in Housing

Low Community
Organization & Leadership

Substandard, Overcrowded &
Deteriorating Housing

SOCIAL PROBLEMS
Low Participation in
Community Affairs

Lack of Observation of
Community Standards

Racial & Ethnic
Prejudice

Lack of
Availability of Credit

Inadequate
Commercial Services

Inadequate Community
Medical, Education, Social,
Legal Services & Facilities

Inadequate Public &
Private Transportation

Inadequate Public
Information System

Lack of Choice
in Housing

Lack of Opportunities
for Participation &
Self-Expression

Lack of Ability to
Obtain Goods & Services

Lack of Access
to Opportunities

Lack of Job
Opportunities

Lack of Motivation;
Feelings of Frustration,
Powerlessness & Isolation

Racial & Ethnic
Concentrations

Excessive
Internal Mobility

Poor Police Relations

Crime & Violence

Drug Addiction

Alcoholism

Juvenile Delinquency

Illegitimacy

Lack of Family Stability

ECONOMIC PROBLEMS

Low Income

High Unemployment

High Dependency

High Debt & Low Savings

HEALTH PROBLEMS

High Illness

High Infant Mortality

Low Life Expectancy

EDUCATION PROBLEMS

Low Educational
Attainment

High School
Drop-Out Rates

Poor Communication
& Understanding

Changing
Production Methods

Lack of Low-Skilled
Jobs Available

Lack of
On-The-Job
Training Opportunities

Job Restrictions from
Union Practices,
Industry Hiring Practices
& Minimum Wages

Poor Job Skills

Low Work
Performance
& Absenteeism

THE CONTINUING PROCESS OF POVERTY AND NEIGHBORHOOD DEPRIVATION

The Poverty Cycle *Courtesy: Arthur D. Little, Inc.*

THE PROSPECT FOR HUMAN HABITATION

The prospect for human habitation is filled with contradictions—technical and human. No period in history has had greater reason to design the environment nor better means for doing so. Yet the world stands at the edge of so many crises that neither philosophical theory nor political tract adequately present the dimensions of decay and discontent. Both are exacerbated by an awareness that the distance between those that have and those that have not grows at a pace that threatens natural and social chaos. The resulting dilemma can only be explained as temporal conflicts, whose eventual resolution will be universally recognized as a necessary step for survival, growth, and a pervasively better environment. Compounding the crisis is the tyranny of technology, an outpouring of information and invention at a rate that seems faster than Man's ability to use it.

No society has spent more willingly for human welfare than the United States. Yet, while the causes of poverty and social disorder are known, the action to break the vicious cycle is left in the limbo of disjointed political debate. No other country has made greater contributions through medical research for universal physical well-being and at the same time produced a higher incidence of mental disorders. No other nation has had a better distribution of the by-products of technology, such as the individual house and automobile, nor suffered more for it. National policies and practices in home building have abused the land and wrecked central cities. The automobile is the instrument for a thousand deaths a week and incalculable environmental disorder. As the great blackout of 9 November 1965 dramatically demonstrated, the continuance of urban life is increasingly vulnerable to imponderable strains on technological networks of power, communication, and movement, where the severing of a single strand can bring widespread physical paralysis. But the idea of an optimum size for urban places and for environmental standards that eliminate or at least reduce the possibilities of natural and technical disaster is publically decried as impractical and idealistic. Yet private corporations envision the time when human habitation in all its dimensions will be marketed like cereals and toothpaste and accordingly are planning important corporate efforts. At least one hundred private new towns are in various stages of planning in the United States—each designed to provide an optimum environment and all planned to limit their size in population and economic activity. In this panorama one conclusion stands out. The largest contradiction, the biggest paradox, the base of indecision is Man himself. The environmental crises cannot be fully explained otherwise.

Few groups have been studied in as many different ways as the American Nation. Out of these studies anthropologists have been able to glean and identify the American value system—the attitudes underlying the gyroscope of individual and community action and reaction. This value system illuminates much of our present-day predicament, suggesting deep historic conflicts which have motivated when they have not hindered American achievement in the past and produced the climate of aspiration within which further progress may either take place or be aborted. Uniquely, the American Nation has risen from accommodation of disparate, individual values: liberty versus conformity, spending versus saving, love of bigness versus the beatification of the Little Man, simplicity versus extravagance and ostentatious display, success through dehumanizing competition versus brotherly love, novelty versus constancy, theory versus practice, concern for others versus individualism, playing safe versus taking risks, the stimulation of needs versus the possibility of satisfaction. To these values Clyde Kluckhohn would add the following more general characteristics: the ethic of responsibility (wherein consequences matter) and the ethic of intention (wherein motives matter), organizational skills, and hatred and distrust of bureaucracy.

If the contours of American society and the resulting environment can be explained by these pluralisms, divisions, and conflicts in values, what are the prospects for improvement? Predictions can be entertaining, edifying, and helpful in making a point. As we approach the Year 2000, prediction is a popular activity. In the spirit of the times but without suggesting prophecy or scientific pulse-taking, I suggest a significantly improved environment will be created without messianic intervention or doomsday fanaticism. Strung like beads on a necklace of hope, three words suggest a better future: commonality, altruism, survival. What does commonality mean here?

What passes for social disorder—dissention between the young and old, rich and poor, semiliterate and educated, scientist and artist—is constructive anarchy. Sides are being chosen and the battle joined not to subvert progress but to encourage it. The concept of "now" is not so much a quarrel with ends as it is with means, procedures, and techniques. A common purpose is taking visible characteristics. The fouled, polluted, crowded air, water, and land and the impoverishment of spirit, body, and opportunity are being rejected—sometimes with humor, other times with savage civil disorder. The mocking pen of a middle-class cartoonist in Marblehead and an enraged arsonist in a Detroit ghetto are similarly inspired. Both see their communities as victims of uncontrollable external forces. Each

Development Pressures *Reprinted from: Marblehead (Massachusetts) Reporter*

American in Australia *Photo: A. V. Jennings Industries, Ltd., Australia*

responds with the tools at hand. In Marblehead, where an unusually fine town design is threatened by intensive development, and in Detroit, which requires, as far as environment is concerned, resuscitation of large urban areas for any real improvement of human habitation, the basic issues are similar: priorities in environmental design and the organization and management of the factors of increase, change, and continuity.

Out of this commonality, altruism has to appear eventually. While acceptance of an improved life for everyone is not yet universal, the signs of altruism are increasing. Industrial combines are now recognizing that their own satisfaction depends on society's survival. Formation of the Urban Coalition, political action for the restructuring of educational and welfare institutions, the active participation of the poor in determining the purpose and method of changing their lives are additional signs of the times.

Survival is implicit in the ideal of advanced environment planning for increases in population and simultaneously the improvement of sizable amounts of existing substandard environmental conditions—conditions which affect the everyday life of individuals and communities. One hundred million additional Americans are expected to be born in the next two generations. To provide for their needs at our present level of environmental design, America will have to construct the equivalent of a physical place the size of Hartford, Connecticut, once a week, every year, to the end of the century. To the impetus of growth is added the demand for renewal and replacement. About nine million homes are considered substandard and obsolete in the United States today, as well as half our mass-transportation system, twenty-five percent of our schools, and almost all our urban recreation areas. Every city over a hundred thousand in population needs to have its urban space brought up to current ideas of environmental suitability. Levels of expectations rise each year. Vast acreage of natural resources require rehabilitation to serve an urban nation—four-fifths of the total population may live in three dozen metropolitan areas. Today, none of those metropolitan areas has sufficient clean air and water, nor adequate means for disposing of waste. It is said that we reach for the moon standing knee-deep in garbage.

Any degree of completion of this task of improvement and development could profitably engage all our national energy, our abilities, our vision, as well as all known forms of public and private finance. The design of this future environment, could unquestionably bind many interests (personal and communal) into a single national purpose and at the same time accommodate the conflicts in American values, which have motivated our progress to this point in time. If there is a semblance of hope in the most lugubrious of eras, it lies in this direction.

The efficacy of the effort could have import to other continents and nations. Only by demonstrating how well we can help ourselves can we expect others to use our techniques and standards. We cannot hope to impose our models of success, but we can display them. Perhaps the worst aspect of the American environment in the last ten years has been the exportation of the worst American environmental designs. Seen in Australia, shoddy subdivision and shopping-center designs are painful experiences—especially when praised Down Under as a successful American fashion in community design. The American press decries the depopulation of wild beasts in the Great Plains of Africa and the destruction of natural landscapes never again to be seen by Man, but our own recent past is filled with similar losses. Righteous indignation at home should be replaced with designs for emulation more worthy in spirit as well as physical form than those we now send abroad.

Herbert Read sees ahead: "There will be lights everywhere except in the minds of men, and the fall of the last civilization will not be heard above the incessant din." There is a moralizing tone to his remarks that is not unjustified. The events that shaped the modern world—in science, industry, agriculture, transportation, communications, art, and education—have not heralded a millennium of world peace or individual comfort. Other revolutions lie on the horizon, caused either by further advance and application of knowledge or the lack of it. Yet there is nothing inevitable about the coming state of affairs unless Man chooses to have it that way. However uncertain the national mood, the options and choices for changing and bettering the environment are as good now as they have ever been.

Two examples of design trends for everyday environmental elements should suffice as examples: the schoolhouse site and urban open space. Around each of these the prospect for human habitation can pose a challenge to environmental designers as they work with commonplace elements that fill the urban scene.

Early Education Park C. 1880 *Photo: Library of Congress*

New Values, New Designs Experimental elementary school, St. Louis, Missouri. New programs and new concepts of how to teach lead to new design forms. *Architect: Shaver & Company Photo: Banbury Studios*

LARGE SCHOOLS/SMALL SITES

While the United States could accommodate five times its present population—a billion people—at a density less than that of pastoral Switzerland, land for school construction is not plentiful, and the difficult task of accommodating a large school on a small site has become an everyday challenge to administrators and architects in suburban communities as well as central cities. The reasons behind the problem are worth brief comment, for they illuminate trends, hint at immediate solutions, and identify one of the impulses for new environmental-design standards and long-range planning techniques in community development.

Concepts of site standards are changing in many ways. In selecting a school site, these factors are typically considered as having an effect on the eventual size of the school property.

Land cost
Location
Accessibility
Topography
Soil Conditions
Shape of the land
Presence or absence of physical
  nuisances
Adjacent land use
Proximity to utilities
Microclimate

In the end, however well it meets the ideal criteria above, a site is large or small in accordance with the amount of land available to accommodate five elements:

The buildings
Outdoor activity areas
Parking
Site circulation
Amenity

The most universally accepted standards as to site size are those suggested by the National Council on Schoolhouse Construction. The Council was founded in 1921 "to promote the establishment of reasonable standards for school buildings and equipment with due regard for economy of expenditure, dignity of design, utility of space, healthful conditions, and safety of human life."

The Council's present guidelines shown in Table I have been incorporated in almost all the individual states' rules and regulations on school development and with real effect. Recent statistics on medium-size school sites show that the national averages are approaching, for the first time, these standards. (See Table 2.)

Some states, however, already consider the Council figures as only minimum requirements. North Carolina, for example, explicitly equates quality of site with size of site—larger equals better. Others are moving in this direction. In New York State, the accessibility of any schoolhouse site to the pupils it serves is being down-graded as a site-selection factor in favor of site size. Thus, both legislated standards and values are exerting pressure for increase in the size of the site.

Compounding the problems of the small site today is the strong evidence that the trend in the size of the five school elements cited above (building, outdoor activity areas, and so forth) is definitely headed upwards. It is also apparent that the reasons are largely qualitative, sometimes go back many years, and are thus perhaps irreversible, especially as regards the school building and functional outdoor areas. A review of the pressures, historical and current, for this increase in size of school elements will make clear the magnitude of the problem.

In a nation as diverse as the United States, one must be cautious in generalizing about educational matters. Regional differences are sharp, and even in singular metropolitan areas, individual communities are not mirror images of each other in aspiration or attainment. Nonetheless, one can say that while the daily operating expenditure per pupil for public, primary, and secondary education has not significantly risen since 1900, the annual per-capita capital outlay in physical plant in the same period has almost tripled. All this is measured in constant dollars.

Since the turn of the century, expansion in investment in physical plant has occurred in two stages. In the 'twenties, important advances were made in ventilation, lighting, heating, safety, and other utilitarian aspects of the schoolhouse.

The 'thirties were largely a stagnant period with low birth rates, a reduced school population, and a national economy unable to organize itself to materially improve the physical environment.

From the late 'forties to the early 'sixties, there were dramatic increases in population and subsequently in school enrollments. The rate of classroom construction doubled between 1948 and 1958—from thirty-five thousand to seventy thousand classrooms a year. This growth was accompanied by events that fundamentally influenced environmental design decisions, to wit: urbanization patterns which affected the location of schools, community interest in new methods of teaching, administrative reorganization of the educational enterprise, and search for new architectural concepts. During this period about a third of all existing new school buildings in the United States and additions to older structures were built.

Need, desire, and opportunity were nicely matched. For the most part, the new school plants reflected new educational values and markedly so in those communities which were most affected by post-war com-

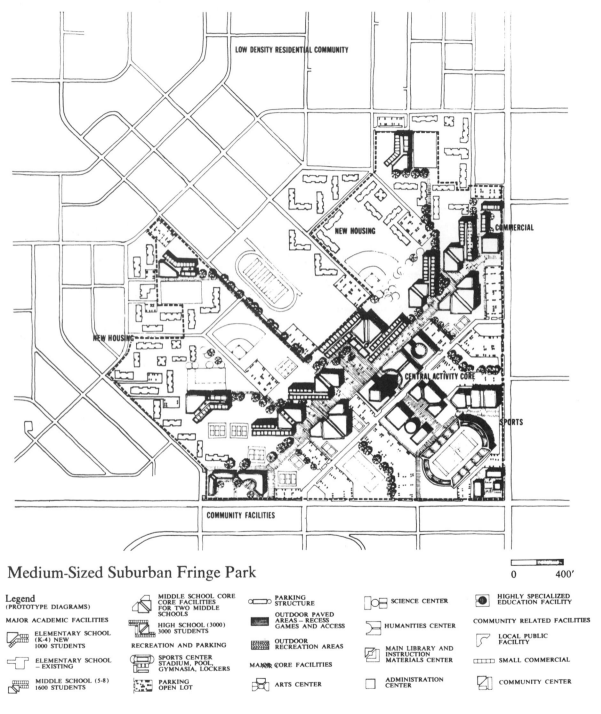

Medium-Sized Suburban Fringe Park

LOW DENSITY RESIDENTIAL COMMUNITY

NEW HOUSING

COMMERCIAL

NEW HOUSING

CENTRAL ACTIVITY CORE

SPORTS

COMMUNITY FACILITIES

0   400'

Legend
(PROTOTYPE DIAGRAMS)

MAJOR ACADEMIC FACILITIES

ELEMENTARY SCHOOL
(K-4) NEW
1000 STUDENTS

ELEMENTARY SCHOOL
– EXISTING

MIDDLE SCHOOL (5-8)
1600 STUDENTS

MIDDLE SCHOOL CORE
CORE FACILITIES
FOR TWO MIDDLE
SCHOOLS

HIGH SCHOOL (3000)
3000 STUDENTS

RECREATION AND PARKING

SPORTS CENTER
STADIUM, POOL,
GYMNASIA, LOCKERS

PARKING
OPEN LOT

PARKING
STRUCTURE

OUTDOOR PAVED
AREAS – RECESS
GAMES AND ACCESS

OUTDOOR
RECREATION AREAS

MAJOR CORE FACILITIES

ARTS CENTER

SCIENCE CENTER

HUMANITIES CENTER

MAIN LIBRARY AND
INSTRUCTION
MATERIALS CENTER

ADMINISTRATION
CENTER

HIGHLY SPECIALIZED
EDUCATION FACILITY

COMMUNITY RELATED FACILITIES

LOCAL PUBLIC
FACILITY

SMALL COMMERCIAL

COMMUNITY CENTER

Contemporary Education Park *Designs: David A. Crane and Peter Brown. Source: Corde Corporation*

munity development. Middle-class America demonstrated its aspirations in the architecture of the suburban school building. Contrastingly, in the central cities they abandoned schoolhouse design and fell further behind in innovation and change.

Current pressures for change underline the qualitative factors which impinge on and force upwards site standards, especially outside central city. Educational improvements, more than anything else, have tended to increase the size of recent school buildings. Team teaching, specialized materials and resources centers; educational television, ungraded learning spaces, improved laboratories and libraries, and improved accommodations for simple back-up services outside the classroom (such as mimeographing, storage, and lounges) have altered and enlarged the conventional floor plan. Though the amount of space per student has increased, the most important physical change is in the variety of specialized rooms and the general level of improved equipment and furnishings. The desk has been liberated from a fixed place on the classroom floor, and from this change alone, new spaces, new designs, new shapes, and new forms emerge.

The modern school building aims to be a community of spaces, not a series of boxes. Adaptability and flexibility are key words, for the definitive educational program and curriculum are not considered to be in sight. Many school administrators feel that the container should not inhibit a change in contents under the superintendent after next.

Community use of the building after hours is increasing and may continue to do so as the school becomes the center of community leisure-time activities, physical recreation, and social and health programs. These opportunities have not been fully grasped and are likely to further increase the size of building and site.

School consolidation is another cause of growth in the size of school buildings. At the elementary level, the United States still has over twenty thousand ineffective one-room schoolhouses. In the public-school system at the senior-high level, educators have pointed out that three out of four schools are not big enough to provide an adequate range of courses and facilities. As a result, there has been considerable effort to eliminate the small school, an effort that has been successful as the one-third decrease of the number of school districts since 1962 in the United States attests.

Also, community attitudes about school design have influenced site and building-design standards in the suburbs. The appearance of the schoolhouse, as well as the kind of education carried on therein, are among the few public actions directly subjected to voter approval. Public referenda determine many a capital appropriation and operational budget for schools in suburban areas. The vote is often as much affected by the architect's renderings and cost estimates as it is by curriculum and pedagogical objectives implicit in the design.

To some extent the voter's aesthetic preferences are arbitrary, though the issues presented at the polls may be phrased as rational choices. For example, two out of three recently approved school buildings have been one-story designs, though two-story schemes cost no more and use less land. For example, a one-story building for an eight-hundred-pupil middle school requires one acre of land more for the building than a two-story design. The one-story building however does allow internal change, and being more easily freed from a rectangular structural grid, induces different and often more ambitious designs than the tra-

ditional box-like schoolhouse usually required for economic reasons in multistory designs. Even if the latter designs are presented, for many communities the one-story school plant represents a clean and clear departure from tradition and is favored at the polls. Parenthetically, central-city voters rarely get a chance to voice their opinions directly on specific school designs, and this lack of opportunity is bitterly recognized by the disadvantaged and disenfranchised.

Expansion of the building and its site for educational and aesthetic reasons finds its parallel in the outdoor-activity areas. Co-education requires duplication of some field spaces, if not in kind, at least in acreage. Enlarged varsity sports and intramural sports programs affect the size of the site at the secondary school level. Provision for teaching after-school-years sports such as tennis and golf can be found on high-school grounds. In the grade schools new insights as to the kinds of outdoor environments that best serve younger children have in turn led to an increased variety of outdoor play spaces.

Some school sites now accommodate driver-education courses and camping facilities for overnight recreation. Schools are also establishing outdoor nature-study zones, either by preserving and enhancing the natural situation on the existing site or by designing and constructing new ecologies and habitats, especially for studying plant, animal, and insect life.

Car-ownership trends and "metromobility" also play their role in enlarging the school site. Per-capita automobile ownership in 1968 is almost double that of 1945. In California, Texas, and New York there is more than one car for every two people living outside central cities. Teachers do not live in the parson's attic and walk to work any more. They may drive considerable dis-

tances, and proper provision for parking is an absolute requirement for recruiting and holding staff. In some communities the students' parking lot is larger than the space occupied by the school building. Safe and proper access to and through the school grounds for automobile drivers—and school buses as the school districts are expanded— also means additional acreage.

Finally, many people now appreciate that the landscape teaches. The physical environment outside the classroom is getting deserved respect and attention. Social and activity spaces are landscaped and the setting for the buildings treated with community pride. The margins between structure and boundary of site are growing, rightly, as a demonstration of public aesthetics in a time when much of the visual world is a demeaning experience. Some educators believe that such amenity, though defensible on the grounds of appearance alone, also affects student behavior and that the softening qualities of a handsome and generously-proportioned site subdue the urge for senseless destruction and marring of the physical plant.

In many communities—for reasons educational, functional, and aesthetic—the desire for the large school and the large site seems here to stay. In those cases where school sites are below standard, many different kinds of measures, procedural and physical, can be used to reduce the size of the school building or the site requirements for that portion of the land not occupied by the building. Maximum utilization of space can help. School day and school week may be lengthened and classes staggered. Not all pupils have to arrive and depart at the same time. Computers can aid in the scheduling. The floating-period concept can be applied, that is, seventy-minute classes with the last half hour set aside for independent study.

This eliminates separate study halls and does away with special supervision. As a general rule, the larger the school in population, the better the opportunity for high utilization, so the application of these measures varies.

Supporting services can be centralized. These include storage spaces, maintenance shops, spectator athletic facilities, certain types of clerical and secretarial activities, and food preparation. Daily food delivery can include meals, dishes, and tableware; each school does not have to run its own warehouse. These ideas can result in a relatively small building.

Departure from the typical double-loaded corridor (classrooms on each side) can result in a more efficiently designed building. Open-plan schools allow teaching and service spaces to serve as circulation spaces; in this way as much as a ten-percent saving in gross square footage can be effected. Multipurpose rooms can save money as well as space. An average New York City elementary school costing $2.4 million includes $400,000 for the auditorium, $300,000 for the gymnasium, and $300,000 for the lunchroom. A combined auditorium and gymnasium can be built for $350,000. Some other alternatives: lunchrooms can be used for study halls, and auditoriums can be designed with movable walls for creating small lecture halls and classrooms.

Outdoor space requirements can be reduced by such measures as changing the physical education and recreation program from an outdoor to an indoor experience. The additional indoor space required may be accommodated on rooftops or in underground gymnasia.

System-wide sports facilities, practice areas, and nature-study areas can be designed for groups of schools. For some grade levels the school month can be divided by

Table 1 National Council on Schoolhouse Construction

STANDARDS FOR SITE SIZE (1964)

• Elementary School	10 Acres*
• Junior High School	20 Acres*
• Senior High School	30 Acres

PLUS ONE ACRE PER HUNDRED PUPILS FOR JUNIOR AND SENOR HIGH SCHOOL SITES

*100 percent increase since 1953.

Table 2 Number of Acres—Median-Size School Sites, Education Statistics, Department of Health, Education and Welfare (1965)

	ELEMENTARY	SECONDARY
Before 1920	1	3
1920-1929	3	6
1930-1939	3	8
1940-1949	4	10
1950-1959	8	20
After 1959	9	26
Under Construction Spring 1962	10	27

Rooftop Playground *Photo: Library of Congress*

Open-Court Schoolhouse *Source: Library of Congress*

attending a "tight-site" school for several weeks and then for several weeks a more generously landscaped environment. The study of nature can be scheduled.

Edu-vans—mobile, outsized trailers fitted out for unique purposes such as a music-listening room, art studio, science room, or health-examination station—can be brought to the site at periodic times in the school day. Students can be bussed to specialized facilities.

On some sites ceiling heights may be reduced and yet multistory buildings designed with more adequate light and ventilation than that provided by conventional floor-to-ceiling standards. Ziggurat and step-like building profiles can produce platforms for outdoor use or, in reverse, sheltered open space at the ground level.

The necessary amenities around the periphery of the building may be reduced to the extent that the environs themselves are attractive and comfortable. Constructing schools next to neighborhood parks will produce multiple-use opportunities. Parking lots may be designed and landscaped so their hard-surfaced areas are available for after-school play.

Most of these ideas are expensive, respectable, tried and true solutions; and some have a history longer than many will remember. They can be revived, revised, and made better by contemporary technology and design.

As described so far, the options for handling the large school/small site problem are these: modifications in school designs, changes in programs and procedures, and advanced planning to assure adequate sites in the future. Ironically, the most exciting environmental design concepts are not of these, but rather the fresh new look being taken at what constitutes an appropriate so-lution to the schoolhouse in the central city. These new designs not only respond to the question of large school/small site but also suggests ways and means to improve the general environment in which they may be situated. They offer alternatives to an increasingly large site. Their application in the future in suburban areas is as relevant as in any urban place. The new designs are especially effective where planning has been neglected and sites either have not been reserved for future schoolhouse construction or are poorly located.

School sites in central cities offer a distinctive design challenge, because the purposes of central city and suburban school sites differ. Essentially, this is because as a physical environment the central city went out of style after World War II, was socially rejected a census period later by the middle class, and now seems almost economically untenable and ungovernable. Central city's difficulties include housing conditions, medical and public health problems, lack of job opportunities as well as comparatively low-quality educational services. The inadequacy of the size of the urban school site—especially when compared with its suburban counterpart—is only one of many inequities, and perhaps not the most significant.

As noted elsewhere in this book, central city's restoration as a humane place will require large-scale reconstruction, and a physical solution without a social and cultural dimension will only postpone the necessary and eventual reconstitution of neighborhoods and districts. New techniques for public and private investment, perhaps new forms of government, and certainly elemental shifts in values and attitudes about community life are required. However, within this context the school and school site can fruitfully occupy central roles in city designs yet to come.

As natural centers for community life in at least thirty-seven cities which have densities of over eight thousand people to the square mile and millions of school children, school sites can accommodate the para-educational activities now in critical demand: social, health, and welfare programs. A school site can be a twenty-four hour environment rather than a formidable and singular building that stands unoccupied afternoons, evenings, and weekends. Sharing of school site may be extended to sharing of school facilities by placing educational activities in nonschool buildings. The new school forms may physically as well as symbolically weave their way through the urban fabric-like seams, recognizably urban—meeting no present-day acreage standards—but marked with those design qualities that our technology can produce and our economy command. Some of the possibilities are illustrated on adjacent pages and further described below. In important ways the resulting designs reflect searches for an urban equivalent to the quality produced in recent schoolhouse construction outside central cities. Many of the central-city designs stress the community context and the role of the site in both school and city development. Some of the solutions are singular themes; others represent ideas extended as required by program needs and land-use patterns. The range of solutions represents a variety of grade groupings and school types. Taken together, the following examples cover site and community conditions that might be found in a cross section of a typical central city and its immediate adjacent urban areas. They are: educational park, high-rise school, schools on shared sites, schools on rooftops or underground, schools on air rights, schools on waste land, minischools fitting into the dense urban environment, and block connectors which tie pieces of the urban fabric together.

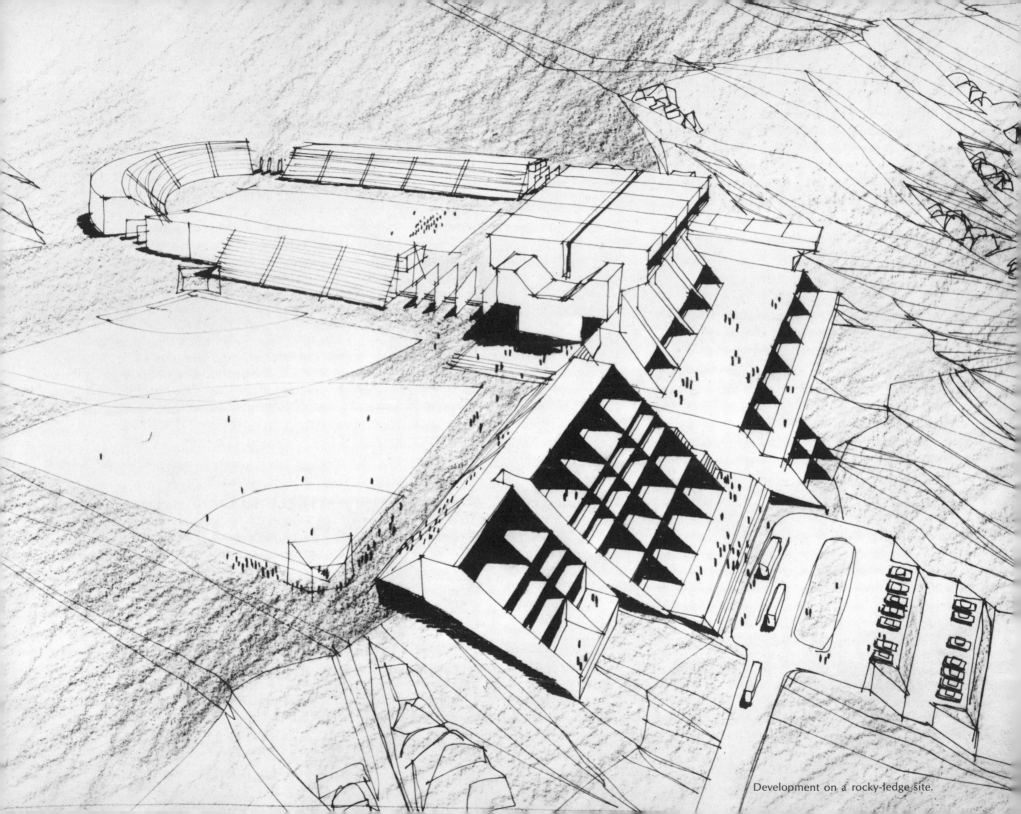

Development on a rocky-ledge site.

Air-rights development over major expressway in which the school-house connects two neighborhoods together.

High-rise development in a high-density area.

Threading the schoolhouse through a renewal site.

The mini-school.

Designs: George F. Connolly, Richard P. Dober, and Earl R. Flansburgh. Source: Educational Facilities Laboratories, Inc.

The education-park concept is gaining interest across the country both for educational reasons, social purposes, and to save land. Typical solutions bring together different kinds of schools on a single site. They may range from pre-kindergarten facilities for working mothers to community-college buildings. These clusters of schools have the advantage of sharing open space, site circulation, and play areas, as well as interior facilities such as auditoriums, gymnasiums, libraries, art and music rooms, remedial centers, language laboratories, and support services. In addition to the obvious economy of sharing, the size of the educational park tends to insure high-quality specialist help in each of the common facilities. Physically, the park may be organized as a series of high-rise classroom areas and low buildings for the common facilities; or these may be spread throughout the site, with each educational grouping having a distinctive location and setting.

There are instances when a location in a high-value land area deserves a school. Proximity to a downtown cultural center or business or trade district, for example, may have intrinsic benefits for the academic programs carried on by the school. A high-rise school is one solution. A functional zoning principle can be applied: laboratories, studios, and classrooms can be economically stacked together, facilities used by large numbers of students can be kept to the ground levels so as to reduce movement by stairs from floor to floor. As a variation on this theme, classrooms and laboratories may be designed as a tower with gymnasiums, auditoriums, dining facilities, and library remaining in a separate but connected structure at ground level. Parking can be handled in several decks. The remaining land on the site can be landscaped and used for playfields.

The schoolhouse can share a site with some other use, placed for example in an office building or at the base of a super-block of high-rise housing. In the latter instance, land may be gained by using community open space for school playfields. The school facilities, clustered around the base of the building, can form after-school and weekend pedestrian precincts free from intrusion by the automobile as well as encourage use by community members because of proximity to the residences. There can be parking for school staff in structures or on lots used by the residences, on the assumption that the schoolday and the workday do not overlap. Again, land requirements are reduced, and a smaller site is required.

Rooftop concepts fall into two categories. Most common are those designs in which the school roof itself is used for maximum utilization of space. Inasmuch as they are open to the sky, rooftops are typically used for recreation facilities. Though long accepted as a suitable solution to urban-land problems for public schools, the rooftop opportunity has been neglected in recent school construction. The provision of canti-levered space platforms extending beyond the building structure (multideck roofs stacked to face the best sun conditions and the least wind) and relatively inexpensive domes that give all-weather protection are technically feasible.

The second kind of rooftop use involves schools built on the roofs of other buildings. Large-scale construction of industrial parks, shopping centers, utility and power plants, transportation terminals, and office complexes affords the opportunity for coordinated design enabling roof space to be used as the platform for school buildings and related outdoor space. For example, a regional shopping-center site could be used for a junior high school. If the retail areas are one-story buildings framing a series of outdoor courts and malls, the educational facilities can be placed on the second level and sometimes extended upwards to a third and fourth floor. A pedestrian bridge can connect a section of a junior high school with an adjacent playfield across the road. Inasmuch as the peak shopping periods are outside the typical school hours, teachers, staff, and visitors can use shopping-center parking.

The idea of a continuous building with stacking of functions, one over the other, has intrigued architects from da Vinci to Le Corbusier. It is conceivable that in the future the implementation of automated production and distribution systems may result in large-scale, completely enclosed, machine-controlled environments. In turn, these structures may produce the bases upon which areas for human activities can be constructed. The words roof and land may then need new definition.

An underground school may be a practical solution to land scarcity, especially in areas where noise is a nuisance, such as near a jetport, or where extreme climatic conditions justify a subsurface solution in order to produce tolerable temperatures with relative economy and ease. In Scandinavia studies have shown that a well-designed artificial, underground environment need not be inimical to good health and good spirits. A totally subterranean school is an extreme solution of course, but it is feasible to place auditorium, gymnasium, library, shops, and dining areas underground. Their size, special purpose, and structural requirements lend themselves to such solutions. The activities inside can be interesting enough to overcome any monotony and boredom that a sealed space might induce. General teaching space can run the entire length of the site, under the buildings (which can be set on stilts) and over the subterranean rooms.

Air-rights development over expressways, railroad tracks, and canals can not only save land but also take advantage of optimum accessibility and help tie together parts of the city otherwise divided. Air-rights designs are not inexpensive, however. Land prices have to go above $200,000 an acre before decking for buildings, playfields and parking becomes a cheaper solution than purchase of land.

Ravines, steep hillsides, ledge areas, and wetlands have traditionally been considered waste areas and have been left undeveloped in many cities. Yet these pockets of land are often well located for community use. With new building and site technology now available, many of these problem areas can become community assets by serving as sites for new school construction. In a rocky-ledge area, precast concrete forms can be set up to serve as the base for a small elementary school, with a small playground on the rooftop and parking kept to the edge of the site. The natural landscape can surround the school and run under the building. A ravine can be bridged by a school building. Selected clearing can open the site at either end for playfields. A steep hillside can be terraced so that a school building can follow the new contours. The structure can be entered at several levels, parking can be placed at the base of the hill, and playfields can be wrapped around the upper levels. Where it is ecologically beneficial to do so, a wetland site can be engineered so as to allow a school building and playfield to be developed and at the same time preserve some of the natural conditions that sustain the natural landscape. In a variation on this concept, clusters of classrooms can be sited at some distance from one another yet connected by a path system that avoids the most difficult building areas.

Minischools are portable shelters erected for short-term school needs. For example, four to six classrooms may be created by erecting walls and ceilings in locations not typically considered school sites—a plaza in a shopping center, a utility easement, a street surface closed to through traffic. Compact, mobile, reusable, minischools give immediate relief to pressing needs, but to work well they have to draw on all the visual arts and transform a shelter problem into a stimulating architectonic landmark. They have to be more than outsized trailers brought to ground in an isolated part of the community.

In some instances, when permanent school solutions have been found, some minischool structures may remain in place to serve other community needs. These needs might include child-care clinics, after-school recreation programs, winter-storage areas for outdoor play equipment and park furniture, library rooms, post offices, telephone booths, and other enclosed public spaces.

A block connector is conceived as a continuous structure, perhaps sixty feet wide, that passes through several high-density urban residential blocks. It is an activity spine largely devoted to education, community health, indoor recreation, and leisure time programs. It might also contain the local post office and a few convenience shops. Parts of the structure might be designed for multideck garages. The roof might be used for outdoor athletics and appropriately planted with trees and flowering shrubs serving as a "pocket park."

The ground level would be punctured to allow local streets to pass through. The second level would be a pedestrian terrace along the total length of the building. At appropriate crossroad locations, stairs and towers would be provided down to the ground. The pedestrian terrace would serve as a protected pedestrian way through the

district as well as the point of entrance to the facilities inside. The structure would have "plug-in-as-needed" flexibility. Spaces could be reserved in the structure for accommodating school needs. As local conditions change, the spaces could revert to some other use, commercial or community service.

These many solutions may at first glance seem radical measures, but as such they are not inappropriate, for a long-range look shows that all our invention will be needed to cope with future conditions in education. With the rate of family formation likely to double by the late 1970's with the maturation of the post-World War II baby boom, sizable schoolhouse construction can be expected. In addition, numerous building replacements can be anticipated, for continuing qualitative advances in education are likely to make obsolete any building constructed prior to 1950. While advanced planning may have reserved large sites in certain suburban communities, other communities will have to fit their new school requirements onto limited sites. Finally, as illustrated earlier, the reconstruction of central cities will continue to offer numerous reasons for new school designs.

As an everyday activity, education occupies one out of four Americans full-time. The environmental design that accompanies education is as important as it is pervasive. We can see that new environmental designs need not spring forth haphazardly but can rise from common necessity, common sense, and concern for joining social purpose with public aesthetics, yielding reasons for wide environmental improvement. In the provision of urban open space we will again see how an everyday requirement for urban life can produce diverse, delightful, and functionally appropriate designs.

The Garden Of Pleasure And Play At the end of the Middle Ages, nature enters the city to serve as the site for games, recreation, and urban pleasures. *Source: Fogg Museum Library*

LANDSCAPE AND OPEN SPACE

Man's attitudes towards the land are expressed in art as well as usage, and nowhere can this be seen more strongly than in his concepts of open space—that lovely greenery of poets and painters. Historic perspective discloses an aesthetic impulse to create gardens and other landscaped open spaces, visionary towns and garden cities, greenbelts and green environments. The prodigious efforts to use nature as an art form differ from place to place in scale and enduring qualities according to the cultural temperament.

Greenery as a civic art seems to rise and fall with the intensity of civilization, gaining a central position in man's design for human habitation seemingly at those points in urbanization when architecture and technology begin to fail to satisfy communal requirements for a pleasant and ordered environment. This pattern seems as true today as it was centuries ago, as a brief glance at past attitudes and events will reveal.

Public gardens as such were not known in antiquity, the Garden of Eden excepted. The sacred groves of Babylon, Egypt, and Rome were private precincts and came into being when there was sufficient wealth to create them and leisure time for their enjoyment and appreciation. Horticultural design began as the province of the rich, evolving from accidental embellishment, as extensively seen in the suburban villas of emperor and merchant and intensively implanted in city courtyards that filled the atrium with verdure and fragrance. Where the space was confined, the Romans used mural paintings of flowers and pastoral scenes, duplicating in paint and plaster what the ground itself could not sustain, as in the row houses of Pompeii. In a similar fashion the Islamic gardens grew from by-products of conquest and leisure into a major art form, as the splendid remains of Moorish Spain beautifully testify. Buildings, plant material, and sculptural effects were joined together. But as a meeting place for all the arts, it is the Renaissance garden—an enclosed space—which reached the zenith of formal design in sixteenth-century Italy, seventeenth-century France, and eighteenth-century England. Writes Derek Clifford:

Painters, architects, sculptors, poets, and philosophers gave their minds to the comprehension of its nature and the perfecting of its practice. Men who excelled at it became the confidants of statesmen and the friends of kings.

The Italian Renaissance gardens and parks were not, however, central features in community design. The magnificent public spaces in Florence, Sienna, and Rome admitted little plant material. Architectural historians have suggested that any attempt to harmonize plant life with architecture must have invariably resulted in a disagreeable conflict in style. The rigid rules of proportion, perspective, modeling, and relief were appropriate only for the "hard" landscape; thus statues, railings, stairs, and fountains were used to embellish urban spaces. The interconnected buildings and the contents of the space between buildings were designed to read as symbolic language; nature was not part of the story. Cities were small, the countryside readily at hand. The city was intended to be all that the surrounding woods, forests, and fields were not. With the exception of house gardens, plant life was largely walled out. Political and social forces brought about changes in these attitudes.

As the classical codification of design principles and design objectives was extended into the Baroque forms, the security that comes through power and accompanying wealth lead to greater public display than previously of formal architecture among the ruling classes. In France, particularly, the art of garden design was melded into classical architecture, at first for palaces and châteaux. The pleasant amenities were later extended to urban places. By the late seventeenth century, much of Paris looked like a garden. For the first time nature was intentionally established in generous areas inside a city.

Christopher Tunnard theorizes that the new fashions in urban greenery in France came about with changes in social behavior, especially recreation. The descriptive vocabulary of civic design mirrors his comments. The mixing of form and function produced tree-lined avenues (allées and malls) originally designed for popular forms of cricket and promenade, in which bourgeoisie and royalty all participated. An early land lease in Paris shows the impact of play on open-space design. The lease commanded the owners of the ramparts between the St. Honoré and St. Denis gates to plant the land with elms, with branches no less than ten feet above the ground, and to roll the surface beneath them to ensure proper playing. Public appreciation of public open space for public use spawned a number of similar landscaped civic improvements in Paris, such as the Place Royale and Place Dauphine, which remain today.

Due to land-tenure systems and communal prerogatives, England's development of open spaces differed from the Italian and French experience, although the design fashions of both latter countries in time found ready acceptance in British cities and towns.

The recreational use of open space was a right long recognized in England. The celebrated judgment in the case of Abbot vs. Weekly (1665) stated

that all the inhabitants of a vill (sic), time out of memory, had used to dance on a certain close at all times of the year at their free will for recreation . . . [it] was a good custom . . . as it is necessary for inhabitants to have their recreations.

Despite popular belief, English greens and commons were never "no man's land," belonging to the public and free for any use. Rather, every piece had an owner, whether individual, government, or a group of trustees holding the land for public benefit. The common aspects were the *rights* to use the area for specific purposes and sometimes for specific times. These rights included: pasture (animal grazing), turbary (digging for fuel), estover (taking wood), piscary (fishing), and soil (using minerals excavated). Common rights were attached to both person and property and as such could be exchanged and rented, but the exercise of that right could only be done for personal use. The wood and fish gathered, for example, could not be sold.

By the fourteenth century, Lords of the Manor realized the potential agricultural value of the commons and systematically began to enclose the land for their exclusive use. Enclosure was accomplished in several ways: by fencing in the common land, by abandoning Medieval strip-cropping for compact farm plots, and by acquiring land for residential estates and parks. The latter provided the opportunity for the magnificent three—the English landscape architects William Kent, Humphrey Repton, and Lancelot Brown—to take denuded land and impoverished agricultural holdings and create memorable, pastoral landscape designs.

Brown especially was the genius of his age, never fearing, in Repton's estimate, "to plan largely, to execute extravagantly, and charge extortionately." Molding green hills, placing still water in juxtaposition with an occasional single tree specimen, clumping trees in a natural ecological zone, Brown's designs were as simple as they were effective. The typical English landscape scene in calendar art and picture postcard is sheer invention. Hugh Walpole rhapsodized: "We have discovered the point of perfection. We have given the true model of gardening to the world."

Through enclosure, agricultural productivity rose as savings in labor and transport were effected, the wasteful features of open-strip cultivation were removed, selective animal breeding was rendered possible, and the sole possession of the land encouraged owners to invest additional money for even greater yields. As a result, capital accrued, relative leisure was made possible for a growing upper class, and styles of life and architecture were engendered, satisfied at first in manor house and country seat but gradually centered in London, where political and social power accumulated. The impulses of modern life took early physical form in London through the work of Inigo Jones and Christopher Wren, in which a sense of geometric regularity—Vitruvian and Palladian—was applied to organize urban spaces and urban scenes.

The Great Fire of 1666 in London wiped out four hundred streets and thirteen thousand houses and left seventy thousand shelterless. The first contemporary opportunity for a planned new town was lost when Wren's scheme for rebuilding London was not accepted by its citizens, who were too impatient for slow and orderly redevelopment. In its place, speculative development began to fill the old city and its environs. This did not mean, however, an end to hopes for a pleasing environment with an appropriate amount of green space.

Some of the London precincts developed by speculators, aided by their inhabitants' wealth from agricultural surplus, set fashions in community design that would be copied for the next three hundred years. In Grosvenor Square, Belgravia, and Bloomsbury, urban life advanced through architectural order and hygienic improvements in housing and community design. The Georgian style of brick architecture in both single and connected houses surrounding an open square with tree-lined streets, suited both merchant and clerk, royal consort and squire. Carried to America, this style served as model for early Philadelphia and Salem, as well as for innumerable college campuses from Harvard onward. Even as more ornate façades were imposed in the late Georgian expressions such as Cumberland Terrace, Carlton Gardens, and the Circuses of Bath and Paddington, urban greenery continued to hold its place and that of nature in the city.

About the same time in France, the visionary architect Ledoux designed perhaps the first Greenbelt town. Inspired in part by the Republicanism of the Revolution, his plan for Chaux called for an eliptical enclosure of trees in place of the traditional city walls. The landscape around and in the city defined the urban form. Never built, Chaux was a progenitor of later English garden cities.

The environmental pleasantries and art that stemmed from agricultural and commercial prosperity in England were not universally shared, however. While the juncture of the eighteenth and nineteenth centuries saw generally stable times and increasing prosperity, the conditions of the poor and disenfranchised were abominable, and a notably callous state of indifference to prison conditions, child labor, and adult indenture prevailed. The continuance of enclosure decreased the demand for agricultural labor. The consequent unemployment drove large numbers of the unskilled to city life. As Charles Dickens captured the turning years: "It was the best of times, it was the worst of times . . . it was the spring of hope, it was the winter of despair."

Surplus capital, surplus labor, invention and innovation, exploration and exploitation, and industrial revolution all led to large

The Picture-Card Landscape Moving earth and plant materials with an intuitive sense of ecological fitness and design effect, Lancelot (Capability) Brown created memorable landscape designs, whose picturesqueness is commonly accepted as a natural as opposed to a man-made creation. *Source: Library of Congress*

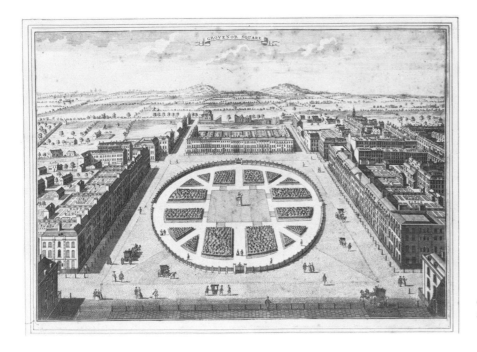

Letchworth Town Center One of the earliest garden cities constructed along the principles laid down by Ebenezer Howard. However laudable the design intentions may have been, the town center is memorable as a historic event but not as a cohesive design. *Photo: Aerofilms Limited (Copyright reserved)*

Grosvenor Square Urbanity disclosed: New values and new designs were imposed upon agricultural land as London grew in size and style. *Drawing: The British Museum*

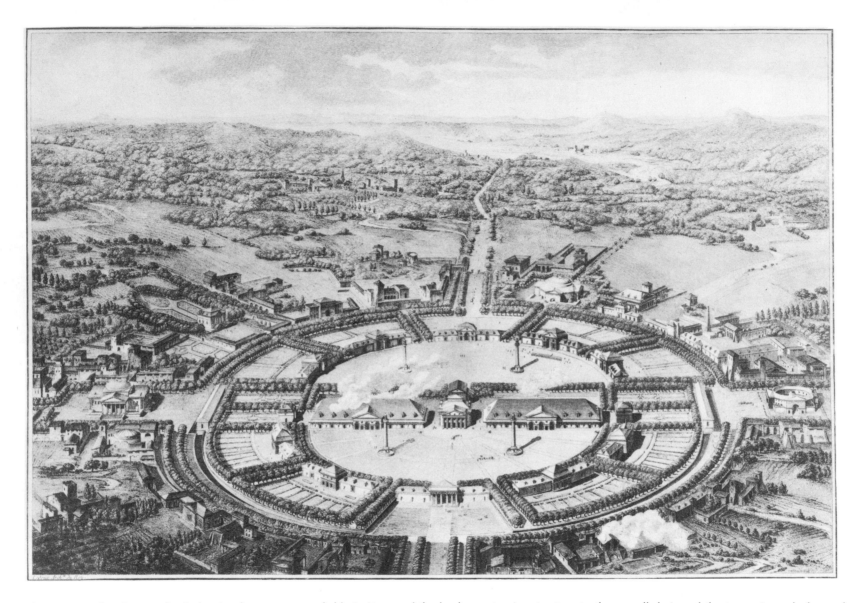

Chaux: A Utopian Community Ledoux's scheme was remarkable in its use of the landscape to give structure to the overall design of the town. An early form of green belt is used to establish the limits of the community. *Source: Library of Congress*

cities, and then the metropolis—unplanned, undesirable, unloved. As rapidly as people moved in to fill the jobs of town and mill, others moved out to the periphery to avoid the aggravations of industrial development, the overloading of old streets and districts, the nascent slums. The amenities of older communities were swamped, and new residential districts in the nineteenth century failed to capture the urban spirit the environmental arts had heralded a century earlier. Those sensitive to environmental decline reacted strongly with pleas for social revolution and new types of community design. The pattern was to repeat itself in countries other than England, as they, too, were subject to industrialization and city growth. The chronicle of need for, aspiration toward, and execution of new environmental designs became worldwide and can be fully sensed in two English experiences, green towns and greenbelts, as well as in the urban park movement in the United States, which will be described subsequently.

In England these visionary schemes for green towns and garden cities were not so much products of aesthetics as they were utilitarian devices to improve individual and public health through improved housing, working conditions, and social order. Where the Renaissance city was considered the sanctuary for well-being, now the countryside was looked upon as the refuge from urban distress. Pamphleteers, philosophers, radicals, and eventually industrialists joined in the crusade, not only to avert the invasion of country by town, but to promote also the return of country to town. For necessity, convenience, and pleasure "the wretched slums will be pulled down, and their sites occupied by parks, recreation grounds, and allotment gardens."

Contempt for the industrial city rose as fast as the distress it induced. Social philoso-

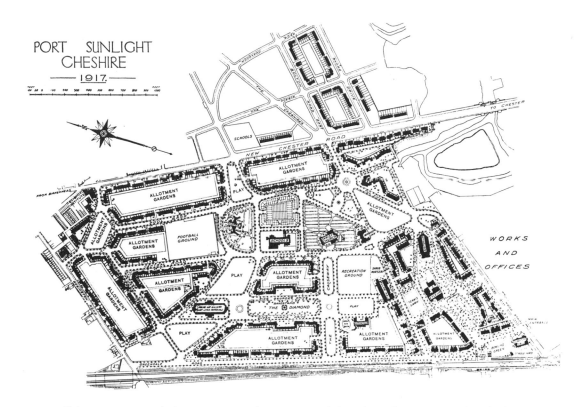

Port Sunlight 1917 Port Sunlight as it appeared in 1917. English industrialists concerned with the impoverished conditions of the working class constructed several utopia-like communities in which generous amounts of open space and recreation strongly contrasted with typical environments in major cities. Founded in 1888 by the Lever Brothers, Port Sunlight, gained world reknown as an example of garden villages and nineteenth-century paternalism. *Source: Harvard Graduate School of Design Library*

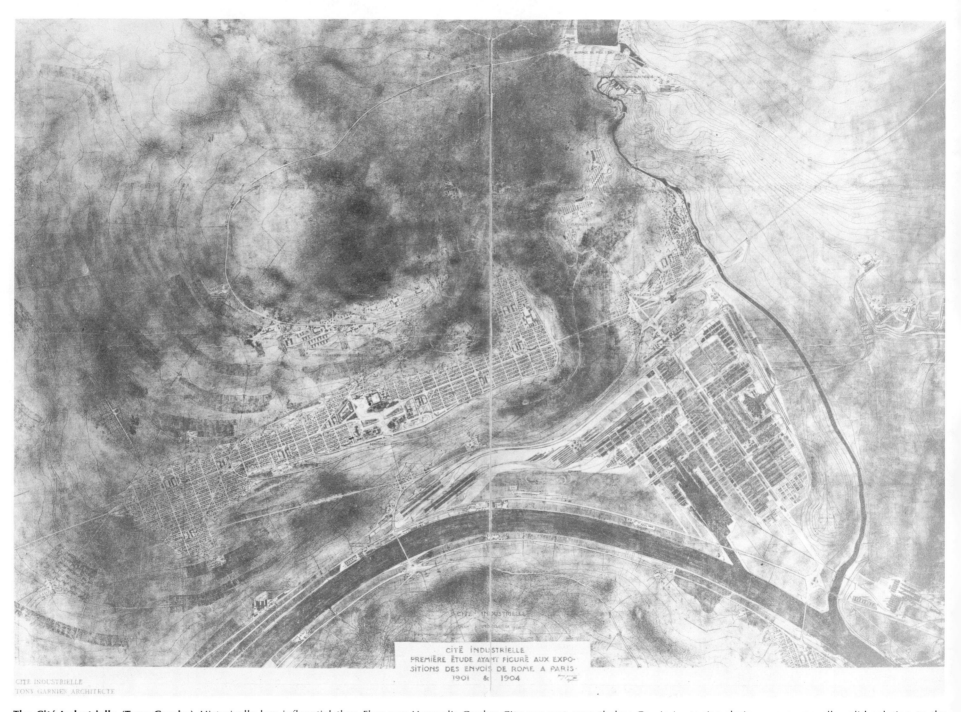

CITÉ INDUSTRIELLE
TONY GARNIER ARCHITECTE

CITÉ INDUSTRIELLE
PREMIÈRE ÉTUDE AYANT FIGURÉ AUX EXPO-
SITIONS DES ENVOIS DE ROME A PARIS
1901 & 1904

The Cité Industrielle (Tony Garnier) Historically less influential than Ebenezer Howard's Garden City concept, nonetheless Garnier's utopian design was an equally valid solution to the environmental deficiencies of the nineteenth-century city. *Source: Harvard Graduate School of Design Library*

phers Spencer, Locke, and Engels cited examples of overpopulation, overcrowding, and economic exploitation to propose both gentle and drastic political and economic reforms. The fears aroused speeded Parliamentary action in England, though the Victorians seemed to progress by inches in tasks that covered square miles. Paternalistic entrepreneurs such as Titus Salt, George Cadbury, and William Lever raised model towns like Saltaire, Bourneville, and Port Sunlight, workable translations of utopian schemes put forward earlier by the French architect LeDoux, and by Robert Owens, and James Silk Buckingham but in the main limited to a chosen few.

The exaggerated qualities of both Owens and Buckingham's plans hold a strong place in planning literature as examples of comprehensiveness and naiveté. Buckingham's prospectus for the town of Victoria was filled with seductive descriptions of what "the latest discoveries in architecture and science can confer" as well as visions of the "highest degree of health, contentment, morality and enjoyment." Buckingham promised hygienic conditions, a stable community life, and a green setting. More than anything Le Corbusier could ever envision a half century later, Victoria was a machine for living, allowing no freedom of movement, nor choice—a cleaned-up version of the ills the new town would replace, different only in form from the demeaning districts which Charles Dickens, again beautifully chronicalling his age, captures in *Hard Times*:

. . . large streets all like one another, and many small streets still more like one another, inhabited by people equally like one another, who all went out and in at the same time, with the same sound on the same pavements, to do the same work, and to whom every day was the same as yesterday and tomorrow, and every year the counterpart of the last and next.

In contrast to Victoria, the garden cities of Ebenezer Howard four decades later, though struck from a similar metal gave off a different tune. London-born, forever frightened of the "Great Wen," evangelist, moralizer, "[neither] masterful type, brilliant, learned, nor obviously clever," Howard managed to construct two cities which demonstrated in enduring form an epitome of early twentieth-century environmental design.

Influenced by Edward Bellamy's utopian novel, *Looking Backwards*, Howard conceived a scheme for community development, which he expressed in *Tomorrow: A Peaceful Path to Real Reform* (1898), reissued as *Garden Cities of Tomorrow* (1902). Favorably received, attracting public attention, and acquiring influence, Howard's book gave him a platform from which to expound his view on urban ills and their solution. His idealism was tempered with a semblance of the economic certitude beloved by the Victorian mind, "If you do well, you'll do good." *The Edinburgh Scotsman* (1906) editorialized:

The Garden City is a fascinating experiment, and it cannot certainly at the present stage be treated as fantastical. It is being conducted to all appearances on business principles, under the guidance of engineers, builders, sanitary authorities, and capitalists who know the value of money, as well as philanthropists and idealists. It can hardly fail to give valuable help towards a solution of some of the most pressing social problems of the twentieth century.

Howard noted the defects and virtues of both town and country environment. The closing out of nature in city life is thus balanced by social opportunity, murky sky by well-lit streets, high wages by high rents and prices. But the country also had its disadvantages. Bright sunlight was not sufficient society, and the beauty of nature did not make up for the lack of amusement. Low rents

were paid for by long hours of work at small wages. Proposing to combine the "advantages of both [city and country] with the disadvantages of neither," Howard called for garden cities of about thirty thousand people, in his words, "securing healthy and beautiful houses and conditions of life and work for all classes of people." (Howard's estimate of self-sufficiency—thirty thousand people—is no longer an economically or socially viable number, and the advance in scale from his time to our own may be taken as one measure of environmental change. Planners today believe a community of 250,000 people is needed to provide a range of services, goods, and amusements.)

In sight of the ideal but in touch with the practical, Howard and his followers incorporated First Garden City Limited (1903) to raise money to build the first garden city, hoping that accomplishment would inspire others to do the same. Capital of £300,000 was authorized but only half the amount raised—enough, however, to begin the construction on 3,818 acres at Letchworth, a rural hamlet on a small branch line from London to Cambridge. In 1920, a second scheme was started in Welwyn, twenty miles from London along the same line.

Both Letchworth and Welwyn met many of its founders' objectives and served as a testing ground for ideas and insights that would allow larger and more impressive undertakings in both England and elsewhere later in the century. Design innovations in Letchworth and Welwyn included the use of small neighborhoods to break down the total town into smaller districts, and zoning and density limits to keep manufacturing and heavy transport from the residential areas. The towns were surrounded and contained by agricultural lands. The corporation was set up so that the profits that accrued from the increase in land values (as farmland became town land) were used—in reflection of Henry George's philosophy—for community benefits.

As the twentieth century came into view, increased respect for nature, distrust of the existing industrial city, and utopian visions of a better environment excited designers, dreamers, and artists worldwide, evoking, with varying degrees of precision and prescience, visions of a better life. In words and pictures it was the age of the noble diagram—among others Howard using his famed concentric circles to represent the future "since the plan cannot be drawn until site [is] selected."

About the same time as Howard's garden cities were taking physical form, Prix de Rome winner Tony Garnier (1901) created the broad outlines of his *Cité Industrielle*. It would occupy him for sixteen years as he filled in the details. Garnier was utopian in his inspiration, especially in his desire for reforms in government and public health and the distribution of the benefits and profits that technology offered. Dora Wiebenson finds threads of socialist thought throughout his work. His scheme made no provision for law courts, jails or churches, for Garnier believed the moral force of a new society would eliminate these needs. Howard had no such illusions.

Yet the *Cité* also looked to the future in a practical way. It anticipated regional development, being designed for a decentralized government and self-sufficiency in all ways. Syndicates would rule and administer food production and distribution, transportation industry, and public services. Garnier emphasized creating a distinctive place through encouragement of local arts and crafts, universities and museums; he gave each of these facets of life a prime location and a special building in his plan.

Garnier envisioned his *Cité* constructed on a plateau, with highland to the north, valley and river to the south. The terrain near the city was ideal for a hydroelectric plant which could use the dammed-up waters and furnish power for factories and mines, all below the city. Noxious fumes would be avoided and the dirt and noise of the nineteenth-century factory town overcome by careful zoning and the use of electric energy. Where Howard left nature largely in the care of household gardeners and greenbelt farmers, Garnier infused his design with parks, playgrounds, and public tree plantings. The shape of the *Cité* was linear, emphasized by rectangular arrangement of residential blocks; and the open spaces which separated them from the nonresidential quarters were generous in size and vegetation. Highly organized, almost inflexible, Garnier's plan nonetheless had a semblance of reality that neither Buckingham's Victoria nor Howard's diagrams could convey. Garnier's plan leaves the viewer with the feeling that it could easily have been drawn from life rather than an architectural fantasy. This is especially true in the buildings themselves. Utilizing concrete as a common material, deservedly praised as harbingers of coming design form, stripped of decoration, these buildings are nonetheless as familiar as mid-twentieth-century fashion. Only the dates of the sketches reveal that the Bauhaus was still ahead.

Garnier's work marks the beginning of a twentieth-century compulsion among architects to design cities, the draftsman's utopia, in which community life can be captured, entire and complete, in a single bold stroke. It is a tradition that has merits for other reasons. Cities cannot be designed that way, but the technique is useful for arousing public interest and stimulating political action—as we will discuss later in this book, in a review of current planning in Los Angeles.

Garnier's designs were not universally acclaimed. Intellectuals and artists were divided about the advantages of twentieth-century technology and the place of nature in the city. In his novel *When the Sleeper Awakes* (1899), H. G. Wells describes a fearful world, a future city enclosed by walls, domed, with subterranean factories and high bridges connecting tall buildings. But the city which Wells lugubriously defined in words was seen in a different light by Sant' Elia. Calling themselves the Futurists, he and his cohorts celebrated the arrival of the machine age with exuberant manifestos and an aesthetic philosophy that called for a new art and architecture.

The Futurists pleaded for a "city like an immense assembly yard, dynamic in every part." Partially a counter-movement against art nouveau and historicism, anti-nature in concept, in some respects as much a political credo as an affirmation of modern technology, Sant' Elia's designs now seem closer to reality than one would have suspected two decades ago. Where Le Corbusier grasped the significance of the million-people city and the scale of the automobile age, Sant' Elia envisioned the impact of the airplane and mechanical movement systems, some robot-like and untended by man. In his architectural forms (circa 1910)—which Helen Rosenau aptly labels "secularized Gothic"—there are images and themes so current that perhaps no other designer of his time can be looked upon as the progenitor of the architecture of motion.

Dead at age twenty-eight (1916), Sant' Elia paradoxically now sits in the shadows of architectural history; because what he drew could not be divorced from what he said. His mechanistic attitudes—as rigid as Buckingham's and as insensitive—command no audience today, and his contributions and

City Without Trees In prescient work of Sant' Elia, the influence of movement systems and building technology are recognized in his designs for the modern city. To the horror of many, the importance of nature in the city was ignored. *Reprinted from: Lacerba, Vol. II, No. 15. Source: Houghton Library*

Frank Lloyd Wright's Broadacre City Several of Wright's drawings for Broadacre City — often criticized as a rural solution — show imaginative anticipation of current city development trends: Broad highways, individual air transport, high-rise buildings, and the use of landscape to set the pattern and context for the location of housing and industry. *Source: Frank Lloyd Wright Foundation*

The City Is a Garden Air view of Tapiola Town Center. *Photo: Suomen Matkailulutto Finish Tourist Bureau*

insights are thus ignored.

When, in a life such as Sant' Elia's, philosophy overcomes technique, the process for the critic of selecting what is useful for environmental improvement is complicated by the age-old problem: What an artist does and what it means to a later generation may be two different things. Thus, Frank Lloyd Wright's Broadacre City and Le Corbusier's *Ville Radeuse,* both appearing the same year (1935), are generally coupled as opposing concepts. Le Corbusier proposes immense skyscrapers that conserve land and place many people in immediate touch with each other. Wright's scheme is a three-dimensional pontification of low-density urban living. Yet both designers accept nature as a binding and enveloping context, not just useful, but necessary in their designs. The appreciation of greenery, open space—it is this commonly held value that sifts down through the many historic precedents and antecedents and connects the century's early visionaries to contemporary environmental design.

These values—and the fact that his visions were built—brought planners, architects, and government officials from all over the world to visit and see Howard's Letchworth and Welwyn. Yet despite genuine achievement, the Garden City enthusiasts had little immediate influence on government policy in England itself. Though housing reforms were begun in mid-nineteenth century, a general planning bill was not passed until 1932. Significantly, Patrick Abercrombie's *Greater London Plan* (1944) divided London into four rings and called for a greenbelt to surround the metropolis and eight satellite towns to be constructed on their far side. The New Town Acts of 1949 and 1950 gave further recognition to Howard's idealized form of urban life, almost a century in evolution.

While the British waited, Radburn and greenbelt towns were completed in the United States, Romerstadt in Germany, Magnitogorsk in Russia, and satellite towns proposed in the Netherlands, Scandinavia, and France. What joins these later schemes to Howard's original plan and the post-World War II English new towns is the idea of predestined size in urban communities, a full range of urban activities and densities, and the generous use of open space to girdle the town and fill the interstices with verdure and landscape—so much so that in the latest version (Tapiola, Finland) the city *is* a park. Ludwig Hilberseimer's prediction comes to mind. Writing about his proposed design for a new community in Chicago (1944), he noted that all designs move toward one concept: "The city will be within the landscape and the landscape within the city."

GREENBELTS

Containing urban growth by means of permanently keeping land open through building restrictions has long been a feature of British planning theory. To ensure an abundance of cheap food within easy distance of the city and to mitigate ill effects of the plague, Queen Elizabeth I ordered "all manners of Persons . . . to desist and forbeare from any new buildings of any house or tenement within three miles of said citte of London." A century later James I similarly prohibited building within ten miles of the city gates and required that new houses be sited on at least four acres.

However, these controls were little exercised, and no firm plans for separating urban growth from rural setting were laid in England until the late 1920's. By then the original deplorable conditions of London life that gave birth to the Garden City movement were somewhat abated through various legislative reforms, but what Patrick Geddes called conurbation (the coalescence of one community into the other) continued. Electric power freed many industries from dependence on sites next to collieries. To save transportation costs on moving manufactured goods to the largest market, many factories relocated in the London metropolis. The unemployed they left behind soon followed. Thus by 1930 London grew into the largest urban region ever known in the world.

For reasons of public health, safety, and economy, Minister of Health Neville Chamberlain called for an agricultural zone that would "form a dividing line between Greater London as it is and the satellites or fresh developments that might take place at greater distance." Raymond Unwin, Howard's designer of Letchworth Garden City, further urged the idea, so that recreation space would be available close to the city center and reserved from future building. Legislation in 1935 and 1938 allowed land to be bought or declared part of the London Greenbelt, and by 1969 over thirty-five thousand acres had been acquired—outdoor land the equivalent of the size of a living room for each person in Greater London today.

These minscule amounts of space have had enormous impact on the quality of the environment and planning. The green girdle has checked "endless urban sprawl," keeping neighboring towns from melding with one another and thus preserving their distinctive qualities. The greenbelt has provided environmental relief from urban congestion and noise as well as amenity and recreation. Irregular in shape, broken into seven sectors, each with a character, coherence, and scenery solely its own, the London Greenbelt combines natural features with historic spaces such as Runnymede. However, not all of the Greenbelt is public land. Some of it is privately owned and predominantly farmland, occasionally private estates, schools, research centers, and colleges. Ten to fifteen miles deep in places, the London Greenbelt stands out in strong contrast to the central city, forming a design at the largest discernible scale.

Enthusiastic acceptance of greenbelt proposals has highlighted recent English town planning. Although less than a handful have full official approval, over six percent of the country's land is included in current greenbelt proposals. Daniel R. Mandelker believes that if national-park, landscape, and other open-space provisions are included, twenty-nine percent of the English countryside is restricted from building development.

A sampling of urban plans in other sections of this book further underlines the importance of greenbelts, green wedges, blobs, and strips of open space in environmental design and demonstrates that the utility of greenbelts in metropolitan growth has been recognized in the United States as well as in Europe. For example, in putting together the Bay Circuit, the beginning of a greenbelt around the outer fringes of Metropolitan Boston (circa 1958), landscape architect Charles W. Eliot II ingeniously used great ponds, wildlife sanctuaries, conservation areas, town forests, agricultural lands, golf and country clubs, institutions, parks, wetlands, and cemeteries to patch together enough land to structure a significant green crescent and thus avoid "the shapeless, scattered, formless urban sprawl."

A recognized need is not, however, always followed with rational public action. Though the work of Eliot and others has alerted civic consciences, and Federal and State legislation has been prepared for the acquisition of open space, it is probably too late to wait for formal metropolitan planning and development to coordinate local government action in order to secure and hold the greenbelts necessary for America's metropolitan cities. What is needed, with dispatch, are vigorous programs among private groups and public bodies to secure now as much fringe land as possible, regardless of shape, condition, and location, and leave tó the next generation the subsequent horse-trading and sorting out necessary to realize a large design of local and metropolitan park systems.

English Green Belts Drawing showing the location and extent of the English Green Belt proposals. The effectiveness of the London Green Belt in controlling urban sprawl ensuring recreation open space close to the centers of population has encouraged local county and national governments to plan for similar green belts around the other major English cities.
Reprinted from: The Green Belts, Ministry of Housing and Local Government, London, 1962

Megalopolitan Sprawl Air view of Northern New Jersey suburbs showing the impact of a linear green belt on the design of the community. *Photo: Richard P. Dober*

Escape A small part of the London Green Belt. *Source: Greater London Council*

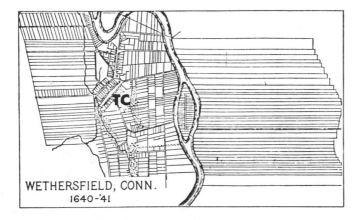

Early New England Villages Agricultural pursuits determine the form of the early New England village. The town center (TC) served as the location of the common lands and meeting houses — convenient and close to all. *Source: Harvard Graduate School of Design Library*

Savannah, Georgia, 1734 An early and successful attempt to organize the design of new American cities. *Source: Savannah, Georgia, Housing Authority*

Mt. Auburn Cemetery A classic example of early nineteenth-century romanticism in cemetery and park design. *Source: Library of Congress*

Frontier Town The main street of Kansas City, Missouri, about 1870. The reasons for the town's being are clearly seen, as well as the beginnings of a town architecture, largely expressed in the size and materials of buildings. *Source: Archives, Native Sons of Kansas City*

THE AMERICAN EXPERIENCE

The size, character, and variety of open spaces in the American scene arose from many different actions, at different times, for different reasons but the threads of purpose and concern that link tot lots to national parks are beauty and utility. The former lies in the eye of the beholder, as well as in the establishment of archetectonic effect through landscape design. The latter includes the conservation and preservation of natural resources. Admittedly, from the beginning the threads are interwoven, for in the earliest New England village and town, the greensward in front of church and meeting house was equally important for grazing cows as it was for providing a setting for buildings and a place for assembly and play. What planners call multi-use is a tradition as old as the European inhabitation of the continent.

As wilderness changed to city, the customs and fashions of continental city buildings were introduced, though not slavishly copied. In his encyclopedic work, *The Making of Urban America,* John W. Reps traces the influence of plaza, place, piazza, and town square in the planning and design of early American communities. Spanish towns and fortified sites, William Penn's design for Philadelphia, Dutch settlements like Albany and Manhattan, and the French outposts at Mobile, Alabama, and St. Louis show an orderly arrangement of major and minor streets, public open spaces, and gardens forming a design skeleton for the community. Unfortunately, the reasoned designs established by the original planners were often abandoned in the interest of real-estate speculation; the spirit of order and control was often ignored as population increased; utility and beauty were largely forgotten during a hundred years of subsequent urbanization.

James Oglethorpe's plan for Savannah,

Georgia, is thus notable in several respects: the articulation of method for incremental growth, the relationship of building lots to open space, and, along the original lines laid down in 1735, the continuity of town design for over a century.

Oglethorpe devised a basic community planning unit consisting of four blocks containing ten house lots each. The blocks were disposed to form a square. At the outer edges of the square were two additional lots designated for churches, stores, or other public uses. At the time of the Civil War, Savannah was the most attractive community in the United States, for Washington, D.C., despite Pierre Charles L'Enfant's bold design, remained in outline form, New York City and Philadelphia had begun to lose the tidiness of their first plans, and the design of Back Bay in Boston had not yet been accomplished.

Savannah demonstrated quite early in American history how a modular system of land planning and open-space design could establish a strong visual order, allow consecutive growth (without diminishing the design strength of the first stages of development), and at the same time allow infinite variety in architectural expression. While Savannah apparently had little impact on the design of later American cities, its influence abroad may be greater than historians to date have suggested. James Silk Buckingham's account of his visit to the town in 1840 and his later proposals for Victoria (a utopian scheme for a new town in England and, as noted earlier, of some importance in the Garden City movement) run parallel in their enthusiasms for the formal order and natural beauty, though Savannah's geometry differs from that of Victoria. An important stop for many other travelers who came to see the New World, well-praised in published journals, diaries, and travelogues, Savannah's amenities and character must have entered the consciousness of anyone concerned with the condition and appearance of city environment in the nineteenth century—as much as the outlines of Radburn and Valingby today are part of everyday professional knowledge and literature.

A rekindling of civic interest in open space and greenery occurs again in the United States during the 1850's. Romanticism, the urban-recreation and public-park movement, metropolitan design, commercial interest, conservationist and inner-city concern, and political support for the "City Beautiful" eventually resulted in substantial designs which in boldness of concept and execution have not been duplicated since.

The new romantic attitudes had their beginning in cemetery planning, where the attractive walks and roads and the embellishments of landscape and stone sculpture drew admirers and visitors from nearby communities as well as distant places. Laurel Hill Cemetery in Philadelphia and Mt. Auburn Cemetery in Cambridge in effect became public parks—so much so that the American horticulturist Andrew Jackson Downing suggested that "the general interest manifested in these cemeteries, proves that public gardens, established in a liberal and suitable manner near our large cities, would be equally successful."

As the eastern seacoast cities grew and population swamped the tiny squares of the earliest planned neighborhoods, the agitation for open space for recreation and relief from congestion became a political issue. In New York City (1853), after his election on a "parks for people" program, Mayor Kingsland instructed the local Commissioners to begin land acquisition and construction for a large central park. Four years later, 840 acres had been obtained, Frederick Law Olmsted had been appointed as superin-

tendent of construction, and the design for the park was opened to a competition. In collaboration with the English emigrant architect Calvert Vaux, Olmsted submitted the winning scheme, which gained him the commission and an additional appointment as architect-in-chief, and a great career and creative influence began.

Olmsted and Vaux's design preserved half the Central Park site in its natural conditions and used the other acreage carefully for outdoor recreation (ball fields, parade grounds, picnic areas, arboretum, horse-and-carriage trails). The outcome was an immense success. Within ten years all the major cities of the country had begun similar undertakings, with Olmsted himself traveling from coast to coast, either actually designing parks or inspiring others to emulate his work. The application of careful site analysis, the attention to climate and topography, the careful consideration of road and path design—all part of Olmsted's technique—were also applied by Olmsted and his colleagues to residential areas and institutional grounds. Open space and landscape came to play a strong role in giving form and content to city design. In Olmsted's work we see the beginnings of contemporary American community planning, campus planning, and site planning.

In some cases the park movement led to the beginnings of metropolitan design. Writing in *Civic Improvement* (1906), Henry Schott describes how Kansas City in the 1880's "found itself a place of 125,000 people, with all its physical features in the rough. In its anxious haste to grow, it had guarded carefully every commercial advantage and had as carefully overlooked the fact that parks, boulevards, and playgrounds are essential in the building of a city." Under George E. Kessler's direction, and with consultation from Olmsted, a large park system

was laid out, totaling two thousand acres and later joined by boulevards to a similar amount of acreage outside the city limits. The city charter had to be changed three times to give legal grounds for establishing a park commission and allowing assessments "to build pleasure grounds with public money."

Large and small landholdings in thirty-eight cities and towns around Boston were similarly organized into a single adminstrative unit, The Metropolitan District Commission, starting in 1893. The MDC pioneered in land conservation as well as recreation planning—and later added water supply, sewage treatment, and policing to its metropolitan responsibilities. The MDC plans for the Back Bay Fens, Middlesex Fells, the Blue Hills Reservation together with Charles Eliot's scheme for the Charles River Basin formed model designs copied worldwide. About the same time, forward-looking city leaders in Chicago promoted the acquisition of large land reservations in nearby Cook County for the city, giving it an invaluable and now highly regarded green preserve.

Commercial enterprises also served the spirit of the day. As trolley-car lines were extended to the suburbs, traction companies catered to new tastes in recreation by constructing amusement parks and picnic areas at the outer termini. River and lake steamships carried day-trippers to similar kinds of parks, while the rich traveled by rail to coastal and mountain resorts.

The appreciation of contrast between city and nature helped conservationists like John Muir, Gifford Pinchot, and Stephen Mather rally political support for better utilization of natural resources and for good cause. By 1890 the Census Bureau reported that there was no more free land left in the United States. The frontier had disappeared. In Peter Farb's words, the "skin of the land showed the abuse—sheer wastefulness had destroyed a quarter of the standing timber wherever logging had been carried on . . . [through other malpractices] . . . two thirds of the United States' original endownment of timber had been consumed."

Fortunately, the national conscience could be aroused before total environmental disaster took place. Yosemite and Yellowstone were saved from encroachment and destruction and the National Park Service created. From 1890 onward Congress slowly increased both the scope and size of Federal interest in conservation and open-space planning. Gigantic reforestation programs were eventually undertaken, most impressively in the southern states, more because of the size of the programs than the tidy, mechanical appearance of the landscape. Today about as much timber is seeded each year as is cut. In the last thirty years about a third of the damaged farmlands, misused for centuries, has been repaired.

But the price of destroying the landscape which filled the eyes of the first explorers is still being paid. Gigantic dams and flood-control works are now needed in the Midwest to do what cedar swamps did naturally: impound the surface water and release it slowly. The farmlands created by unplanned cutting of virgin forest and draining of swamps are now less settled than a century ago. Rural population declines as urbanization grows elsewhere. Land is abandoned and neglected. Patently, this is a needless cycle of changing raw land into a cultivated state and back again that developing countries may profitably observe and avoid.

As these large enterprises for metropolitan and national open space were launched, inner city was not neglected. Boston again pioneered in conservation of natural resources with the establishment of a Park Department charged with developing and operating local playgrounds. Boston aimed (1880) at having a playground for small children within a half mile of every home and recreation grounds for larger boys no more than a mile distant from where they lived. By 1906 Andrew Wright Crawford could report that there were eighteen playgrounds officially opened and other sites were being similarly used.

For older American cities expenses for catching up with decades of neglect were immense. Ten acres for three playgrounds in congested downtown Manhattan (1905) cost more money than the 840 acres acquired for Central Park fifty years earlier. Yet despite little precedent for large public expenditures for city parks and playgrounds, the 1920's and 1930's saw the professionalization of the recreation movement in urban areas and prolific construction, because supporters such as Luther Halsey Gluick successfully interpreted the meaning of play in human life. Others like M. Jane Reaney showed *The Place of Play in Education,* while Charles Platt commented on *Leisure as a Cause or Cure of Crime.* Platt opted for the latter and gave a moral tone to the pleas for organized recreation. Simultaneously, public-health leaders found additional reasons for sustaining green environments in the central city, and the slogan "green lungs for city people" was popular.

Finally, the development of park and playground systems was speeded up by local pride instilled by the City Beautiful Movement. The capital letters, then as now, seem appropriate for a great crusade to give the appearance of the city monumentality, dignity, and order. Out of a set of common values they shared in the design of the Chicago Fair of 1893, architects and planners like Daniel Burnham, Charles McKim, Cass Gilbert, and Frederick Law Olmsted, Jr., were able to propagate the idea that American cities had come of age in city planning. The affirmation of their beliefs by restoring L'Enfant's plan for Washington and producing the Chicago Plan of 1909 (see page 225) symbolized the peak of park and park-like designs for the central cities. For thirty years thereafter the urban park and playground remained a life-giving element in civic design.

Despite growing population and continuing utility, city parks declined in use and public support after World War II. The automobile made older forms of urban recreation obsolete. Many found play and pleasure by leaving central-city housing for suburban sites. Those who filled their places in the central city lacked political leadership for getting funds to maintain and operate the parks and playgrounds. Within the lifetime of one family, valued and prestigious old, urban open spaces fell into neglect and decay from which they have not yet emerged.

The migrations to suburbs and warmer climates that marked post-World War II were inspired in part by a search for open space and amenity, but there is little reason to think that the community development that resulted successfully met these objectives. Events in Santa Clara County, California, dramatize the state of environmental design. especially greenery, at the beginning of the Space Age.

The Cookie-Cutter Landscape The Valley floor embroidered with subdivisions without sensitivity or sensibility. *Photo: County of Santa Clara Planning Department*

COOKIE-CUTTER LANDSCAPE

In a state whose growth is considered amazing, Santa Clara County's urbanization has been phenomenal. Since 1950, it has been building communities the size of Wilmington, Delaware, once a year, every year; and the end is not in sight. Up to 1940 the environment and life of the County were dominated by agriculture. The fortunes of the County rose and fell with the success or failure of the crops, largely fruit trees, and the canneries that packaged them for national distribution. During World War II, however, there were gradual shifts in employment toward manufacturing. Many of the station-stop towns (strung along the railroad built through the Valley to take the agricultural products to San Francisco) became the nuclei for commercially and industrially diversified urban complexes. Yet the environment still remained open and pleasant.

The 1950's witnessed spectacular economic advances; at the same time the roles of agriculture and manufacturing as backbones of the County economy were dramatically reversed. The advantages of climate, excellent living conditions, and land for expansion attracted electronic and aerospace industries. In ten years Lockheed Aircraft alone *added* more workers than were employed previously on all of the farms and orchards. With new job opportunities came homes, highways, and commercial areas. In less time than it takes for a child to go from kindergarten to high school, Santa Clara County changed from a Garden of Eden to a sprawling metropolis.

Some have called the resulting design "scrambled cities" or "slurbs." The historic past—Indian burying grounds, Spanish missions, the homes of early settlers—were physically rooted up. Blossoming orchards,

invigorating air, abundant water and pleasing prospects were lost, and the picturesque qualities of an agricultural setting were not replaced in kind. With sixteen cities in the County fighting for economic advantage over one another through zoning and annexation, the benefits of large-area planning were lost.

Because of topographic constraints, most of the development occurred along the valley floor, all of it marked with ironies that illuminate the environmental dilemma not only of Santa Clara but of other fast-growing American communities elsewhere. Agricultural land in Santa Clara, for example, has been thoughtlessly eliminated by tract housing in just those areas that need open space to balance the man-made environment. In addition to their utility as a design element, these crop lands stabilize the County's economy, provide a close source of fresh food, and neutralize smog.

Paradoxes appear in the quality of environment where people work and where they live. Much of the industry in Santa Clara is viewed as "blue ribbon": the plants and sites are generous in design and landscape; employes are highly paid, highly educated, but their homes and the community surrounding them are quite different. The adjacent air view shows the new "cookie-cutter culture"—housing tracts stamped out on the landscape like cookies on a baker's tray. With ample money for some other choice, Santa Clara has ended up with the largest and most undifferentiated housing tracts in the country. Single-family areas are particularly flat and dull. Streets take up over a fourth of the land. There is little community open space and until very recently no easy means to walk away for a short distance to escape the sound of the automobile and the next-door neighbor.

Visual monotony mirrors social monot-

ony. Santa Clara planners note that any given tract is likely to be an enclave of the same kinds of families, of the same age, the same income, the same racial background, the same collection of consumer goods and time payments. In this sameness lies not just a design problem but a social problem. Public-health workers suggest that mental disease is bred in such monotony. Public administrators worry about lack of contact with other styles of life and the resulting deepening chasms in mutual respect and understanding—these being important to the climate of opinion in which constructive social change can be carried out.

Santa Clara is the grimness anticipated in the worst dreams of academicians and philosophers of environment. The ugliness and inefficiency will be seen for decades, the economic consequences will become increasingly destructive, and a designed future will seem increasingly less conceivable. Nevertheless, the design assets and liabilities of the past fifteen years can be summed up easily: The remaining natural conditions of site and situation plus the excitement and vitality of a booming economy are positive forces; the environmental deficiencies of unplanned and uncoordinated urbanization are negative. Santa Clara cannot replace what it has destroyed, nor is it likely to hold back the forces behind urbanization. The most realistic course of action is to ameliorate the damaged landscape, establish some general design structure for the growth to come, and hope that local initiative may preserve the amenities that remain.

Encouragingly enough, public desire for improved community amenities is quite evident in more than one way. For instance, as part of the 1965 Bay Area Transportation Study, four thousand householders in Santa Clara County were interviewed and administered an Environmental Attitudes Questionnaire asking the respondents about the quality of their present environment and the kind of environment they would wish to live in. Statistical analyses of the resulting data, assisted by a computer, enabled planners to correlate 118 different environmental issues.

The results clearly showed that the direction of public policy on land development should be reversed. Forty-four percent of the households responding stated that no more filling of San Francisco Bay should be allowed. An additional forty-seven percent felt that filling should be done for parks and recreation only. Thirty-three percent felt no further development should occur on any farm and orchard lands, while an additional fifty-five percent suggested that the "best farm lands should be saved." Eighty-five percent of the group wanted the hills to remain with their present natural contours and landscape. On a four-point scale (very good, good, fair, poor), not one of fourteen general descriptions on community services and amenities rated a "very good." "Public Transportation" and "Cultural Activities" were rated "poor." "Freedom from Traffic Problems," "General Appearance of Downtown Areas," and "Purity of Air" were rated "fair."

In many parts of the valley, furthermore, local home owners and citizen groups, well-educated by now through national media in what is wrong with the life they lead, are increasingly vocal about environmental quality. Fortunately, the existing local governments in Santa Clara County are strong, representative at the grass-roots level, and fully empowered to control local destiny. But unfortunately, many environmental questions go beyond the boundaries of local jurisdiction. Because of the scale of the problems, no one municipality can independently resolve such issues as flooding, air pollution, traffic and transit, water supply and waste disposal.

Thus, Santa Clara County points up the interdependency of local communities sharing a common environmental district, and the need for a governmental process that will effectively address itself to area-wide problems without impairing local prerogatives. In any case, a comprehensively designed environment for Santa Clara is not likely to emerge until jurisdiction conflicts between competing governments are settled.

Under these conditions, the most feasible starting point at present for ameliorating the environment through design is the County's plan for parks, recreation, and open space. Because of the current quantity and rapidity of growth of urbanization in Santa Clara (two million people are expected by 1985, an eightfold increase since 1950), the elements that compose the design deserve comment, for the Santa Clara Plan reveals signs of quality that are still within reach.

A range of attractive landscapes is perhaps the Valley's remaining unique natural feature. Hardwood and conifer forests, oak savannas, grasslands, brushlands, farmland, stream banks, and shorelines are not only present but within driving distance of each other. Each has its own special ecologically related wildlife and land forms. Each shows differing textures and colors as the seasons change. Preserving significant acreage of each kind of landscape is, in itself, a move against visual monotony.

The County Plan proposes to organize these landscapes into three groups: agricultural open space, functional open space, and parks-and-recreation open space.

The agricultural open space is the most vulnerable part of the Plan. Agricultural lands are used for greenbelts around communities to prevent urban areas from visually and physically connecting together. Because urbanization does affect the quality of the crops, large acreage holdings are necessary. Lying at the edge of urban areas, such land is continually subjected to economic pressures (rising taxes and land values) for conversion into subdivisions. The soil conditions are typically excellent for buildings, and the large size of the parcels means easy land assembly.

Since 1954 the County has tried to abate these pressures on agricultural lands through zoning. Fifty thousand acres have been designated "exclusive agriculture." This protects the farmer from involuntary annexation by an expanding city or town. He can, however, part with land voluntarily, and farms and orchards continue to change ownership and use. As a countermeasure, local planners would like to obtain development rights on strategically located parcels. The public would purchase the owner's right to build urban structures on the land, just as it purchases easements for freeways and flood control. In this way prime land can be kept agricultural.

Functional open space includes all land which is to be kept relatively free of building because the land either acts as a setting for architecture or is needed to carry on specialized activity. Such land is well used. The largest and most important examples are the watershed preserves. Mountains rim the Santa Clara Valley, and if the covering forests are leveled, irreparable losses will follow: modifications in weather, damaged water supply, and land erosion. To rule the mountain is to design the valley, and fortunately, public-health and safety regulations may be used to keep the highlands inviolate.

Other sizable tracts of functional open space in the Plan include golf courses, airports, cultural and educational institutions, facilities used by the public such as civic centers and shopping centers.

It should be noted that public purpose in Santa Clara is served by keeping these sites for functional open space as large as possible. With housing divisions so fine-grained, large and munificent sites for community uses are visually desirable. Furthermore, public intervention to meet this objective in residential as well as public-purpose land may be accomplished in two ways: directly through public investment in large sites for public facilities and indirectly by writing subdivision and zoning controls that encourage private developers to use generous standards in site design.

The County's third category of landscape includes all parks and formally designated recreation areas. It intends to increase the public supply of such land from twenty thousand acres to sixty thousand acres in twenty-five years. The 1985 population would then have about thirty acres of park and recreation land per one thousand people—about the same as that proposed in the Baltimore Regional Plan. This is a modest goal for a County whose major visual attraction is what's left of the "outdoors." The County's Plan also calls for 140 miles of freeway, 340 miles of recreation roads, 280 miles of riding trails, and thirty miles of bicycle trails—all landscaped and tree-lined.

Since Santa Clara County does not yet have the equivalent of Golden Gate Park (San Francisco) or Central Park (New York City), parks on the Valley floor (150 to four hundred acres in size) are urgently needed. Seven sites have been spotted. Each will be designed for urban-oriented activities. They will contain picnic areas, informal playfields, museums, outdoor amphitheater, and organized spectator sports. Outdoor camping and water sports will be developed in selected spaces in the mountains and watersheds. The County's shoreline, along South San Francisco Bay, will be used for marine parks. Now an inaccessible, neglected area of salt ponds, marshland, and garbage dumps, this shoreline will be improved for swimming, boating, golf courses, and playgrounds.

Santa Clara is innovative in two design areas: the use of utility easements and the stream-side park concept. The latter, as proposed by the Santa Clara planners, is both practical and aesthetic. The Valley is rich in streams that run from the mountains to the Bay. Though many are waterless in summer, the vegetation along the banks (alder, sycamore, and oak) makes them attractive and comfortable places for year-round recreation. Many streams are already under the jurisdiction of flood-control and water-conservation districts and therefore afford access and development rights more easily than purchase of private land. These stream-side preserves can link together the large public parks and in this way create the skeleton for a distinctive valley design. Pleasant to see, walk along, or cross over, these green and wet lines may serve as edges and boundaries to subparts of the community, as well as connectors. Proposed horse and bicycle trails and landscaped freeways will have similar design effect.

Santa Clara's use of utility easements also shows how environmental designers can turn a liability into an asset. The Valley is crisscrossed with narrow strips of land set aside for gas lines, high-tension wires, water and sewer pipes, percolation ponds, flood-control channels, and fire control. They vary from sixty to 160 feet in width and total about one thousand acres. By planting appropriate trees and laying gravel, the County will create safe, pleasant, and attractive walkways out of the weed-growing, trash-collecting easements. A test of the idea in an elementary-school district revealed that eight miles of continuous easements could be landscaped, and there were several intersections where the design could be expanded for a local playground and pocket park. It is this kind of incisive action that shows what can be done through environmental design in seemingly discouraging circumstances.

OUTDOOR RECREATION #1

Walt Whitman wrote in his *Autobiographia* that American democracy must be

fibred, vitalized, by regular contact with outdoor light, air, and growth, farm scenes, animals, fields, birds, sun-warmth and free skies, or it will morbidly dwindle and pale.

Whitman and others notwithstanding, the impetus for recent action in acquiring and developing open space for recreation and, in turn, further interest in conservation, is not due to a national philosophy, but rather the shortening of the work week, increases in personal income, widespread use of the automobile for getting to and from recreation areas, and the construction of roads and highway systems to make the journey as safe and as quick as possible. Because of these conditions, the character of urban recreation facilities began to change in the early 1920's. Up to that time, mass transportation carried people to urban parks, commercial and cooperative enterprises satisfied much of the adult leisure time, and the playground-and-park movement was making progress in meeting the daily needs of family life. Then regional, eventually national, large automobile-oriented facilities became a new goal.

The shift in demand was recognized as early as 1922, particularly in studies leading to *A Regional Plan of New York and its Environs*. Encompassing all the area in which New Yorkers earned their living, made their homes, and played, the Regional Plan was a landmark in comprehensive planning. The survey and recommendations on *Public Recreation* (1928), extensive and detailed, required a separate volume.

The automobile was viewed in the Plan as a force greatly increasing

the need for more playgrounds and neighborhood parks in central areas . . . Until the coming of the high-speed motor vehicle, streets and highways were much used for walking and children's play . . . There is no going back, of course, to the time when the vehicle took a place second to the pedestrian in the use of the highway. The position of the motor car as the predominant user of street surface is impregnable, and every other use has to give way to it.

But, as the regional planners noted, the automobile was not "a wholly destructive agent in regard to recreation." The automobile

has modified the character and location of the facilities that are needed and has increased the demand for special kinds of recreation spaces. It has made the distant places accessible, perhaps, to the majority.

Thus, the Regional Plan report called not only for attention to local playgrounds but for regional recreation areas as well.

Acquisition of large public reservations for weekend and holiday use in country areas should be continued coincident with the effort to obtain sufficient space in the urban areas for everyday use.

Several of the Regional Plan's proposals were later executed by Robert Moses while he was simultaneously administering programs for City, County, and State park and transportation agencies in New York City and vicinity—with varying degrees of success. His development of Jones Beach and Bear Mountain were harbingers of a quiet revolution in recreation needs, of which the economic consequences were hardly anticipated. Recreation changed from a local effort to fill constructively leisure time to a business of first-rank importance.

Post World War II urbanization was stimulated by significant rises in birth rate, income, and automobile ownership, and a one-third reduction in the average work week. The pressures on outdoor recreation places thus continued to build during the

1950's and stimulated widespread activity at the Federal and State levels. Accordingly, the Federal Park Service's *Mission 66 Program* (1956), designed to overcome the deterioration caused by overuse of facilities and to develop new recreation areas, was politically successful and nationally popular. Despite increasingly large Federal and State appropriations for recreation by the end of the decade (1959), the demand for outdoor recreation generated by the urban population still exceeded the supply of land and water resources readily available to them. As a result of public pressure, the Congress authorized a bipartisan Outdoor Recreation Resources Review Commission (ORRRC)

to make comprehensive information and recommendations

so as

to determine the types and locations of such resources and opportunities which will be required by present and future generations.

It backed the Commission with a $2.5 million appropriation.

The Commission investigated all aspects of leisure-time pursuits in contemporary America, measured the probable volume of recreation by the Year 2000, and proposed widespread improvements for managing, developing, and operating outdoor recreation areas. The statistical profiles emerging from the Commission's work (1962) illuminated how often and how well America plays. Driving and walking for pleasure accounted for a third of all recreation time, with travel by automobile slightly favored over travel by foot. The nation was not entirely possessed by sedentary activities as social critics had feared. Less than a fourth of the recreation time was devoted to spectator events or pleasure driving. Recreation turned out to be a $40 billion business and, for parts of the nation, the predominant economic asset. In

the span of a single generation, recreation had changed in emphasis, location, and importance and had become a major service industry.

The year 1962 marked the advent of an increasingly strong national response for meeting recreation needs and relating these to environmental quality. The Bureau of Outdoor Recreation (BOR) was established within the Department of Interior to coordinate Federal planning, programming, and policies. The Bureau continued to track the dynamics of recreation, following the lines of inquiry established in the 1962 ORRRC study. Surveys taken by the Census Bureau in 1965 showed a fifty percent increase in outdoor recreation between 1960 and 1965. Walking and swimming took over the top places in the nation's popularity poll of favored recreation activities.

The same census report showed that the race for recreation space was engendered by high land costs. Prices of acreage suitable for public recreation were rising faster than the general land market itself. BOR projections indicated that The Land and Water Conservation Fund, the prime source of moneys for Federal and State recreation land acquisition funds, would be $2.7 billion short of needs if even the minimum amounts of new acreage for outdoor recreation in 1978 were to be met. BOR recommended (1968) that income received by the Federal government from mineral leases on public land and on the Outer Continental Shelf be added to the land acquisition fund, but unfortunately for the country, Congress was not disposed to pass such controversial legislation in an election year.

Though 1968 was marked more by planning than by action, Federal and State interests in environmental improvements through recreation design were commendable and eventful. BOR began studying the feasibility of using the entire Connecticut River Valley as a national recreational area, running from the hills of New Hampshire and Vermont- through Massachusetts and Connecticut to the sea. The use of the Valley would place a new National Park within a half day's driving distance of thirty million people. Similar planning began for the Missouri and Hudson Rivers. The concept of multi-use (recreation and conservation) was strengthened in these studies, as well as a broad attack on water pollution. The National Park Service's Division of Environmental Design, set up in 1966, began a search for

a silent, efficient way to transport people unobtrusively through the national parks without making it necessary to carve nature's grandeur into more highways and parking lots.

The Division was also instrumental in having twenty miles of power lines placed underground, rather than strung on power poles, across the Florida Everglades Park. Costing $250,000 more than a conventional solution, the underground cable preserved the silhouette of cabbage palms against the sky. To preserve but increase access to the visual effect of canyon walls around the Rainbow Bridge National Monument, the Division designed and constructed a complex of inconspicuous man-made "islands" near the shore for boat landings, ranger stations, rest rooms, and supply stores.

Urban dwellers in the Chicago metropolitan region could look forward to enlarged State appropriations for outdoor recreation and natural resources. At the University of Illinois/Chicago, faculty and students in the School of Art and Architecture demonstrated the feasibility of constructing polders in Lake Michigan for new towns with densities up to sixty thousand people to the square mile in order to pre-

vent further encroachment into the green environment that lay behind the existing towns and cities. Once again, provision for recreation and open space came into focus as a major element in city and environmental design.

Open-Space Diagram, Santa Clara County *Plan: County of Santa Clara Planning Department*

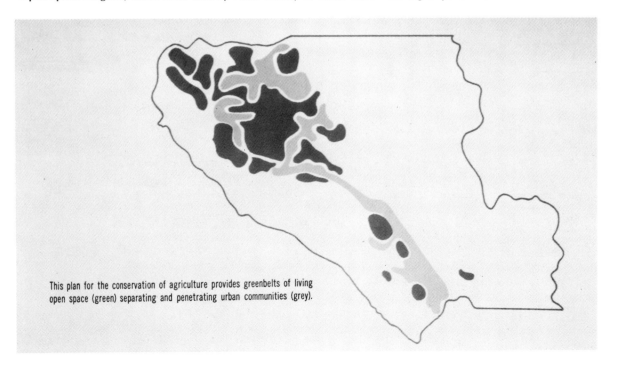

This plan for the conservation of agriculture provides greenbelts of living open space (green) separating and penetrating urban communities (grey).

IMPLEMENTATION

PURCHASE

Outright purchase of streamside lands
by City or County,
with assistance through open space funds.

Purchase of flood control right-of-way
by the County of Santa Clara
Flood Control and Water District.

Grants from state funds
for fish enhancement
along the Los Alamitos-Calero streams.

DEDICATION

Voluntary dedication by the owner.

Dedication of adjacent stream lands
for lot reduction in subdivisions
on a percentage basis.

Use of State Bill #1150,
which would require the subdivider
to dedicate land for park purposes.

Dedication of land through
"Planned Developments" adjacent to the stream.
The dedicated lands would be in lieu
of the required open space
in the Planned Development.

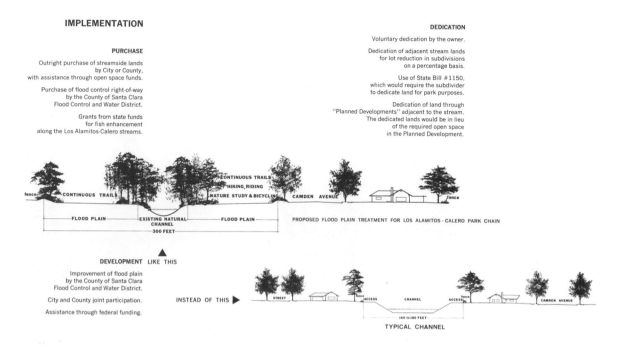

PROPOSED FLOOD PLAIN TREATMENT FOR LOS ALAMITOS - CALERO PARK CHAIN

DEVELOPMENT LIKE THIS

Improvement of flood plain
by the County of Santa Clara
Flood Control and Water District.

City and County joint participation.

Assistance through federal funding.

INSTEAD OF THIS ▶

TYPICAL CHANNEL

Opportunity Grasped Design proposals for using flood-control rights-of-way for linear parks and public open space. *Design: County of Santa Clara Planning Department*

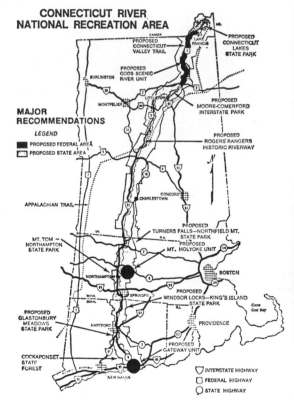

Proposed National Park Along The Connecticut River
Drawing: National Park Service

CENTURY OF CRISES: DECADE OF CONCERN

The events that took place in Santa Clara, California, and the national response to outdoor recreation are all part of a resurgence of concern for public beauty that began in the mid-1950's and continued through 1968 with deep-felt resolution but little genuine accomplishment. For the first time basic research in urban aesthetics was produced in a respected and forward-looking academic setting—under Gyorgy Kepes and Kevin Lynch at the Massachusetts Institute of Technology. Harvard's Graduate School of Design started in 1956 an influential series of annual meetings that continues to draw together practitioners and theorists. The words "urban design" became a useful term to describe a wide variety of viewing points and interests in the physical form and appearance of regions, communities, and groups of buildings.

Environmental design gained a place on the national agenda when writers like Peter Blake and Rachel Carson (respectively) described *God's Own Junkyard* in a stunning photographic essay on environmental decline in the United States or touched on survival in *The Silent Spring* while Edward Higbee showed *The Squeeze: Cities Without Space*. On the hopeful side, Christopher Tunnard illustrated what could be done in *Man-made America*. Jane Jacobs' polemic on *The Death and Life of Great American Cities* was balanced by Paul D. Spreiregen's *Urban Design: The Architecture of Towns and Cities*. Sponsored by the American Institute of Architects, skillfully produced and publicized, Spreiregen's book resuscitated the City Beautiful Movement, instilled it with a larger vision than it possessed before, and gave incentive for countless meetings and proposals for urban beautification. Nationally respected newspapers added fluent and knowledgeable environmental design critics to their staffs. Ada Louise Huxtable of the *New York Times*, Grady Clay of the *Louisville Courier*, Wolf von Eckhardt of the *Washington Post,* and George McCue in the *St. Louis Post Dispatch* kindled local interest and national conscience with timely stories and descriptions of what was wrong and what was right in the architecture and development of American communities.

The contemporary concern with aesthetics is well reflected in more than one effort; for example, the laws enacted by the California Legislature (July 1963), which designated 4,900 miles of State highway routes "State Scenic Highways." In a rare contemporary pronouncement on aesthetic matters by a public body, the Legislature declared that the program was

a vital part of the all-encompassing effort which the State must make to protect and enhance California's beauty, amenity, and quality of life.

While preservationists and design groups supported the legislation for aesthetic reasons, the bill and the subsequent appropriations to carry out the measure were rationalized on economic grounds, as if beauty of itself were an inconsequential matter. The official publication of the California Department of Public Works thus noted that

the quality of the living and working environment is playing a significant role in industrial plant location and in long-range investment decisions.

Whether declared in the name of art or regional economy, the California Scenic Highway Plan is a substantial effort in environmental design. The highways will lace together all sections of the State, facilitating day trips from dense urban cities to distant areas of unusual scenic beauty. Highways that cross the state's border will be designed and improved to give immediate visual satisfaction to visitor and resident, and it will be possible to bypass truck-crowded expressways when transversing the state. Though not yet fully implemented, the California act is a model for other states and perhaps even Congressional action, just as California's earlier legislation on scenic easements and billboard control served as an example for national standards in the Federal Government's Interstate Highway Act.

The culminating event of aroused concern for environmental aesthetics in the 1960's was the Presidential message on *Beauty For America*. Sent to the Congress on 8 February 1965, it contained Lyndon B. Johnson's statement:

For centuries Americans have drawn strength and inspiration from the beauty of the country. It would be a neglectful generation indeed, indifferent to the judgment of history and the command of principle, which failed to preserve and extend such a heritage for its descendents.

Johnson called for increased appropriations for land and water conservation and open-space acquisition. The latter program had proved to be popular country-wide, and he sought to extend the program `to include matching grants for "small parks, squares, pedestrian malls, and playgrounds" as well as

landscaping, installation of outdoor lights and benches, creating attractive cityscapes along roads and in business areas, and for other beautification purposes.

Five months later, a special advisory panel to the President reported on national priorities. Chaired by Edmund N. Bacon, the panel on Townscape stressed that beautification funds should be commingled with provisions for the development of neighborhood centers and "particularly directed to the poorest areas of cities, to insure that all American families share in the effort to beautify our townscapes." The panel recommended the establishment of a National Urban Design Center, improved measures for preservation of historic areas, the integration of comprehensive design into all aspects of planning, and a massive tree-planting effort, including the improvement of the technical process of large-scale mass moving of big trees and the reduction of costs of such operations. Shortly afterwards, other urgencies and priorities pushed beauty off the political platform. A century of crises did yield a decade of concern, although it was not immediately followed with widespread environmental improvements. Thanks to Ladybird Johnson's consistent efforts her husband's successor was inaugurated in a capitol city aesthetically improved, but at the same time several National Parks were scheduled to be closed several days a week—like a bankrupt museum in a provincial backwater town they hadn't enough money to stay open.

National Park, 1920 As fast as the parks were opened to people and automobiles, they were well used by the public. *Photo: National Park Service*

Preserving The Natural Scene Man-made islands at the foot of Glen Canyon provide viewing stations and service points for tourists and park rangers without cutting into the natural setting. *Source: National Park Service*

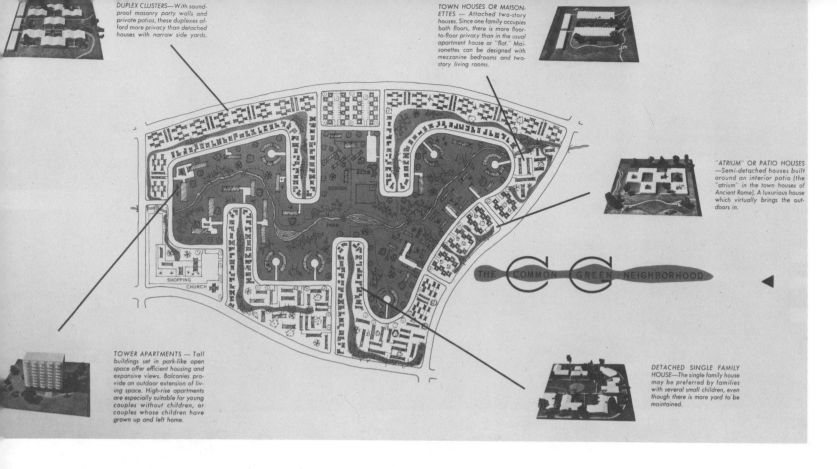

DUPLEX CLUSTERS—With sound-proof masonry party walls and private patios, these duplexes afford more privacy than detached houses with narrow side yards.

TOWN HOUSES OR MAISON-ETTES — Attached two-story houses. Since one family occupies both floors, there is more floor-to-floor privacy than in the usual apartment house or "flat." Maisonettes can be designed with mezzanine bedrooms and two-story living rooms.

"ATRIUM" OR PATIO HOUSES —Semi-detached houses built around an interior patio (the "atrium" in the town houses of Ancient Rome). A luxurious house which virtually brings the outdoors in.

THE COMMON GREEN NEIGHBORHOOD

TOWER APARTMENTS — Tall buildings set in park-like open space offer efficient housing and expansive views. Balconies provide an outdoor extension of living space. High-rise apartments are especially suitable for young couples without children, or couples whose children have grown up and left home.

DETACHED SINGLE FAMILY HOUSE—The single family house may be preferred by families with several small children, even though there is more yard to be maintained.

The Advantages Of Clustering For Open Space *Source: Santa Clara County Planning Department*

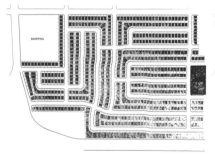 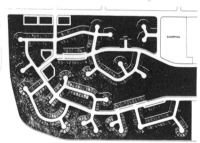

	CONVENTIONAL SUBDIVISION		CLUSTER SUBDIVISION
	32	ACRES IN STREETS	24
	22,500	LINEAR FEET OF STREET	16,055
	29	PERCENT OF SITE IN STREETS	19
	80	ACRES IN BUILDING SITES	41
	590	DWELLING UNITS	604
	0	ACRES OF USABLE OPEN SPACE	51

STAGNATING SALES

The typical tract house has been over-produced. Buyers are tired of:
Peas-in-a-pod uniformity
Gimmicks in place of real values
A forest of overhead wires
Weed growing, trash collecting, useless side yards.

BUYER APPEAL

The dwelling oriented to the Common Green is a fresh concept with a great potential. Buyers are looking for:
Variety in housing
Nearby open space
Sky views uncluttered with overhead wires
Minimum yard maintenance.

CLUSTERING FOR OPEN SPACE

The "race for inner space"—the national urgency for acquiring land for parks and recreation and conservation of natural and scenic resources—is so well documented that no thoughtful person should be unaware of this special environmental crisis. As indicated earlier, however, there are reasons to suspect that, at least in terms of our cities, rhetoric and exhortation have not been and will not be sufficiently effective. Strictly in terms of square miles of open space available, the national stockpile can meet known needs to the end of the century and beyond, as less than ten percent of the land available is needed for all urban development and outdoor recreation for the next three decades or so. But this is *no* comfort in face of the fact that where open space is needed most, it is available least, and, paradoxically, many of the great nineteenth-century urban park systems are neglected and decaying. Since land occupied by human habitation is the most difficult to return to some natural state, the chances for significant additions to the open-space stockpile in central cities and the immediate communities around them are severely limited by cost and the attendant problems.

Cities with conscience, neighborhood-improvement groups, and dedicated individuals have certainly made valiant attempts to meet the open-space requirements, and Federal and State Governments have promoted programs for acquiring and preserving open space. Nevertheless, the action taken to date has not reduced the backlog of current needs. A sampling of recent land-use statistics in the thirty largest American cities shows that few meet the generally accepted planning standards for public open space. Public administrative policies for responding to scarcity in these cities have been timid and uncertain—understandable perhaps in light of requirements made on the public purse for education, health, and welfare programs in recent years. For similar reasons little has been done in anticipation of future requirements. It was estimated in 1960 that Metropolitan Washington would require two hundred square miles of park and recreation land by 1980. In 1968 the Maryland-National Capital Park and Planning Commission, one of the most vigorous open-space acquisition agencies in the country, though preparing plans at a fast pace, was far behind schedule in acquiring and developing recreation land.

The Washington situation is not unique, and nationally great efforts must be made to increase the stock of open space now. Several approaches come to mind. First, land-absorbing urbanization must use the everyday construction of housing tracts and community facilities as a way of enlarging the stock of open space by making it an integral part of the development cycle: vacant land to occupied land. In this effort the quality and location of open space should take precedent over the shape and quantity. Lacking a public purse to match public policy, government locally and nationally may have to try to shift the burden of open-space development to the private sector more extensively than it has in the past. Finally, the opportune time for developing public open space in suburban communities is before the value of land has risen because of new construction, not afterwards; and cluster housing is one technique for achieving this goal. As described below, this simple measure can dramatically influence architecture and site design, as well as affect community design. The expectations that come with population increase support an optimistic view of cluster housing and similar designs.

Large-scale housing construction is forecasted for the early 1970's when the postwar "baby-boom" population is expected to mature and form families. Corporate efforts in housing construction are anticipated. With the advent of systems analysis and the application of management sciences, the small builder may be replaced by a consortium or special corporation. Economies in the process and product can then be realized, though many observers believe that because of labor practices and popular tastes, revolutionary new forms of housing are not likely to appear. Efficiencies in construction may be paralleled by improved site and community design, especially if large amounts of land are developed through corporate or cooperative efforts. Thus in the very context of massive housing development, it may be possible to write subdivision regulations to encourage the application of land-conserving devices such as cluster-housing development. Cluster housing allows the builder to group his houses closely together maintaining the density implicit in subdivision regulations but producing unfragmented open space because all the land need not be divided equally among individual house lots.

From the developer's viewpoint, the economic advantages are substantial. A study of the Ville du Parc cluster development near Milwaukee showed that for a plot of land costing $360,000, the cluster scheme returned 104 percent on the investment, a conventional scheme about twenty percent. Further, half the savings in development costs (because of shortened utility runs and roads) were put back in the project in the form of recreation areas, thus increasing the sales potential of the design. From the community's point of view, a significant amount of open space was produced at little public cost, and the amenities of this green space throughout the housing area improved the appearance of the community.

Clustering has its problems, such as community suspicions about innovative designs, the setting up of administrative devices for maintaining the open space, and the possible prejudice and bias that comes with small precincts of one-price and one-class housing districts. In the hands of hit-and-run speculators, the cluster idea could break down the development controls afforded by traditional zoning without replacing them with something as good. In addition, it is possible to honor the idea of cluster development without producing a substantive design. For this reason, environmental planning and review procedures well beyond the everyday subdivision regulations are needed to ensure that the paper plan filed by the developer is well executed and the burden of maintaining and operating the open space is not passed on to the general public.

Towards clustering, hope (from the profit motive) and fear (from existing zoning which builders feel handicap their business) may be manipulated to reach several kinds of desirable environmental benefits simultaneously. For instance, communities may invite a consortium of small builders, or a large-scale builder, to use the cluster concept with generous open-space provisions in exchange for a variance on local regulations. The persuasion which comes about from discreet application of local regulatory powers may be extended in other directions. For example, requests can be made to include a reasonable amount of housing for low-income groups in any overall scheme.

Further, communities that can foster a development plan going beyond land use, zoning, and the location of community facilities, will see a constructive partnership between private developers and community materialize. The tug-of-war that went on in the late 1940's and 1950's as private market and community fought for the control of land need not be duplicated. Neither builders nor community officials won long-term gains, as the disrepute of the cracker-box subdivision so eloquently proclaims. The spector of an even larger housing boom lies just ahead. The older ways do not encourage many to believe that the answers to environmental improvement lie in continuance of the traditional disorganized system. Dramatic alterations in land law and procedures cannot be depended upon but the strategic application of existing regulations is a reasonable expectation. In this light the cluster concept is a good example of enlightened self-interest.

To reiterate, the cluster concept is important for a country that is not meeting its self-proclaimed needs for acquiring open land independent of the community building process, for clustering encourages a housing type that generates its own open space. Also, clustering works for high-rise apartments as well as single detached houses. Even so, the idea of clustering is not new, the design being recognizable in the medieval village and early New England town. In today's setting cluster housing constitutes an application of an old technique that can have many benefits. Anticipate fifty million new homes in the next three decades, and the utility of the cluster concept in obtaining some of the needed open space is evident and clear. Some of the other advantages have been reviewed; a possible indirect benefit, the restoration of row housing to design prominence, deserves elaboration.

THE ROW HOUSE AND COMMUNITY DESIGN

The relationship between row housing, open space, and community development is worth exploring in terms of past and present attitudes and future possibilities. Out of such relationships emerge the community designs of the future. In the American experience, row housing is an aborted form of urban shelter, traditionally cramped by cost and divested of status by the Great American Dream—the free-standing, single-family home. Yet despite the lack of consumer preference and prestige and the small amount of construction to date, row housing may become the most significant type of shelter in what we hope will be the brighter half of the twentieth century.

History illumines the present predicament. Row housing—repeated units horizontally connected by a party wall—appears in history whenever people congregate in dense urban areas. Many residential sectors in ancient, medieval, and modern cities fit the definition. Pompeii, the Vicar's Close at Wells, Bonito Pueblo, and sixteenth-century Amsterdam manifest the tradition. The Georgian elegance of Bloomsbury and Bath and Beacon Hill (Boston) are superb examples of the building type. It is not these masterworks, however, that constitute the public conception of row housing, but rather the environmental poverty of nineteenth-century industrial England.

Drawn by industrialization, the new urban masses sought shelter within an easy journey to work. The tenement served some, the row house attracted the others.

The popularity of the row house was due in part to its middle-class image and in part to its economic advantages. The image came partly from the romantic notion of Charles Dickens, the "patron saint of the

The Radburn Concept A large idea carried out through a small project having world-wide influence on community design. This view shows the site design before the landscape had grown to maturity. *Photo: Harvard Graduate School of Design Library*

Contemporary Row Housing Plans showing the variety of interior layouts that can be accomplished within a modular construction system. *Architects: Geddes, Brecher, Qualls, and Cunningham. Source: Design Proposal for the Eastwick Project*

English home." He depicted the happy domestic life, replete with comfort and good cheer, for those living in their own domicile, however modest it might be. For the rest the image came from a dream of social welfare nicely entwined with the profit motive and all well-described by Henry R. Aldridge in the *National Housing Manual*, London, 1923. "One house, one family" was, he reports, the considered solution to crime, bad health, and poor work habits. Groups such as the Improved Industrial Development Company put up housing in rows at thirty to forty units to the acre. The rising middle class and the small investors were attracted to this kind of speculation and were accommodated through the formation of building societies. At a time when industrial stocks were hardly worth the paper they were printed on, "safe as houses" was more than a slogan. Building societies prospered, if the environment did not.

What began as a desire to do good ended with land sweating. What was meant to secure order achieved monotony. To be fair, the "brick box with the slate roof" of the end of the nineteenth century was an improvement over the dank, dark hovels of the beginning. The (English) Health Act of 1875 checked the jerry-building tendencies. Structures were sound. A minimum of sanitary provisions was ensured. But the growing ability of the average citizen to buy good shelter at the close of the nineteenth century was not matched by improvement in design. There was little amenity and no sense of site development or communal aesthetics. The "estates" of yesterday became the slums of today.

The ideal and the reality were also evident in the United States.

There are advantages of an independent home for every family no matter how humble it is, in the country or in a district where each house is entirely separate from all others, and has its own ground; and (with) a small garden, so much the better.

Small Homes Within the City for Unskilled Workers (1910)

Such housing conditions were impossible for the unskilled worker to achieve by himself, so companies such as the Sanitary Improvement Corporation attempted to supply a satisfactory substitute. The favorite technique was to buy a vacant city block that had water and sewage and split it down the middle with a new right of way. Houses would be erected with fifteen-foot frontages and a small backyard. "Mantelpieces and wallpaper do not add to the tenants' comfort but do add to the rent he must pay," wrote the incorporators. Again, the results are minimum shelter, minimum design, but inappropriate cost to the resident and fine profit for the building. For example, Sanitary Improvement Corporation paid over five percent yearly to its investors and, in addition, had accumulated within a decade a surplus of $211,000 on a capitalization of $500,000.

Not all row housing was so poorly devised, impoverished from economies, and exploited. Boston was land-shy. It continually grew by land filling, which in turn limited high-rise structures and encouraged side-by-side housing. But the Back Bay of Boston mixed speculation and design to produce a lively urban scene. The clientele, however, was affluent and accustomed to this manner of urban living.

Since 1900, despite some promising beginnings for row housing among the middle classes, housing construction has shown growing preference to detached units. This pattern seems to follow the reduction of immigration and the speeding up of acculturation in which the single-family house becomes a symbol, a value, and a social goal. The decline of the multiple-family dwelling and the fulfillment of the Dream (every family its own house) were postponed by the Depression and World War II, but afterwards a dramatic building surge took place. The causes were: an increase in variety of financing schemes for individual ownership, new journey-to-work patterns made possible by the automobile and highway development, and consumer antipathy towards the choices of housing and community environment available in the central city.

Numerically, single-family housing now preponderates in the United States—forty million out of fifty-eight million units in the 1960 census. About thirteen million units are in multi-dwelling structures and about 3.6 million units are one-unit attached houses. The latter category includes row housing, for the Bureau of Census does not have (in its own words) a "definitive definition." The category *one-unit attached* includes double houses and houses attached to non-residential structures and sharing a party wall.

Currently, row housing in the United States is as under-represented as it is significant. Since it is less than five percent of the existing inventory, row housing obviously does not represent a common form of national shelter, as it does in Denmark or Great Britain. Regionally, it is important. In Philadelphia and Baltimore sixty-six percent and fifty-seven percent respectively of the total housing stock is row housing. These are cities with a row-house tradition. Yet scan current literature, and the very words *row housing* in America seem an anathema, especially in areas where it has been a historical shelter type. Connotations of poverty and slums are veiled by such phrases as town house, cluster house, and patio house. The imaginative plans for Metropolitan Balti-

Variety And Modularity In Reston (Virginia) Town Center, high-rise apartments and row houses are combined to create a distinctive design. See also Tapiola Town Center (page 25), for an example of a similar design effect. *Source: Reston Corporation. Photo: Blue Ridge Aerial Survey*

Regional Expression An excellent example of regional architecture using the principles of row-house design. (Glover Landing, Marblehead, Massachusetts.) *Photo: Chapman and Goyette, Architects*

more use the euphemism "group house."

Build to minimum standards a pseudo-Georgian house on a swamp, call the development Powdermill Hollow, and sales will boom. Construct decent shelter in pleasant groupings, and the development's proper name may be an impediment to sales. Still the fact of the matter is that in concept and execution, row housing is the best buy in the market today—in terms of shelter and the costs and benefits of community development.

This historical and statistical background presents pointed questions. What are the reasons for optimistically assuming a change in attitudes among builders, investors, and consumers? Why row housing? The answer lies in the potential it has for open-space planning and, in general, the designed environment.

An absolute requirement for higher-density development plus a repugnance of jerry-built and inadequate subdivisions will be the basic incentives for row housing. The world outside the academies and planning offices has been aroused to the surrounding environmental grimness, the quandaries no longer of tomorrow but of today, and the possibilities (but not inevitabilities) of worse to come. Ugliness and inefficiency, seen for decades, are becoming more intensified. Escapes are increasingly hard to find, excuses unconvincing, and economic consequences inexorable. A future without a move towards openmindedness in housing attitudes and practice is hard to conceive of— again if for no other reason than the fact of a great increase in the number of people on a diminished supply of land.

Total land area in the continental United States today is about 9.6 acres per person. Around the Year 2000, this is expected to drop to 6.3 acres per person. In a general sense, as previously noted, the United States is fairly well supplied with land when compared to parts of Western Europe; but the location of the open land is the central factor, and by the end of the century, ninety percent of the population may live in metropolitan areas on less than an acre per person.

Furthermore, the citizen is now in a position to see that properly controlled metropolitan growth of this magnitude need not be a continuous sprawl, lacking design structure, undifferentiated in content, and abusive of the landscape. Assuming effective planning at all levels of government (existing and yet to be devised), a metropolitan existence can result in:

Wider and more diverse cultural and social choices as to ways of life, residential environments, and community facilities than heretofore available.

Maximum use of urban facilities in all parts of the metropolitan area in accordance with appropriate development policies on land use and transportation activity. The latter is a key matter. Highways and transit systems are public enterprises. Private land development depends on public access. The location of transportation elements and the timing of their construction can thwart or encourage development.

Sustaining of the core city and especially those activities that have metropolitan-wide significance. This would be accompanied by an increase in the provision of housing for all those linked to the central city by occupation or choice.

As described later, all those conditions are central to any macroview of environmental design. The implementation of such broad-scale planning objectives should afford design opportunities at a scale previously considered utopian. Within this context, row housing can help furnish a satisfying environment in the renewal of existing urban places, in better subdivisions, and in the construction of new towns.

The same conditions that gave rise to row housing in the past are present today, but this time around, the techniques for good design are more readily available. The necessary technical competence is of a high order and includes a level of architectural design that melds site, structure, materials, mechanical systems, and interior and exterior spaces into a single unit which can bear repetition without regret, together with a firm grasp of the requirements of urban development, a sense of community-scale design, and simultaneously, a sensitivity to the varieties of living patterns that a pluralistic society will continue to encourage. Not an easy task, to be sure.

In the direction of community development, the very term "row housing" implies community design and the provision of common space. By definition, row housing is more than the single house on the single site, and the automobile must be handled as part of the housing unit. Site development can then provide private open spaces and at the same time allow easy access to large public areas. The public areas can be connected together so as to help give structure to a larger scale design.

In many ways, the Radburn diagram, America's first contribution to row housing, (see page 47) is an ideal plan and achieves the kind of open-space environment in the city that many families are seeking. Using this plan, parking, servicing, and kitchens are on the automobile side of the house; living areas, private gardens, and common spaces are on the other. An accompanying respect for site and situation will then yield gratifying land design. The items to be considered should be well-known by now, having been set down in 1803 by Humphrey

Repton in his book *The Art of Landscape Gardening.* They are:

the natural character of the surrounding country . . . the style, character, and size of the house . . . the aspects of exposure, both with regard to the sun and the prevalent winds . . . the shape of the ground near the house . . . the views from the several [dwellings] . . . and the numerous objects of comfort [arranged outdoors with their] proper space.

One added design difficulty in contemporary developments that Repton never faced is the boundary line between the individual units and common areas. Several recent projects disclose that a strong, sophisticated architectural arrangement achieved in a $40,000 town house can easily give way to clutter and a lack of privacy between units at the $15,000 price level. In the latter case, the selection of an economically priced enclosure for the private spaces is a prime requirement for visual order and also for privacy. The lack of either one mars the environment.

Common spaces are the key to improved community design. To be effectively used, they must be maintained. Planning boards in the past have rejected some development plans because responsibility for the public spaces was not clearly assigned to owner, developer, or community. William H. Whyte's recent study for the American Conservation Association (*Cluster Development*) points out that new procedures for common-space ownership are now working well.

"There are three basic methods," Whyte writes:

One, favored in New Jersey, is to deed the space to the local government . . . The second method is to set up a special government district, the boundaries of which coincide with that of the development, and deed the land to the district. Such districts are empowered to levy assessments on the residents for maintenance and development of the open space. The third is basically the same except that the vehicle is a non-profit corporation consisting of the homeowners.

Row housing also promotes a community scale of design through facilitating cluster development, and Whyte's comments on the economies of good design, as discovered in his survey, are worth noting: Per-capita costs for sewage and storm drains can be halved in comparison with typical subdivision. Land development costs for roads and other site improvements can be reduced $1,000 per lot. Preserving the existing landscape proves to have strong sales appeal. Utilities can be placed underground. In site planning, "good aesthetics, to repeat, make good economics," Whyte concluded.

Thus social purpose and good design can be mutually served through row housing. As America shifts from a production-oriented society to a consumer-oriented housing market, support for community facilities can be built into the housing areas. Then the recreational, leisure-time, and cultural resources of the common areas will be seen as part of the price of shelter, not something extra but something essential. Reflecting, as it must, seasonal activities and regional preferences in recreation, the opportunity is great for a rich design expression forestalling a cookie-cutter chaos from coast to coast.

Following this line of reasoning, a description of other housing types (high-rise and detached units) is relevant. Variety in physical form must be ensured, heterogeneity in community formation encouraged. High densities can favor these goals by making it feasible for volunteer community associations or private enterprise to support the satisfying environment economically through, for instance, large-scale development corporations which, of course, are possible under current new housing policies.

Reston, Virginia, and Columbia, Maryland, may be the beginning of a trend which heretofore has been largely a California phenomenon: A house and a lot are superseded by a house and a way of life. This concept of housing and environment seems reasonable for a pluralistic society as long as there are many different environments to choose from and everyone is free to make a choice.

Within one type of row housing, the design problem of diversity of content with continuity of form is intense. To avoid the monotony of repeated units, a rich plastic expression is called for. It must be more than an arbitrary change in the color of the paint or a common drapery material. All the elements of architecture must be skillfully exploited: changes in elevation, setbacks, heights, wing walls, balconies, doorways, lights, window treatments, color, and building materials. The individual expressions of the private spaces through landscape treatment add to the total effect, especially when contrasted with the (usually) less embellished public areas.

As site planning can engender heterogeneity and togetherness simultaneously, architecture must secure privacy. Placement of visual barriers between adjacent units is a prime requirement for a successful development. The extension of party walls, fencing, and heavy planting can achieve the desired effects outdoors.

The immediate environment also needs special care. Design of the individual unit cannot be considered apart from design of the site. Marginal land, steep slopes, and unusual terrain can excite solutions which otherwise would be pattern-book architecture. In such cases ingenious circulation systems are discovered, especially to avoid vehicular trespass in pedestrian zones.

Inside the house the sequence of spaces, interior circulation patterns, and location of different activity areas offer as varied a set of solutions as any kind of housing. Remembering the historic aura of poverty that row housing bears, it appears ironic that some designers may be compromising current solutions by skimping on circulation space. There is also need for better design of kitchens, bathrooms, and utility systems.

Row housing can be improved in all ways. To be viable in an urban age, it must continue to advance itself as a key form of shelter for urban life. It should generate design opportunities not otherwise possible, such as ample public open space. In this fashion, environmental design can serve society as an art larger than architecture. And this we must continually remind ourselves is an essential objective: to make the whole better than the simple sum of the parts.

RENEWAL, RESTORATION, RECLAMATION

The stockpile of strategically located open space needed for known urbanization trends may be obtained in part by reclaiming land through urban renewal, by restoring derelict land, or by creatively using foreshore, marshland, and water surface. Each of these measures not only increases the quantity of land but simultaneously affords an opportunity to improve the quantity of the environment—if enough imagination, care, and comprehensiveness in assessing goals and combining professional skills and methods are used.

Typically, urban renewal aims at replacing obsolete land uses. The market for re-use determines the eventual character for the project and Federal funds are usually the means for making it possible. About a thousand Federally-assisted urban-renewal projects are in various stages of execution, each adding large and small increments of open space to the urban fabric. Some, like the Charles Center Project in Baltimore, through exemplary public and private enterprise, result in functional and symbolic civic open spaces—grand, imposing, and enduring. In renewal plans open space can serve as settings for high-density buildings, buffers between land uses, and space for general recreation and amenity. Nevertheless, the fullest opportunities for environmental improvement are not always obtained.

The time span between the technical identification of a renewal project and the final execution of the scheme may run as long as ten years. Often after the urban blight has been cleared away and while final designs and disposition of the land are being carried out, the renewal area has a desolate appearance. Because of legal and funding restrictions, temporary landscape is considered unfeasible, though the very presence of cleared land in raw condition may accelerate further blight in the adjacent area. Perhaps temporary open space designs should be established during the waiting period when complicated, elaborate, and lengthy administrative procedures have to be completed.

Karl Linn's pioneering effort in Philadelphia is a good example of what can be done to find and develop open space, even on an interim basis. Local legislation permitted responsible neighborhood groups to lease tax-delinquent properties, acquired by the City at sheriff's sales, for recreation purposes. Using student designers from the local university, building materials that remained from demolitions, and household junk scoured from basement and attic, Linn established imaginative and unusual playgrounds in which local pride was quite evident.

In the derelict areas of the hinterlands significant measures to repair the damage done by strip-mining and other unconscionable exploitations of natural resources are cause for celebration—however late the party may have started. The application of scientific methods for working extractive industries should prevent further degradation of these special environments. Because they lend themselves to dramatic explanation, such methods attract public notice and support, and rightfully so. In the older urban areas of the country, similar efforts are needed to bring new life to similar if smaller dead areas.

Derelict lands are of two kinds: those passed by in the first waves of urbanization and those abused by industrialization; in neither case do they have any immediate economic return in sight. Most central cities and their fringe areas are pockmarked with acreage whose natural terrain, proximity to obnoxious land uses, or uncertain ownership discouraged development. Weed-covered, dank, and dumpy, they are obvious targets for improvement and reuse. Unkempt, unclean, the derelict land disfigures the landscape and depresses the environmental values all around it.

Most derelict land in and about cities, however, is left over from nineteenth-century manufacturing and processing. This land includes disused quarries and surface mineral workings, abandoned industrial buildings and railroad rights-of-way, spoil heaps of mineral waste, slag-heaps, chemicals, hollows due to subsided earth and filled with stagnating water and refuse. Such derelict lands stand as monuments to indifference, memorials to an era in which land could be easily abandoned after abuse because there was more land elsewhere. No commuter travelling daily by train to any large city could be unaware of the large acres of

New Urban Open Spaces The thirty-three-acre Charles Center urban renewal project added three major plazas to downtown Baltimore. *Source: Rogers, Taliaferro, Kostritsky, Lamb, Architects*

abused and underused land that lie along the rights-of-way—and what they see is only a fraction of what exists just beyond the weed-choked horizon.

Reclamation of derelict land is not inexpensive, but with current earth-moving machinery, hydraulic and soil engineering, costs have reached the point where restoration is as reasonable as it is desirable. Burying wastes, regrading, and detoxifying the soil can be accomplished in a relatively short period of time. In some instances the restored land may be used for rebuilding; in others, only landscape materials can be implanted.

True, derelict land is often spotty rather than pervasive, often removed from everyday environment, bypassed, and left neglected because there is no one to care. A systematic restoration of these lands, however, may produce open space close to the heart of the metropolitan area, relieve the pressure for mindless extension of urbanization into the rural areas, and stimulate ideas about how to avoid the same cycle of deterioration in the future.

Capturing land from the edges of the sea has been practiced since antiquity. In early times the reclamation was done largely for the purpose of gaining agricultural land. Two methods have been used: enclosure of tideland and filling. The first is the oldest method, the Netherlands being the outstanding example. The foreshore is surrounded by banks, seawalls, or dikes made from earth, concrete, or other stable materials. The land enclosed is then pumped dry, and mechanical means are introduced for keeping the area permanently drained. The second method is reclamation by filling. Low-lying land may be overlaid with mud, shingle, or similar material. (Robert Moses used garbage to create the sites for his New York World's Fairs.) When the surface is brought above the highest tides or water table, compaction and drying-out takes place.

As nearby drawings show, much of Boston and East Cambridge was developed on such reclaimed land, including the Back Bay, the Charles River Basin, and the site for Massachusetts Institute of Technology. Further expansion of the city is expected to take place by reclamation of the Fort Point Channel and adjacent land: Designs prepared for the 1975 Boston General Plan illustrate how industrial, recreational, and educational buildings, stadia and outdoor areas would be sited next to the existing expressways on land now underused, obsolete, and presently otherwise unsuitable for construction.

Projects reclaiming land from the water have a way of inciting audacious and visionary urban plans. When serving as planning director in Chicago, Charles A. Blessing proposed a new city fourteen miles long on artificially created islands about a mile off the Lake Michigan shore. Branded as a utopian fantasy in 1952—"Atlantis in reverse" —the idea was justified in part a decade later by Northwestern University's sixty-five-acre reclamation project further up the coastline. Supporting Blessing's scheme, architect Harry Weese described how rock coffer dams could be built to form the perimeter of the islands. Fill could be pumped from the lake bottom as far as five miles offshore. He estimated that land could be made ready for building for about $1.65 a square foot, including streets and utilities—one sixth the cost of inland territory for apartments and row housing.

Far to the south in San Juan Bay, Puerto Rico, a one-hundred-acre market and port facility was developed in two years from swampy and low-lying land. The technical feat was impressive, including as it did rerouting of the river that drained the area. In Florida the drying out of mango groves and swamps to create subdivisions has proven profitable, though occasionally dangerous ecologically.

Wherever ecological areas are subject to reclamation for human habitation, extreme care must be exercised, especially in wetlands. Not only do wetlands serve as feeding grounds of biotic species that maintain the cycle of fish and fowl life, but these precious lands also reduce and prevent erosion of the seashore and serve as run-off reservoirs and as natural stabilizers for ground-water storage and flood-water accommodation. Without these natural stabilizers the extent of flood peaks in particular hydrologic areas could double and triple. Spring floods in New England communities in 1968 were particularly severe in those districts that allowed developers to fill wetlands. Front-page stories carried photographs of worried home owners with flooded basements and little insurance protection. The elected officials who allowed the housing on wetlands were no longer in office, the builders no longer in the community, their profits taken. Banks, householder, and town were left with recourse only to recrimination and new laws to prevent future ecological encroachment.

Thus it is clear that land reclamation calls for more than lip service to good planning: it requires great care and collaborative efforts of scientist, engineer, and designer very early in any project. When the questions of feasibility and justification have been settled and the project is ready to proceed, the same care and variety of disciplines and skills are needed to shape the final design. Without comprehensive planning even expert advice is not enough.

Because it appears to the unknowing as a waste area, shoreland is continually subject to dumping and filling. San Francisco Bay has lost one third of its surface area in fifty years because of indiscriminate reclamation. Population growth in nearby cities led to land filling for industrial sites, subdivisions, recreation facilities, and garbage dumps. Such filling promised large profits to contractors who had to find a way to dispose of large amounts of excavated earth, to truckers who carried the materials, to investors who lent money for the building, and to politicians who benefited from the haphazard jurisdictional control over the entire Bay Area, for as long as a single-interest group could do what it wished in one area, no overall environmental design was possible, and favor-giving and favor-taking was rampant.

Each succeeding fill, revealed reporter Harold Gilliam, threatened "to turn the Bay into a biological desert, to curtail navigation, to diminish the natural beauty of the region, and to affect adversely the weather conditions of the shoreline cities." Responding to public outrage over the environmental destruction, the State Legislature established in 1965 the San Francisco Bay Conservation and Development Commission and gave it responsibility for developing a master plan for the area. The Commission had authority to grant or deny any fill project during the four years of master-plan work. For the first time in American history, a potentially strong environmental-planning commission was backed with funds and power to confront the issues and pressures head on.

The success or failure of the San Francisco experiment will be keenly observed by groups in other parts of the country who face similar situations. The most dramatic of these is the twenty-five thousand acres of tide marsh immediately west of New York City, the New Jersey Meadows. This is the largest and most strategically located open space adjacent to any major city in the world but also the most misused, underutilized, and enigmatic. Crisscrossed by highways and railroads, radio towers, and small industrial sites, the New Jersey Mead-

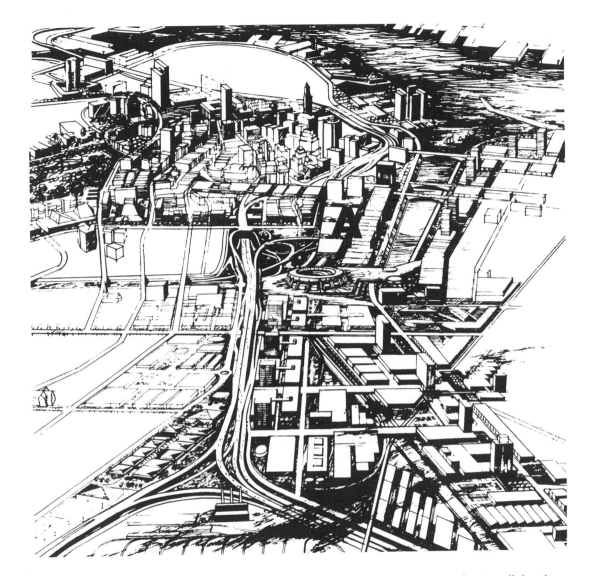

Land-Filling In Fort Point Channel Plans prepared by the Boston Redevelopment Authority call for the filling in of an obsolete waterway and the construction of a sports stadium; commercial, warehouse, and industrial building south of (A) and adjacent to a major existing expressway. *Source: Boston Redevelopment Authority*

ows represents a political and design challenge of unusual proportions.

Urbanization to the east of New York City has filled two thirds of Long Island in two decades. Further expansion is limited. The Meadowlands, always desired but relatively untouched, now take on new importance. Forty square miles of sparsely occupied space within twenty minutes of Manhattan represents the New York Region's last uncommitted land area. It will be either nibbled away in small pieces (and thus lost), or subject to intensive, rational, and comprehensive planning. The size of the opportunity is such that it probably cannot be realized without a special urbanization authority with strong development powers, long-term funding, and highly skilled management. Assuming that each acre can eventually represent $40,000 of construction, a billion-dollar effort can be anticipated. Only at this scale of effort can the appropriate environmental designs emerge, replacing present and averting future pockets of auto-wrecking and junkyards, used-car lots, diners, cheap motels, and third-rate industry, with verdure, intensified human habitation, and urban activity.

How practical? What might the area look like? There are few extant models, but in the visionary work of designers such as Kenzo Tange, a glimpse of the future may be seen. In his plan for Tokyo, Tange proposed that a linked system of transportation and street elements, "the civic axis" be extended into Tokyo Bay. The engineering construction would be the "long cycle" investment. Various structures and groupings of buildings on filled land, or *pillotis*, or in air rights above would be attached to the civic axis. The size of these added components would be determined by environmental conditions and market needs.

A similar concept applied to the Jersey Meadows might include a series of micro-cities, industrial parks, recreation areas, education centers, or health complexes surrounded by parkland. Easy access to the southern New Jersey shore resorts and to New York City would place any resident within a half hour's travel time of excellent outdoor recreation, indoor culture, and the largest choice of goods and professional services in the country. Tange's concept may not be readily applied in New Jersey, but it does expose the timidity of known schemes for the Meadows and the requirements for organizing a massive effort to reach a suitable environmental design for the region's last land frontier.

Parenthetically, the efficacy of any plan for the Meadows is absolutely dependent upon clean air and clean water in the New York Region. The Meadows site is an industrial cesspool today, below and above the surface. However ingenious the engineering of the polders may be, only the replacement of yellow skies with blue, and the oily, grey ooze that passes for Newark Bay with potable water will allow human habitation.

The very size of large projects such as the Meadows raises the pitch of public interest, attracts able professional people and funds, and in turn engenders those options and alternatives that come about with the participation of many skills and talents. But everyday community-scale problems do not seem to attract audacious and aggressive designers and political leaders. Such problems typically must be handled within the context of the community-planning process and accordingly must find a place on a crowded priority list. In this situation, nevertheless, the same elements of care and combination of method again enable environmental design to be useful in two ways: first, to alert the community to what has to be done and, secondly, to illustrate the benefits that may

Portion Of The New Jersey Meadows Looking East To New York City *Photo: New Jersey State Department of Public Works*

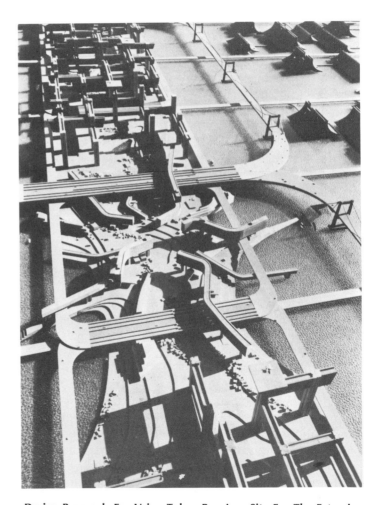

Design Proposals For Using Tokyo Bay As a Site For The Extension Of The City *Source: Kenzo Tange, Architect*

accrue from thoughtful action. The Oahu Development Conference's plan for the Manoa community in Honolulu is a good example of how open-space planning and modest land reclamation enlivened and improved a district's design and helped counteract development pressures that could further diminish the neighborhood environment.

The Manoa District runs from the mountains to Ala Wai Canal near Waikiki. It is a memorable place now subject to intense settlement, especially at the head of the valley where agricultural land is being turned into house lots, and the area around the University of Hawaii, where high-rise structures have threatened to cut off the traditional views and vistas to mountains and seas. With the support of local planners, the community met the problem head-on. Design proposals by the local planning group will avert further destruction of the farmland, allow only low-rise buildings in the center of the valley, and restrict high-density facilities to an urban spine (possibly developed through urban renewal) that follows the course of an existing mass-transit line. Tying all these proposals together is a linear park (the Manoa Stream Trail) which will extend from the rain forest at the head of the valley to Waikiki. The Trail will connect several existing community open spaces as well as reclaim public property, now fenced and underused, for flood-control purposes.

The Manoa plan sums up in one planning effort three ways of gaining significant additions to the stockpile of urban open space: renewal, restoration, reclamation. There are additional opportunities, such as using the everyday wastes of civilization for some purpose other than mere disposal, and perhaps in so doing continuing an art form as old as civilization itself.

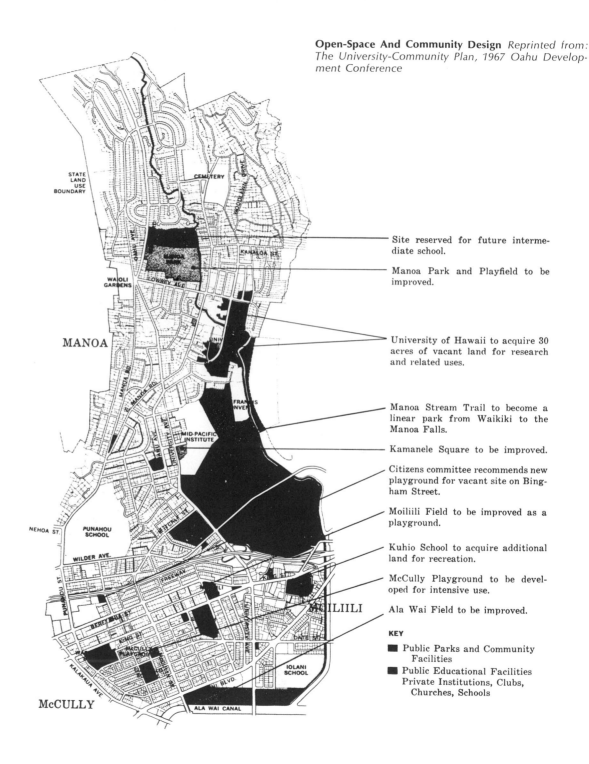

Open-Space And Community Design *Reprinted from: The University-Community Plan, 1967 Oahu Development Conference*

Site reserved for future intermediate school.

Manoa Park and Playfield to be improved.

University of Hawaii to acquire 30 acres of vacant land for research and related uses.

Manoa Stream Trail to become a linear park from Waikiki to the Manoa Falls.

Kamanele Square to be improved.

Citizens committee recommends new playground for vacant site on Bingham Street.

Moiliili Field to be improved as a playground.

Kuhio School to acquire additional land for recreation.

McCully Playground to be developed for intensive use.

Ala Wai Field to be improved.

KEY

■ Public Parks and Community Facilities

■ Public Educational Facilities
Private Institutions, Clubs, Churches, Schools

*The Community Facilities Plan proposes
a standard complement of school, park,
and playground facilities. It also proposes
a Manoa Stream Trail as a new
regional facility.*

STREAM TRAIL

The Manoa Stream Trail would be a
linear park which would extend from the
Manoa rain forest to urban Waikiki.
Beginning at Manoa Falls, it would pass
the University Arboretum, the proposed
new aviary, Manoa School and Park, and
Manoa shopping center. It would join
the new University lands (former war
housing site), St. Francis Convent, the
East-West Center, and the lower campus of
the University. Continuing makai, it would
provide a pedestrian connection from the
University to Waikiki by way of Moiliili,
Iolani School and Ala Wai Park.
Very little private land would be required
for this new combined flood control
and recreational facility.

RECREATION

Two district playfields are located in the
area, Manoa Park and Ala Wai Park. These
playfields also provide playground facilities.
A new playground is proposed for the
former McCully Language School property
and another is proposed by the citizen
planning committee for a site on Bingham
Street at Alexander. Moiliili Field would
be converted to a playground.

SCHOOLS

Most of the schools in the area will find
it necessary to expand as the population
increases. An enlarged site is indicated for
Kuhio School. Land is available for a
new intermediate school adjacent to
Manoa Elementary School.

OTHER FACILITIES

The Plan includes two libraries,
one near the Manoa shopping center
(recently completed) and one to be built
in McCully-Moiliili on King Street.
After the Honolulu Stadium is relocated,
the present stadium site is indicated for
possible University use.

Stream trail at the Manoa Falls

Stream trail through the University Campus

Stream trail through Moiliili to Waikiki

Högdalstopper An inspired design constructed from waste material. *Photo: Bengt Almquist. Source: Stockholm Department of Planning and Building Control*

LANDSCAPING: SHAPING AND SCULPTING

Creating refuse at the rate of 251 tons a minute, the United States approaches the point wherein, as the U.S. Public Health Service warns, "The capacity of the modern city to drown in sewage is more than matched by its talent for smothering itself under a blanket of garbage and refuse." Rather than view remnants and remains as the end of a cycle—production, use, disposal—the public could consider waste as raw material for environmental improvements, especially in light of the continuing scarcity of open space in many metropolitan areas and the risks of further environmental decay because of haphazard dumping and waste disposal.

Meat packers have made money by packaging everything but the squeal. Petroleum companies extract many by-products when changing crude oil to highly combustible fuels. Similarly, experimental devices have been invented to compress, atomize, and burn garbage to produce salable gases and residues. Other kinds of bulk waste cannot be treated and transformed by known technology, and it is this stuff that challenges the environmental designer.

A commendable example of creative efforts to handle this problem is Stockholm's use of material excavated in the preparation of sites for urban construction, and wastes created by building demolitions. Rather than spend sizable sums to cart this material miles away or to spread it insensitively over the landscape, Swedish engineers and landscape architects used the debris to form man-made hills. The sites chosen were those closest to the area of greatest building activity and least useful in their present form for future building, agriculture, or recreation. The first of these hills, near Hogdalen, a district of rapid urbanization, covers fifty-five acres. Present estimates are that 3,250,000 cubic yards will eventually be brought to the site. Those using the site to dump waste material are charged a fee of approximately 40 cents a cubic yard. The income (eventually totaling over a million dollars) is used to prepare the land, operate the dump day and night (which in winter comes early in Sweden), build roads, and complete the landscape when the dumping is finished. Högdal's Hill is designed to become a major recreation center, summer and winter, for the growing suburban communities nearby and can also be reached by the Stockholm subway. Out of waste and at minimum cost, an important open space has been added to the metropolis.

The Stockholm experiment has succeeded in several respects: economy, land utilization, and environmental design. The Stockholm planners did not simply pile the debris into small copies of nearby hills. As landscape architect Karen Anderson wrote in her critical evaluation of Högdal's Hill:

Man has already made such an impression upon the land with his buildings and roads and urban parks that he should not attempt to imitate nature when he builds (such) hills. Instead, he should create a piece of sculpture like Högdal's Hill, a thing of planes and angles, emphasized or accented by trees and shrubs, a positive element in the suburban cityscape.

Högdal's Hill is thus an art form, part of a tradition that shapes, sculpts, and scrapes land into aesthetic and symbolic expressions well beyond simple engineering, an aspect of environmental design that has precedents and antecedents that go far back into history. These arts are as applicable today as they were in ancient times, and our utilization of them can animate and enliven otherwise humdrum artifacts: for environmental design can go beyond the mere safe and sanitary and add a dimension of enjoyment and aesthetic satisfaction when establishing urban landscapes. The distinction between preserving the natural ecology and landscapes of human contrivance is an important one. Environmental design concerns both conservation and development, but the latter is rooted more in art than in science.

Man has continually tried to create an art object out of the landscape. Of the historic examples, molding of the earth seems to have preceded carving at many scales of design; and of things molded, sculptural mounds—often forming defensive positions or places of the honored dead—are among the oldest examples of environmental art. Their remnants can be seen today in all regions inhabited by man. Frazer's *Golden Bough* describes several: the Hill of Lloyd at Kells, the Great Barrow at Pelops near Olympia, and the epic Babylonian mounds of Semiramis. Some have suggested that the pyramids evolved from these simpler forms, as Jean Cocteau writes, ". . . in a murmur that only a delicate ear can catch." There is logic to the assumption that a common primordial instinct produced such archetypes centuries apart, continents removed, in civilizations significantly different.

Technology effectively controls the precision of this environmental art but not necessarily its size. Uppsala, Figsbury Ring, all such earth sculptures conjure up actions willful and art simple. In the United States the Indian mounds in the Ohio and Mississippi Valleys are comparable to those anywhere in the world, both in scale and image. In Wisconsin, totemic effigy mounds can be recognized as mammals, birds, and reptiles. One eagle has a wingspread of one thousand feet. The Great Serpent of Adams County, Ohio, is 1,348 feet long and has jaws seventy-five feet apart. Recent scrutiny of the contents of several earth sculptures disclosed beautiful objects of carved stone, metal, and shell. Carbon-dating and comparative analysis have laid to rest the romantic notion that these mounds were the works of a special breed now long extinct. They are actually the artifacts of Stone-Age Indians, ancestors of Hiawatha and Pontiac, and are impressive because of their primitive origin.

Gutzon Borglum's work at Mount Rushmore and Stone Mountain (1900 A.D.) follows the tradition of carving the earth seen in the cliff temple at Abu-Simbel (B.C. 1300) —an old art but one less primitive than earth sculpture. These carvings and rock-hewn temples around the world—the great ancient architectural panoramas of India and South America, for example—are part of the spectrum of environmental arts appropriate at the urban scale. Other examples of melding earth design and natural environment include the Anacona Pueblo in Arizona and the desert city of Ain-Leuh in the Central Atlas Mountains. The latter city commands attention by the way wind-driven sands are held off by a series of interlocking domes, vaults, walls, and courtyards so well set into the ground that the joining of earth to building and town to desert is difficult to perceive.

In Macchu Picchu, as well as in Mayan and Aztec cities, one sees the art of urban form and earth carving rising from a subtle appreciation of natural conditions. In contrast, contemporary North American technology —with immense energies and machines for moving earth—seems unskilled, insensitive, unsure, doing with thousand-horse-power engines little more than pick-and-shovel exercises. The environmental disasters that hang from man-made cliffs or are plopped on man-made terraces in Southern California are the results. To create salable lots, developers cut shallow shelves into the hills, denuding the landscape of topsoil and vegetation and carving awkward scars. In Garrett Eckbo's words: "The trinity of equipment, efficiency, and economy allowed no wasteful variations that might have salvaged the amenities nature was building up for years." Loose fills, unskilled cuts, and heavy rainfalls have several times resulted in mud floods and landslides that wiped out a number of houses. Those sites least responsive to basic design principles have been the hardest hit, proving again that designs which look wrong in natural settings tend to be wrong in their interpretation of what is practical as well as purposeful.

When properly handled, the art of earth shaping, sculpting, and carving can be applied to all scales of environmental design, from backyard garden to city park to entire cities. The application of technology to a problem need not diminish the possibility of special aesthetic effects, and the resulting forms can be enduring environmental designs, rooted in art as well as in function.

AN OVERVIEW OF OPEN SPACE

The planning of urban open space should be directed to these three objectives: the immediate protection of rare and unusual natural environments which are now subject to damage and destruction because of urbanization or ignorance, the enlargement of the stockpile of open space for recreation and aesthetic reasons but without ecological degradation, and the increased use of open space as a device to give form and animation to urban settlement. None of these objectives is really separable from the others, and together they form the most needed environmental design measures of all. This section elaborates on the need and illustrates new environmental-planning techniques to meet it.

Biologists suspect that Man's demand for open space is a natural instinct, related perhaps to concepts of territorial rights and community formation and colonization, but the pursuit may be a cultural phenomenon as well. The search for open space may lie deep in the American ethos—we may be driven to finding new land when the constraints of existing locale are too great to bear. While geographically the frontier has vanished, Americans today still find ways to satisfy an instinctive desire for open space and the aesthetic satisfaction it offers. As in the past, however, the immediate satisfaction of that desire is not of itself insurance that the environment will have been materially improved or even saved from harm. Planning with science as well as art is therefore necessary. The present state of the New England forests is a case in point.

After centuries of extensive commercial timbering, New England still remains one of the most heavily forested areas in the United States. Historically, corporations and small farmers have cut most of the marketable timber, often leaving a scarred and decrepit landscape. In the past few decades a new ownership pattern has emerged and a new environmental problem in which the impulse for conservation without technical knowledge creates as dangerous an ecological situation as did earlier economic exploitation of natural resources.

Today seventy percent of the total acreage of forest land in Connecticut, Massachusetts, and Rhode Island is held by private owners averaging about fifty acres each. "The new woodland owner," reports the Federal Reserve Bank of Boston, "is a city worker who commutes or a professional man or executive who has acquired land in the countryside for recreational or aesthetic purposes. These owners generally do not have an intense interest in forestry or forest

practices but use their land as a retreat from the pressures of city life or own it for speculative purposes."

Preservation-oriented, a majority of these owners oppose timber cutting, and as a result, less than forty percent of the annual forest growth in New England is now cut. The new owners' attitudes, commendable on the surface, are ecologically erroneous. Faculty research in the Forestry School at the University of Massachusetts indicates that a forest allowed to grow and mature without timber cutting also ends up a biological desert. Ecologically, the values the uninformed owners are seeking can not be met without some form of cutting. As a result, extensive forest lands are in biological jeopardy. Recognizing that aesthetic despoliation by loggers was equally intolerable to the new owners, the University faculty has advocated an educational program for both owners and loggers involving selective cutting for economic, conservation, and aesthetic reasons.

In examining the land resources of future metropolitan areas, Lowdon Wingo, Jr. found that the problems of open space vary with the distance from the center of the region. "Open space will be in short supply only in the inner parts of the region, while the outer edges will be small urban islands in a sea of open space." The actions to be taken inside the metropolis are those commented on earlier—preservation and better utilization of existing open spaces and a creative search for additional acreage—defensive measures. On the fringes, where extensive land is not yet developed, aggressive measures are called for to protect and enlarge the stockpile of open space and use it to enhance urbanization. The most effective of these may be the land-planning techniques used in Ian McHarg's system of ecological determinism.

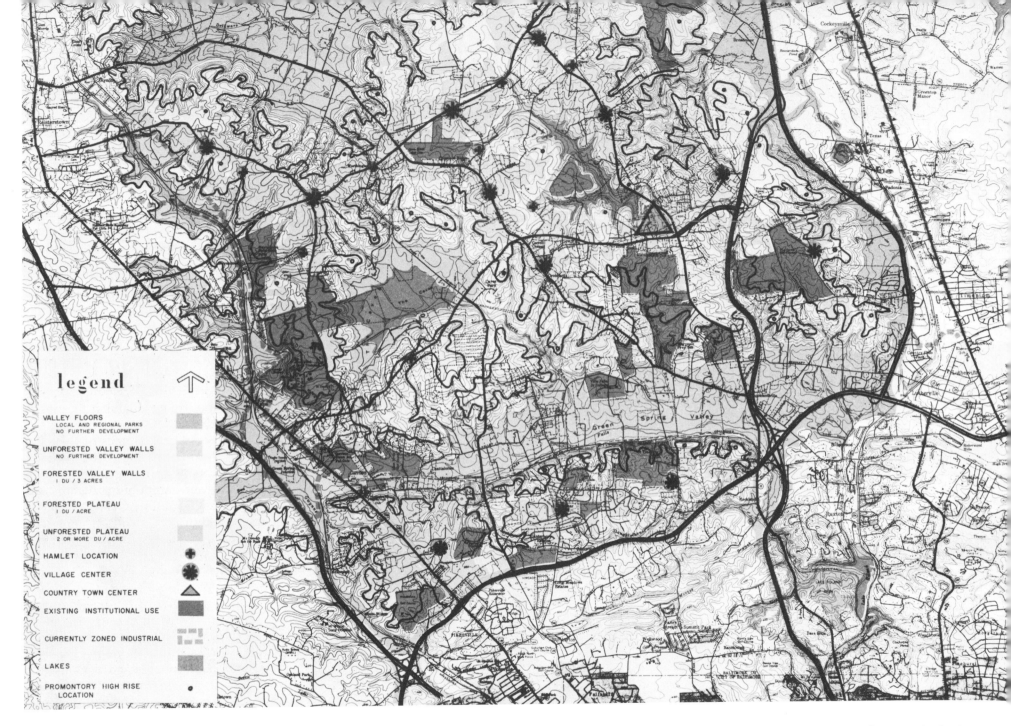

legend

VALLEY FLOORS
LOCAL AND REGIONAL PARKS
NO FURTHER DEVELOPMENT

UNFORESTED VALLEY WALLS
NO FURTHER DEVELOPMENT

FORESTED VALLEY WALLS
1 DU / 3 ACRES

FORESTED PLATEAU
1 DU / ACRE

UNFORESTED PLATEAU
2 OR MORE DU / ACRE

HAMLET LOCATION

VILLAGE CENTER

COUNTRY TOWN CENTER

EXISTING INSTITUTIONAL USE

CURRENTLY ZONED INDUSTRIAL

LAKES

PROMONTORY HIGH RISE
LOCATION

Preserving The Landscape And Guiding Growth A portion of the proposed land-use plan for Green Spring and Worthington Valleys showing the ecological zones and proposed urbanization pattern. *Reprinted from: Plan For The Valleys. Green Spring and Worthington Valleys Planning Council, Inc. Wallace-McHarg Associates, Planners*

63

McHarg persuasively argues that modern ecology can be the basis of city, metropolis, and megalopolis design. Noting how the eighteenth-century English landscape architects rehabilitated the countryside and produced enduring works of art by using and recreating replicas of natural environments in ecologically correct combinations, McHarg suggests that a similar approach today would yield improved human habitations and design forms, derived from a recognition of natural conditions and as aesthetically pleasing. His system involves a three-step planning procedure: inventory, analysis, policy making.

Step One is the preparation of ecological inventories and interpretive descriptions of how the various ecosystems (total ecological areas) are affected by Man and by each other. McHarg maps the large environments such as coastal plain, piedmont, and highlands as well as such discrete environments as sand dunes, pine barrens, and estuarine ecologies. These inventories may be accompanied by identification of factors which can preserve any ecosystem, improve it, or remove it. The detail of observed interdependency among ecologies varies in accordance with the scale of the inventory, but a generalized ecological mapping just by itself advances the traditional land-use diagram from an archaic designation of general open space to a strategic recording of sensitive planning data.

Step Two consists of analysis of the information collected and attribution of value to the various ecologies. The object is to arrive at decisions for intervention or nonintervention and the location, character, size, and timing of such actions. For each ecology it is possible to establish a range of human land uses that is appropriate to the natural environment. In instances where the match-up between Man and nature is un-

avoidable but cannot be made perfectly, problems of environmental damage can be anticipated and ameliorated.

McHarg says there are land-use values discernible after the inventory and analysis: productive, performing, negative, and intrinsic. Productive environments are those which can be intelligently utilized for economic gain: agriculture, forestry, fisheries, extractive minerals, housing, and so forth. Areas with performing values are those lands which are needed for water purification and climate, drought, water, and air control. Negative-value zones include lands subject to earthquakes, hurricanes, floods, subsiding, and other natural disasters where occupancy by man might lead to costly destruction. Intrinsic lands are those which neither produce, perform, nor endanger—such as land having scenic, scientific, and educational value.

Application of these four main values to the inventories brings out what the possibilities and prohibitions for development might be. The costs and benefits of selecting one set of values over another are brought into the community-planning process, Step Three. Both private and public programs and interests can be judiciously balanced against the background of ecological facts strengthening typical procedures for land-planning decisions.

Ecological determinism offers a frame of reference for intensive scientific investigation into the quality of the environment. The relationship between density, toleration of sensory overloads (noise, air pollution, etc.), and human habitation can be expressed with measurable standards and new insights can be gained concerning the desirable relationships between Man and his designed environment. Procedural problems in land-planning can be lessened. For example, using public health as a basis for land-use controls may be sanctioned constitutionally and may prove easier than the usual method of using the arbitration of traditional views of common-law property rights. Thus ecology can give incentive for new kinds of ordinances and controls—strong, rational, and effective—and these in turn can further protect and enhance community development and community design, whether in a relatively circumscribed area or on a state-wide or other large scale.

Highway construction opens new lands for urbanization and often simultaneously subjects such areas to ecological dangers as well as ugly urban sprawl. This was the case in Green Springs and Worthington Valley, seventy square miles of threatened farmland near Baltimore, where McHarg and his associates used the concept of ecological determinism to good effect. Ecological studies pointed out the best locations and densities for a variety of housing types and in turn produced a logical hierarchy of communities—hamlets, villages, and country towns. The emergent design comprised neither man-made aesthetics nor nature and gave clear design direction to the urbanization pattern of the area.

The utilization of ecological factors for design effect is equally valid at a large design scale as in the State Plan for Wisconsin. The origins of the ecological approach to the Plan lie in Wisconsin's historic concern for conservation, the state's strategic location for serving the recreational needs of millions of people, and the sensitivity of landscape architect Philip H. Lewis, Jr.

Less than two percent of Wisconsin's land is urbanized, and the rest contains the most accessible and attractive scenery in the Midwest. Over five million acres of public land are open for outdoor recreation, and sizable additional acreage is privately operated for recreational purposes. Recreation is Wisconsin's billion-dollar industry. Vacationers come to the state from all over the United States, Canada, and Europe. In 1960 over fifty-six million recreation visits were made. By 1980 this is expected to grow to 123 million visits. Most of the nonresident visitors will travel no further than seventy-five miles to reach the state line. With continuing growth of population just outside the state's boundaries in Illinois, Iowa, and Minnesota, improved transportation and highway access, and growth in leisure time and incomes, Wisconsin is presented with unparalleled economic opportunities for its recreation industry as well as an immense obligation to plan for the future.

In placing recreation planning within the concept of state planning, Lewis, like McHarg, set up a method of inventorying environmental resources, projecting needs, and establishing policies to guide growth. Emphasis was given to the qualitative dimensions of the landscape, especially those Lewis thought were appropriate for an American culture that had become more urban than agrarian in its values and outlook.

Recreational resources in Wisconsin were divided into those possessing intrinsic values and those possessing extrinsic values. The former referred to natural elements, the latter to man-made facilities. For example, the topographic-recreational resources inventory included steep slopes and mineral-ore outcroppings under the heading of natural resources; and ski trails, nature trails, shelters, and cabins under the man-made facilities list. Many recreational areas exhibited both intrinsic and extrinsic qualities. Protected and developed wisely as enjoined environments, such areas can give to present and future inhabitants attractive, functional, and healthy environments. The full list contained over 100 items.

After making an inventory of these resources in a test area, Lewis concluded that the important resources that offered special recreational and environmental qualities to the Wisconsin landscape were water, wetland, and strong topographic features. These elements could be shaped to form elongated patterns or environmental corridors. Within these continuous patterns could be found seventy to eighty percent of the other recreation-resources inventory items clustered in sub areas or in isolated locations. Many of these inventory items were clustered together because the corridors paralleled the routes taken by Indians and settlers in opening up the state. History and nature reinforced each other to yield a distinctive image in the quality of the corridors.

Lewis believes that with minimum care and good planning, the corridors can be used to give design structure to the state and subregions. The corridors can separate urban land uses or tie them together or serve as edges and boundaries to natural and man-made environments. Strung together, the corridors form a skeleton with a well-defined purpose and a design character that can be substantially maintained over a period of time.

Visual contrasts and variety may be the most widely sought scenic values in the future, especially as sightseeing, pleasure driving, and the human desire for novelty and change rise in frequency with more leisure time available and increased personal mobility. In anticipation of these trends, the Wisconsin planners proposed "Heritage Trails"—a series of highways and byways that used parts of the environmental corridors for access to and circulation through special tourist areas. These trails capitalize on the economic and other assets of the environment and discourage haphazard tourism.

The Wisconsin environmental-corridor studies at the subregional scale also disclose distinctive landscape patterns. At the micro-scale such things as climate, water, soil, vegetation, and man-made elements show certain affinities for each other, as well as contrasts. A village stands out from the countryside, a tree-lined river from the housing along the banks. The preservation and enhancement of distinctive qualities of the smaller areas do much to overcome the monotony and boredom that fill too much landscape and urbanscape.

The use of open space to constrain and guide urban development for reasons of public health, economy, and pleasure has historic precedent. The recent examples of theory and practice above are part of a continuous search to find ways to design human habitations that satisfy many objectives simultaneously. Open-space designs are especially prized because they help find "a place for man in nature, and nature in man." They create a sense of place and give order and comprehensible dimensions to the environment. But there are other ways to achieve this aim as one can observe in the techniques by which design structure can be established.

Unregulated development eliminates all natural beauty.

Selective development preserves river landscape.

Unregulated development in corridor pattern eliminates natural river landscrape.

Clustering houses preserves river.

Design Concepts For Landscape Corridors A visual interpretation of Philip Lewis' proposals for treating state-wide natural resources in Wisconsin. *Source: Connection Magazine, June, 1965*

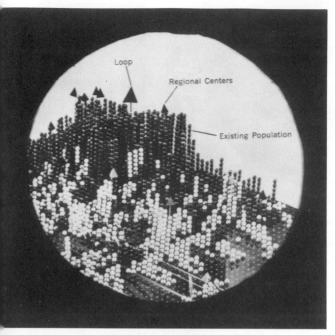

Portion of a Population-density Model. *Source: North-eastern Illinois Planning Commission*

Megalopolis at Dusk. *Photo: Richard P. Dober*

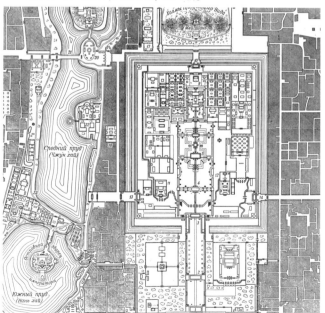

Sculpture by Beppe Sesia. *Source: Art International*

Plan of the Imperial City, Peking, China. *Source: A. V. Bunin and M. G. Kruglova, The Architectural Composition Of Cities, Moscow, no date*

Air View, North African Village. *Photo: George Gester*

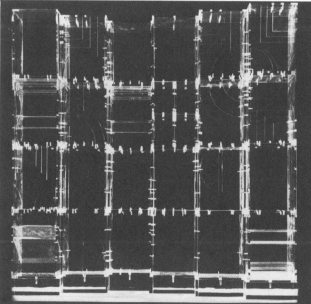

Sculpture by Louise Nevelson. *Photo: Pace Gallery*

DESIGN STRUCTURE

DESIGN STRUCTURE

Observable in nature as in art, design structure is an organizing principle through which parts and pieces are brought together into a unified shape and form. Visual analogies more than words best show the universality of design structure. In environmental planning design structure is art with social purpose. Its objective is the establishment of visual order in response to the technical requirements of developing the physical forms of community life and in a manner that allows, over time, accommodation to diversity, increase, and change while maintaining coherence in the whole.

This section describes the creation of design structure and the consequences that follow. In the examples and discussion herein, there is ample evidence that the idea of design structure is not only a contemporary device for planning the environment but also an approach with historic antecedents and precedents. In the beginning design structure may be seen as adventitious events emerging from natural conditions and symbolic acts. Later, design structure arises from the functional requisites of defense, safety, efficiency, economy in moving goods and people, health and administrative demands, legal controls, and the aesthetic desires of large numbers of people sharing a common environment.

Design structure is capable of being recorded, communicated, tested, and reproduced; and its discernment is an important factor in any analytical approach to theories of urban form. It is apparent in all scales of design. The physical elements which constitute design structure can be related to community planning goals and objectives and thus can be manipulated in size, location, and content to meet these ends. This manipulability is inherent in any elements that may be said to give design structure. While the manipulation may be intuitive, arbitrary, and simplistic as in the past *or* rationalized and deliberate as in recent times, the act of creating design structure involves choosing, shaping, and directing.

These definitions and descriptions raise the question: Is design structure an art form? Hans Blumenfeld has stated that

a city is not a work of visual art. It cannot be perceived from one point and at one time. Its mental image can evolve only as a sequence of perceptions. In this respect it is akin to the products not of the 'spatial' but of the 'temporal' arts, to works of music or literature. But a symphony or a novel is the work of one mind, created in a definite span of time; once completed, it remains unchanged. Not so the city . . . the city is a historical process; change is its very essence.

There are, however, forms of contemporary "spatial" art—painting, sculpture, and architecture—which are sequentially perceived, which change over time, and whose completion is dependent on individual and group participation. It is not inconsistent with Blumenfeld's insights to suggest that new definitions of "spatial" art may be applicable to design structure, making the city a visual work of art. The concepts of continuity, change, and participation in the New Spatial Arts contribute to such new definitions and are observable in the work and attitudes of the Action Painters—Jackson Pollock, Clyfford Still, and Mark Rothko, among others.

In his calligraphic presentations, Pollock constructs a space which has in A. Ossorio's words, "static qualities of perpetual motion." The design form provokes the observer into making for himself images that may or may not have been originally intended. Pollock's achievement in structuring a spatial area, "limited and limitless," is a further advance along paths opened up by the Cubists. Cézanne with his hatch-like brush strokes, unclosed contours, changing perspectives—

elements derived from Pissarro—was able to reduce the deep space of the paintings of his predecessors so that his pictorial design was free from perspective and structurally integrated enough to allow him to merge foreground and background. Cubism broke up the object, just as Impressionism broke up light. Pollock is a link between these aesthetic concerns and the redefinition of time, space, imagery, and permanency as seen in the light patterns of Wilfred Owen, the sculpture of José de Rivera, Yves Tanguy's kinetic designs, and the multiple assault on the sensorium evidenced in experimental sound and motion pictures.

A common attribute of these latter art forms is that, like the city as a visual art form, their physical characteristics cannot be adequately expressed through three-point perspective.

While full description of the parallels between design structure, the city as an art form, and some of the New Spatial Arts, requires proper investigation, other points of similarity seem evident.

Rothko and Still speak of "eliminating all obstacles between the painter and the idea and the idea and the observer." Further, "an art may be readily perceived without communicating anything else but its own being." This admittedly "elusive quality of art" may flow from a single aspect of the daily environment or emerge as an effect whose value is gained by being greater than the sum of the parts. In this sense some forms of design structure may be "pure" and thus have little in common with the traditional ways of making three-dimensional environments.

Wilhelm de Kooning tells of an enchantment with the manner in which all that is within the picture frame "keeps changing in front of us." The structure of design is set, but the completion—the filling in—is left to the observer. In the art of the collage and montage (again forms of contemporary arts), multiplicity without redundancy, organized and defined, gives visual order and opportunity for change through chance occurrences. The literal messages of the individual scraps that Kurt Schwitters gathers into visual order have limited meaning to one who doesn't speak the language inscribed thereon or one to whom the partial pictures are not everyday images, but the design endures.

The inherent aesthetics in design structure evolves in part from these new concepts of spatial art, not from Renaissance picture-making. The charge may be leveled that some of these recent forms are nihilistic, not readily communicable, obscure, and logically inconsistent. Because the burden of explaining their effect is placed on words and static images and not on their perception over time, they may be erroneously labeled nonspatial arts. Whether or not they are merely fad or fashion, they do contribute at least two insights into environmental design: Continuity and change can exist in a spatial art, and art-former and art-user may be part of a single process, which requires both to complete the art form.

Design structure has similarities to organic theories of urban form, planning, and design. Professional literature is filled with theories of environmental aesthetics concerned with contrasting parts changing and adapting yet coherent and sustaining. As Gilbert Herbert reminds us, all such ideas are metaphorical as they apply to man-made urban environments. Each is a variant of a single theme: The designed environment is not a monolith. These organic theories are useful to the extent that they help in the classification and description of physical elements that comprise the environment and their relationship to one another or deepen a philosophical understanding of the promise offered by a comprehensive view of the benefits of design structure.

Lewis Mumford says that organic design may not necessarily be preconceived but instead may be an intuitive seizure of "an accidental advantage," prompting a strong design feature, which a priori planning "could not anticipate and in all probability [would] overlook or rule out." The purposeful blending of site, situation, and mixture of activities through adaptation to sequential opportunities in community development may lead to a design as distinctive and as firm as a plan predetermined in all its critical parts. The same organic happening may produce similarities in form but differences in meaning. A city gridiron set by an ancient priest has heavenly import. The same design imposed by a site engineer is simply a way of creating the largest number of real estate lots possible.

The architects Frank Lloyd Wright, Walter Gropius, and Richard Neutra show a cosmological bias in their use of the organic. Neutra speaks, in almost mystical tones, of a belief "in the unity that underlines the phenomena in which we live." Gropius sees a unity of "land, nature, man, and his art"—an ordered universe. To Wright there "are immutable circumstances" in which "all things are in the process of flowing in some continuous state of being."

Wright, Gropius, and Neutra lack precision in their analogies; but there are ordered natural forces at work, as can be gathered from contemporary sciences of cause and effect in many fields. Regional economists observe forces shaping the form of the metropolis and comment that the pattern of the designed environment is partly a reflection of market values and choices and individual selection. The behavioral sciences are beginning to close in on predicting human responses to environmental conditions, physical as well as psychological.

The biological sciences stand now where theoretical physics stood on the advent of nuclear fission, seminal to the same degree and the ecologists see affinities and analogies between Man and his community and the zoological and botanical worlds. As F. Fraser Darling has stated, the beauty of environmental research in this area is "its bigness and readiness to cross boundaries, looking into less well-understood fields than one's own and finding links, correlations, comparisons, contrasts and differences of exquisitely fine scale and subtlety." Ecologist Pierre Dansereau's Law of the Inoptimum, simple and demonstrable, is of fundamental importance to anyone concerned with the designed environment: "No species encounters in a given habitat the optimum conditions for all its functions." Be that as it may, there is a growing and far-from-complete body of knowledge that hints at a web of life where relationships are persistent and immutable.

On the other hand, fear of a mechanical environment with all the gears enmeshed by some large, ordering, but extrahuman force is ungrounded. Human intervention in the environment still can be exercised and in substantial ways. Innovation and invention also can play a role. Accordingly, there is hope that a unified approach to environmental design lies on the horizon, drawing on many sciences and many arts. This hope conforms to the anticipations of the leading philosopher of organism, Alfred North Whitehead: Differentiation leads towards unity.

Within this context of contemporary aesthetics and all the uncertainties of the organic analogy, the concept of design structure, as defined at the outset, holds up and can be stated in a second way: Design structure is a human artifact composed of willful and random acts finally and intelligently directed to aesthetic social purpose. By examining the historic origins of the elements that make up design structure one can see how and why our urban places took on the form and shape they possess today.

NATURAL AND CULTURAL INFLUENCES

Design structure may be created by establishing strong centers and manipulating lines of circulation and transport, as illustrated later in this section. Underlying these designs are natural and cultural conditions that may constrain the designer's dreams and ambitions.

The available evidence of the earliest forms of the designed environment comes largely from archeological reconstructions of settlements that emerged from post-Pleistocene changes when man began to increase his subsistence efficiency. The means of increase varied in each environmental niche, especially at the points of contact between gallery forest and savannah, land and sea, lake, mountain, and river. R. J. Braidwood and G. R. Willey note that this incipient cultivation and domestication of themselves had "little immediate effect on general cultural development." But without a system of surplus food production and distribution, urban places could never be sustained.

G. Sjuberg lists the salient factors in determining the location of prehistoric urban communities: environment, technology, economic and power structures, and communicable cultural values. He believes environment was not of itself an active agent but an ecological condition that allowed surplus in food supply, exchange, and easy transport. It permitted an accretion of social and technological skills and population growth. In turn, Sjuberg writes, population increases encouraged technical development: large-scale irrigation works, animal husbandry, metallurgy, and other innovations. The exchange of surplus required specialization in occupation, including the management of these skills. The transfer of this knowledge to others engendered class, caste, and a hierarchical society.

The salient physical features of this social phenomenon, reflecting varying patterns of life style and approaches to form-giving are universal throughout most early communities.

The archaic urban place is composed of cellular units and, around the whole, a prominent boundary—wall, moat, or similar device. It serves for defense but also regulates the activity of merchants and has symbolic importance.

In his beautiful essay, "The Idea of a Town," Joseph Rykwert emphasizes the efficacy of this outer boundary in design structure. He touches the quick when he notes that the "structure and configuration of a town is related to the way its inhabitants consider and treat it; and further, that the state of town planning at any given time is connected intimately with the way in which most town dwellers regard their home ground." The ceremonial ditch, the plowed furrow that became the wall, the sacred intersection of planes and coordinates which gave direction to streets, all these added degrees of certainty to the form of the primitive urban environment.

To the ancient Chinese, a city without a wall was as inconceivable as a house without a roof. C. Z. Chen explains that the very word for city, ch'eng-tzu means fortified city wall and a market. Defended or otherwise, any settlement without a wall was not a city.

A Confucian sense of symbolic order determined the design structure in other aspects of early Chinese cities, making some interesting parallels with elements of our own time and culture. In ancient China strength and dignity were measured in terms of the number of chariots a kingdom possessed. The size and decorations of the chariot marked the status of the owner. These distinctions were reflected in the street layout of any city. Peking, as home of the Emperor, had the most spacious streets, which not only permitted the chariots to pass easily but added the dignity that was due a special city. Similar symbolism led to separate walks for men and women, separate precincts for the ruled and the rulers.

Symbolism affected the placement of public buildings. Confucius' *Book of Etiquette* made these specific suggestions: In the middle of the city should be the palace, on its left the Temple of Ancestors ("where lies the way of humanity"), on its right the Temple of State ("where lies the ways of the world"), in front the Court ("where lie faith and loyalty"), and at its back the Market ("where lies profit").

To the Pueblo Indians the surrounding environment, rather than a manual of artificial rules, directly gave spirit and form to the creations of the people. Each artifact created had the intention of finding and fulfilling a harmony with nature. The configurations of river, desert, and mountain found their parallels in an imitative design structure. Paul Horgan describes how: "The cave became a room. The room became part of a butte. The butte joined others like it, resembled a mesa, terraced and stepped back." The Pueblo town resembled the landforms around it. The kiva, or house of worship, was itself like a small butte with a flat top, a typical sight in the New Mexico landscape.

One of the largest of the Pueblos (Kin Tiel) shows a remarkable degree of regularity in outline and form despite the obstacle of topography. A sense of form following function is also clear, for the symmetrical parts are shaped by the location of a wash which, impounded, provided water to the community. Considering the size of Kin Tiel and the primitiveness of its technology, the evidence and power of a managed design seem obvious.

"Man builds what nature permits" neatly summarized the inherent characteristics of design structure in early communities, according to one theory; and geographers and anthropologists have had their own varying notions about how design structure develops. The classical theory of civilization in the Nile Valley holds that primitive man learned to reap before he learned to sow. He observed the annual flooding and the resultant wild cereal crops. From this rudimentary initiation into cultivation grew irrigation projects and towns.

For protection against the floods, Man first occupied the higher elevations in the Valley and for further relief built artificial banks around the village. In time the idea was extended to embanking the river itself, leaving an ingress for water. Later, to help keep water in the basin, embankments were constructed perpendicular to the river; towns followed, then cities.

Natural requisites dictated road locations and settlement patterns. The eventual elaboration of the resulting system throughout the Valley was dependent on a strong centralized government. Thus a civilization was born to protect elaborate civil works, for without defense against surrounding forces, an upstream basin could endanger the riparian activities of people downstream. Thus G. Hamden notes the Latin *rivalus* (rival) is a derivative of *rivus* (river). H. Frankfurt, S. Gidieon, L. Mumford, and others have written extensively about evolving arts and rituals that parallel the above sequence of community development. It is sufficient to say that the many manifestaions of technology and culture developing in continuity through two thousand years of history in ancient Egypt were all influenced by the natural situation.

The French geographer Pierre Lavedan believes that the critical confluence of natural and man-made elements determines design structure in the emerging form of an urban place quite early. In his view there are common-sense relationships between these natural and human elements that are grasped early and lead to self-generating designs. A river sets the location of a main road, either because the road runs parallel to it or crosses it at right angles. The main road sets up a fish-bone street pattern, one branch of which leads to the "monument"—a castle, monastery, town hall, market place, or court. The monument setting becomes the central place, the point of major destination, around which the town grows.

The central plaza, square, or extended main street is large and imposing, if only in contrast with the narrow lanes and streets that converge on it. It serves as an outdoor meeting place for public pronouncements and communication. Mass media being nonexistent, the size of this place is in part a reflection of the number of people whose status and importance require them to come together for the above purposes. In imperial cities, where safety and security are not dependent in part on narrow streets (which impede passage of invaders), the central spaces and connecting streets take the form of large and straight boulevards and thoroughfares.

Whatever political and social forces may be at work, the size and character of the center are further affected by the steepness of terrain, availability of water, and other given site conditions. Central places shift location when natural conditions or growth require this change.

Observing the same phenomenon, another French geographer, Georges Chabot, concludes that functions create structure—that human activities are the dominant form-

Mill Town The plan of Lowell, Massachusetts, 1852, reveals the influence of place of employment and social standing on the design of the community. *Source: Harvard Graduate School of Design Library*

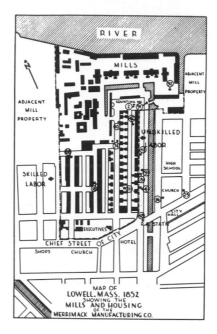

Leisure And Community Design Portion of residential area, Stafford (new town), Virginia. Site planning and building design reflect emerging attitudes about how to shelter a life of leisure in a landscaped suburban setting. *Drawing: Paul Rudolph, Architect*

La Perla Reconstructed A splendid design concept for rehousing the slum-dwellers in Old San Juan. The scheme imaginatively responds to the special needs and desires of the people living there. *Source: Jan Wampler, Architect*

givers. Through a typological study of singular cities—military town, commercial town, resort town, he discovered a variety and diversity in dimensions and locations of such elements as streets and squares. The differences reflected more than anything else the town's reason for being. Thus a style of life created its own unique design forms.

Anthropologist Conrad M. Arensberg notes that the correlation of community form to community culture is one to one. Looking at the American scene, he says that in the United States there is a form for every American culture. These forms include the mill town, New England village, frontier town, and metropolitan conglomeration. Identify cultural characteristics, and form will follow? Quite possible, Stafford, Virginia, a new town planned under private auspices, attempted to build a community around leisure and recreation, as well as urbanity in a rural setting. A distinctive design emerged. If Oscar Lewis is correct, poverty is another culture with a way of life peculiarly its own. Reconstructed cities for the disadvantaged perhaps should not be reflections of middle-class values but rather something appropriate and distinctive to the problem of poverty. Plans for rehousing the La Perla community in San Juan, Puerto Rico, neither intimidate nor condescend but recognize that design solutions for the slums must find a cultural base in the unique environment at hand, as in the design of La Perla's public spaces and streets.

SITE AND SITUATION

The designed environment, a created landscape both geographical and cultural in character, is structured in part by the site and situation. Site characteristics are the precise physical feaures of the ground on which the settlement began and over which it spreads.

Geographers have used the term "situation" as an all-embracing set of general conditions, physical and human, throughout the surroundings of the urban place—conditions that affect its origins, development, and character.

Situation has been interpreted as a two-way relationship between site and society, and in this relationship *nodality* plays a strong role. Until the introduction of modern means of transport, special advantages accrued to river situations and tideland estuaries; because water transport was superior to land transport and the points of contact between land and water, nodal in character, were natural places for settlement. Mountain gaps and the junctions of regional roads were equally favorable situations, serving as natural points for human encounter and the exchange of goods and ideas.

The designed environment is not, however, a process of geographical determinism. As R. E. Dickinson points out, "Man chooses the site prepared by Nature and then organizes it in such a way that it meets his desires and wants." Neither site nor the purely physical facts of situation is a determining force on human choice except to the extent that man's destruction or pollution of the natural environment may force his removal from it. The designed environment is a complicated human organization. Its survival depends first on its effectiveness, on its ability to adjust to new human conditions, and then on site and the physical conditions of situation. This dynamic interaction is well illustrated in the history of Belfast, Ireland, where site, situation, and human chronology have vividly marked the designed environment.

Few cities in the world have had so striking a setting as Belfast, and its natural advantages have allowed the land to fulfill many different functions over the course of history. To the west three miles from the

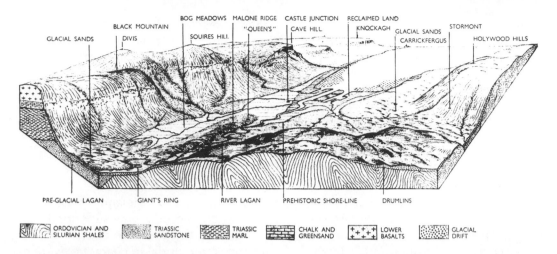

GLACIAL SANDS · BLACK MOUNTAIN · DIVIS · BOG MEADOWS · SOUIRES HILL · MALONE RIDGE · "QUEEN'S" · CASTLE JUNCTION · CAVE HILL · RECLAIMED LAND · KNOCKAGH · GLACIAL SANDS · CARRICKFERGUS · STORMONT · HOLYWOOD HILLS

PRE-GLACIAL LAGAN · GIANT'S RING · RIVER LAGAN · PREHISTORIC SHORE-LINE · DRUMLINS

ORDOVICIAN AND SILURIAN SHALES · TRIASSIC SANDSTONE · TRIASSIC MARL · CHALK AND GREENSAND · LOWER BASALTS · GLACIAL DRIFT

Geological Setting Of Belfast, Ireland From the very beginning the natural site has had a strong influence on urban development. *Source: Greater Belfast Council*

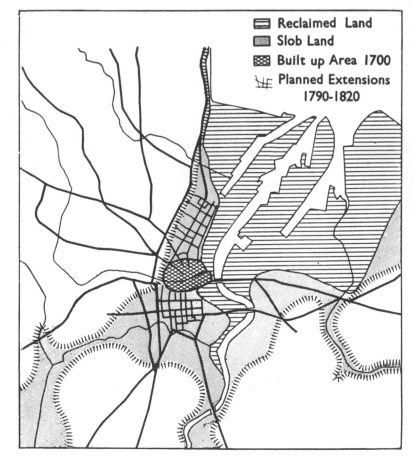

Reclaimed Land
Slob Land
Built up Area 1700
Planned Extensions 1790-1820

The Pattern Of City Growth Belfast planned and unplanned from the early eighteenth century. *Source: Greater Belfast Council*

present city center, a bold, striking escarpment rises fifteen hundred feet above the valley floor, through which the River Lagan flows. Northward, the precipices are broken, and access is given to the valleys behind. Dark basalt covers a bedrock of white chalk, containing abundant flints. The valley floor itself is underlaid with Triassic marls and sandstone, supplying the excellent water and exceptional building conditions eventually to be used for Belfast's best suburbs.

The beauty of site and proximity to river and ground water continuously cultivated over the last thousand years, were not overlooked by primitive man. The Giant's Ring, one of the most famous archeological monuments in Northern Ireland, marks the site's importance in early history. Neolithic and Bronze Age Man hunted and grazed on the hills and exploited and exported the stone, as seen in the burial cairns and flint implements found on the site. During the Middle Ages, agricultural settlements, mostly farmsteads, occupied the lower slopes, burning lime and raising crops.

The Lagan River is the only eastcoast outlet facing Great Britain; thus the juncture of river and sea and the corridor inland along the river valley have long been of special urban and economic consequence to Northern Ireland. On the highest point of the two tributaries crossing the tidal flat, the best defensive position, the first castle was built: Belfast, meaning in Gaelic the ford of the sandbank. The thirteenth-century castle and the church nearby, the Chapel of the Ford, were connected by a track that ran alongside a small tributary stream, the Farset. Of historic origin, the stream-bank path was the easy and convenient way to move between the two monuments.

The extensive tidal foreshore of the Lagan provides level sites for urbanization, and its reclamation for this purpose has pro-

ceeded since the seventeenth century when the town was founded. Then the main street, High Street, occupied both sides of the Farset; and the stream served as an open sewer. It was covered in the nineteenth century, giving the city center a broad and wide street in an otherwise congested area.

Well-situated at the mouth of the Lagan and pinned to high land by its castle, Belfast grew and evolved, pushed and pulled by forces within and without. The Castle itself, standing beyond High Street, gave "an aristocratic air" to that end of town. The castle was rebuilt of brick in the seventeenth century, as were many of the buildings in town. According to E. Evans and E. Jones, the founder, Sir Arthur Chichester, had his workmen "manufacture many more [bricks] than he needed so that his English and Scots followers could also build substantial houses."

Partially surrounded by sea, Belfast was a fortified town, with a conventional ditch and rampart on the land side. High Street was the main axis within the defenses, Commercial activities were divided between a market place on one end and quays on the other; remnants of both remain known as the Cornmarket and Queen's Square, respectively.

Outside the defenses and beyond the castle, the Farset was lined with rows of cottages that led to the linen mills. On this site, the two physical features of the mills and the town walls endured through the history of Belfast, physically and socially. The mills were the first of many that used the outward-flowing stream in manufacture and eventually turned West Belfast into an industrial district. The town walls separated the Protestant English and Scots within from the local Irish Catholics outside. Their habitat along the stream and near the mills became the site of their chapel, around which grew a compact district still there today.

Town expansion was gradual through the eighteenth century, led by advances in transportation and technology. The valley floor, previously wooded and inaccessible, was opened up by the construction of a canal to the hinterlands. In addition a change in vehicles (the solid-wheel Irish cart was replaced by a lighter and better Scottish version) led to improved roads through the valley.

Entrepreneurial skills also played a role in town design. The Belfast merchants began to compete with other European markets in the sale of bleached cloth. The home-crafted brown cloth, made and sold like farm produce, was replaced with white linen. The manufacturing process did not lend itself to small units, and the industry became concentrated in fewer and larger factories. These gravitated to the city where labor and commerce existed side by side. The capital accumulated allowed a major market building to be erected for direct trade with England, bypassing Dublin, making new fortunes, and setting the scene for further economic expansion in the city.

The new wealth, the advent of steam power, the presence of entrepreneurs, and a splendid geographic situation were advantages in the flood tide of industrialization. The town expanded in all directions, including into the sea, through land reclamation. The mud waste of the Town Parks, used for cockfighting and bullbaiting, was reconstructed as a residential area. York Street, the central artery in the design, was said to have been wider than "the avenues of European Renaissance cities." Southward an additional sector of rectangular blocks was raised, centering on the magnificent White Linen Hall though otherwise a dull civic design.

The York Street and White Linen Hall Districts subsequently grew very differently. The north sector was close to shipping, lumber firms, and the port quays. Its outer rim was already a poor district, and in the latter part of the nineteenth century, small terraced housing expanded, vying for sites with warehouses and factories. Into the southern district, however, the presence of the White Linen Hall and the newly rich industrial class brought expensive homes and clubs. Grand mansions and magnificent architecture overcame the unimaginative street design.

Still further south the private land of the Donegall family prevented expansion in that direction. The carriage drives that connected their country seat and demesne were laid out in radiating lines "in the French style," notes Evans and Jones. Later, as the land was developed, the carriage drives were joined to the city center by an extension of roads around Linen Hall. Today, however, the district around these intersections gives little clue to their original reason for being, except that one is called Donegall Pass.

Beyond the confines of the old town, country residences of the rich townfolk further affected the direction of town growth. The Donegall family had preserved the woods of Cave Hill "from unsightly manufacture," and these became the site of several dozen such country estates.

In the 1830's Belfast entered its period of maximum expansion, largely due to further industrialization and lasting roughly seven decades. Railroads now linked the country's interior. Easy import of coal and raw materials plus a labor supply drawn to the city by the Great Famine produced a sevenfold increase in population by 1900. Many of the latest immigrants lived around the three great railroad termini that were built close to the mills.

Not everyone was employed by linen and cotton manufacture. Inventive techniques for building iron and steel ships brought Belfast to the fore among the world's shipbuilding ports. All this technical and commercial development was facilitated by civic improvements that preceded it, including the cutting of a deep-water channel and, from the material excavated, the creation of a large-scale artificial island. Industrial land was thus grafted to the original contours of the city.

As town became city, the design structure of Victorian Belfast was decisively shaped by events that contravened the continuity and quality of its earlier development. The symmetry of Georgian town planning was abandoned. From six key intersections (or asters) which marked the important juncture points along the historic coastline (and thus the established routeways to the inlands), nineteenth-century urbanization pushed outwards; but the radial form and direction were constrained by topography and social and cultural events. With urban growth occurring well beyond them, the asters in turn became the location of important shopping centers, especially to the south and east; because the previously mentioned fashionable suburbs rejected all commercial enterprise.

The resulting diminution of the designed environment did not take long. From a handsome town, especially the distinguished residential Georgian architecture in the southern sector, Belfast declined to "dismal squalor" in three decades. Already by 1834 a traveler had noted, "It is greatly to be regretted that in the improvement and extension of Belfast, architectural beauty has been so little consulted. Anything is better than the liny-hungry look of the modern street."

Though written as a conclusion to a larger history, William H. McNeil's view of past and present well underlines the lesson of Belfast.

... although human intelligence has been used as never before to fire the boilers of social change, mankind has not miraculously shed its tangled inheritance of habit and impulse, of tradition and feeling, of custom and instinct. Human irrationality is as real and as powerful as ever. We face our contemporary difficulties with a psychological nature little different from that with which men faced the world since the beginning of history.

CENTERS

A center fixes a town in space and is a zone markedly unlike the context around it. The differences may come about through contrast in activity, physical dimensions, or symbolic importance, or typically a combination of all three. The very act of creating a center or recognizing a natural one imposes some kind of visual order on the environment, and the stronger the order, the more distinctive the design structure. Furthermore while centers do vary in size, the idea of a hierarchical relationship can be more important than scale. As a living room is the center of a house, a school the center of a neighborhood, a downtown is the center of a city, and a city the center of a region. Hierarchies among early urban places were consonant with the powers of king, princes and knights, and priests. Accordingly, throughout history, city, town and village, district and parish had an architecture and independence of action in keeping with their position relative to some larger social order.

If an egg can be seen as an invention of the chicken so as to perpetuate itself, so the main center of most ancient cities can be seen as a means of the city's self-perpetuation. To the center—and not necessarily the geometric center—of these cities came most of the prominent religious, government, and commercial buildings with their related open space; around these places came the homes of the elite. The center's physical form and spatial position in the city reflected its social organization and available technology. Accordingly, in the art and archeological reconstructions of all early urban places we find a consistent design character. Religious and government buildings dominate the skyline and the ground plane. Essential, significant, resplendent, they are spirited testaments of urban ambitions. The ancient city's architecture is the architecture of these public monuments; and if it is now banal, it is nevertheless consistent throughout the history of civic design to say: Examine the skyline, and it will tell you what is important.

Economic matters have less status than politics and religion in preindustrial cities. The market place is therefore subsidiary in design, though often established in the center and usually next to the major religious buildings. As Sjuberg asserts, "Markets and ambulatory vendors tend to gather about religious buildings throughout the city, apparently to take advantage of the considerable pedestrian traffic they attract."

From the time of the earliest cities, as already noted, the center has been the main residence of the elite. Luxurious dwellings face inward with blank walls to the street side. Thus privacy is heightened and exposure to the other classes minimized—"a pattern still found, most especially in capital cities of countries with widespread poverty and power in the hands of the few." Suburbs, retreats outside the wall for reasons of health, novelty, and escape, also come into being as early as cities themselves; but the first suburbs largely contain summer homes or ancillary dwellings, not permanent residences.

The distribution of classes through the historic city and the design structure that follows (a strong center and many subcenters) take place for many reasons. Spatial mobility is achieved mostly by foot, though occasionally the elite ride on hand-carried sedans or the backs of animals or are in drawn carts. Speeds are nevertheless slow and vehicle movement uncomfortable, while quick access to temple, hall, forum, and stoa is needed for communication and keeping informed. So that they can readily take advantage of news and opportunity and influence decisions, the elite locate their housing close to the center. The strategic locale also has other benefits, including the best protection in time of war, revolution, and banditry. Degrees of comfort accrue, as well as convenience.

An elite center means an outer context of non-elite groups. Nuisance activities and non-elite occupations such as tanning, butchering, and bleaching were thus kept to the periphery of ancient towns.

Special districts come into being sheltered in neutral architectural forms that are pervasive throughout the city. Differentiation of the sectors comes from the broad nature of internal activities, not the buildings themselves. There is a richer variety of life in these environments, especially in their small details, than an air photograph of repetitive building forms would reveal. Often these districts reflect occupational or ethnic characteristics. Several wards, parishes, precincts, and neighborhoods may be as different from each other as they are from the dominant center.

Ghettos are not unusual: Jews in the Middle East and Europe, Muslims in Chinese cities, Christians in the "pagan" lands, Hindus in Central Asia, Catholics in Protestant Belfast have all lived in restricted districts.

Aside from religious and ethnic considerations, the greatest differences in districts are task-generated: Craftsmen with special trades and merchants with special produce keep to their own preserve. Common-sense business principles are the major reasons why this pattern appears in all commercial cities, old and young. Originally, a lack of any form of communication other than the spoken word (except for those very few who could read and write) limited mobility. A highly organized guild system encouraged centrality. Buyers found it useful to compare prices and terms of contracts. Propinquity of similar activities increased the size of the market. Even today in the largest urban markets, speciality shops tend to cluster in one place for equivalent reasons.

Residential and other land uses are not separated in these ancient subcenters, and housing and shops are often combined. Spatial arrangements are further influenced by the sense of time and pace. The dawn-to-dusk environment—a center used by day by all classes and abandoned by some at night—is of early origin. Precise time measurement is foreign. The exacting demands of minutes and hours that standardize routines in contemporary life are missing. The seasons and religious calendar, rather than a clock, mark occasion, event, and passage of time. The design structure is affected accordingly. Urban space is shaped to hold fairs, festivals, and holiday gatherings. There is no where else to go.

The accustomed image of dirt, noise, overcrowding, poor sanitation, and social disease—true in the main for ancient cities—has its important exceptions, brought about by excellence of design. In his excavations of Ur (B.C. 2300), Sir Leonard Wooley found a well-developed city of sophisticated brick architecture. Houses were of two or more stories, with much variety in plan. There were as many as thirteen or fourteen rooms arranged around a central, paved court, which gave light and air. The houses of Ur had internal provision for washing and elimination and a well-engineered drainage system for the town as a whole. Lest this overview of city form from ancient times tends to blur the fact that design has not progressed uniformly everywhere at once, remember that during Ur's ascendency, a continent away the early Englishmen had not yet raised Stonehenge.

Center, city, civilization—the alliteration suggests a poetry of symbols and a natural sequence which brings opportunity and humanity to the lives of many. At the beginning of the list, yet another word must be added: citadel—little city, a place of special prerogatives, and, as we shall see, a center in embryo. Primitive citadels in Western Europe were fortified positions, largely circular and oval earthen embankments that offered protection against inimical spirits, wild animals, and foreign clans. By coercion and magic, temporal and sacred powers were gradually assumed by a few. Accordingly, the citadel grew in function, size, construction, and architectural effect out of all proportion to the defensive needs of the area, which was the original reason for the citadel's being.

Leonard Cottrell visited the site of Sumeria, now in Iraq, in 1957. While the houses of the ancient excavations are almost exactly like those in which the peasants live today (small-roomed, mud-brick dwellings with flat roofs); and the agricultural system (the crops grown, cattle raised, and food eaten) is much the same now as it was five thousand years ago, the dramatic differences between the landscapes of then and now showed in the remnants of the "public buildings, the palaces of the kings, and the high-walled temples and towers." From this citadel and others, by "cultural seepage," to use Mumford's expression, came most of the technological advances and civilizing arts. In the citadel one saw for the first time an army, police force, barracks, foreign office, bureaucracy, law court, temple, astronomical observatory, library, school, theater, private rooms, plumbing, gardens, the best works of art of the time.

The civilizing role of the center today is not unlike that of the ancient citadel. E. B. White writes about Manhattan drawing those "who have pulled up stakes somewhere and come to town, seeking sanctuary or fulfillment or some greater or lesser grail." A city like Boston renews itself periodically by being a center for ideas. Its present plan for survival and continuation depends upon its once more becoming an intellectual citadel for a region. Other declining cities must restore a civilizing quantity to their cores if they and their environs are to counter blight and decay, and redress social and economic imbalances. The concept of dominant and subordinate centers composed into a design structure is a valid one, but it has only those assets and liabilities that come from human intervention. There are few inevitabilities, as the lessons of Sumeria and classical Greece show.

THE AGORA AS AN ARCHETYPE

In Greece there are no navigable rivers, and the mountains come directly to the shores. Early Greek cities therefore share these situations in common: good water supply, nearby cultivable land, a defensible site, and proximity to the sea for communication. In contrast, Wycherley contends that the earliest urban place characteristically shows "no set pattern in shape or structure; this is partly due to the fact that it is usually built on uneven ground but also due to Greek individualism." By B.C. 400, however, similarity of design structure could be seen in many Greek cities and the agora became an archetype center. The Greeks' ability to have formal laws for design of the agora but freedom in exact interpretation can be seen first in the evolution of the street pattern into an ordered environment.

The idea of an imposed design for city growth is attributed to Hippodamus. Aristotle described him as a Milesian "having long hair and holding interesting political views." Knowing the Greek habit for summarizing the culmination of change in the personification of one man, historians surmise that while Hippodamus may have created singular designs for particular cities during his lifetime, the method he used was not original with him.

The imposition in ancient times of a rectangular street system on a site (orthogonal planning) and the advanced reservation of an extensive area for a central place is evident everywhere—China, India, Egypt, Mesopotamia, and Central America—probably in the wake of any system of land tenure that called for elementary methods of surveying. The latter may have involved stellar, lunar, or solar calculations, thus giving the computations a celestial or religious meaning. Ritual and myth guided human behavior in rational and practical matters, justifying choices that could otherwise only be made by intuition. Thus, site planning precepts were filled with references to the ways of the gods. Auguries, divination, and ceremony were the "working drawings and specifications" of their day; but plans—inscribed documents guiding the construction—were not known. Site descriptions found on tablets, coins, and stonework are often after the fact, not preludes. More likely than not, the design was worked out in the field. Whatever the cosmological import, in Greece common sense prevailed. The Hippodamian order, the gridiron, was *adjusted* to, not forced on, the ground.

The division of land, perhaps more than a network of streets, was the first objective in setting the gridiron design in many Greek cities. Several level main streets were sufficient for traffic. As reconstructions show, the minor rights-of-way were often no more than stairways and ramps. The staggered effects of building lines and walls and accordingly the adjacent rights-of-way (now seen as picturesque townscape) may originally have been due to ease of site development and superiority of defense position than to any aesthetic reason.

To R. F. Wycherley and others, the engineering, architectural, and ceremonial ennoblement of the street is more Roman than Greek. The latter thought of streets "simply as a means of access to houses and to public places and buildings." While straight streets at right angles may have been the *sine qua non* of Greek planning, the significant Greek contribution to contemporary design structure is the symbolism of the center.

From caves, grottos, groves, and the backrooms of citadel and palace, the Greeks brought their gods out into the sunlight and openness of public spaces. This detachment from a private place had symbolic meaning and physical importance for the design of the city-state. While the temple became the chief glory and conspicuous element in Greek architecture, the precedence was set for other special activities to be housed in public buildings rather than in private sanctuaries. The process paralleled changes in the political framework from monarchic to oligarchic to democratic forms of government; and these two trends, religious and political, led to the location and grouping of buildings in a distinctive precinct, creating a new kind of center: the agora.

The agora was a large public open space surrounded by, and giving access to, public buildings such as the temples, stoaé, stadia, and gymnasiums. Many classical architectural historians overlook the agora's

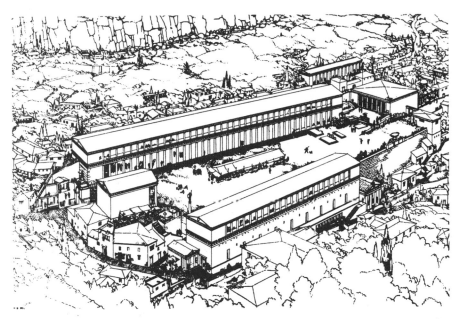

The Greek Agora Shopping center, political meeting hall, crossroads of ideas and action — the beginnings of the humanization of urban life. *Source: Fogg Museum Library*

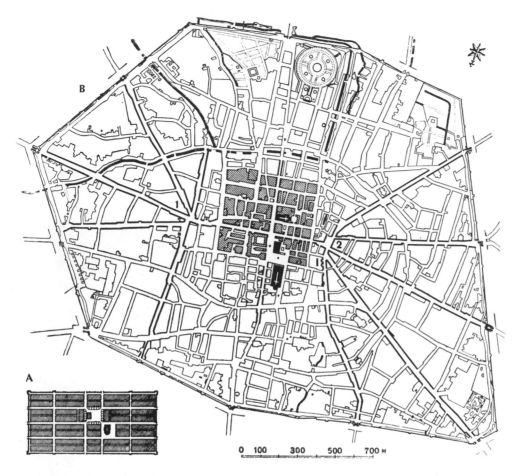

A — Bologna; B — Monpazier Plans of the two cities at the same scale. The Bologna map shows the traces of the rectilinear Roman design on the Medieval city pattern. The intersecting axes, *cardo*, and *decamanus* can still be found in many cities and towns founded by the Romans. Aigues Mortes and Monpazier are classic examples of the application of this principle to fortified towns in the thirteenth century. *Source: A. V. Bunin History Of City Building Moscow, 1953*

value in civic design by stressing the character of the buildings rather than the weight of the space. In his glossary of architectural terms, however, B. Fletcher defines the agora "as the Greek equivalent of the Roman forum, a place of open-air assembly or market."

But the agora was more than such a place. It was the heart of the Hellenic city, the commercial and political locus, the point of exchange between being and becoming. It was the birthplace of much of our Greek heritage in philosophy, theater, science, ideas of commerce, and politics; and perhaps most dramatic in comparison to the centers and citadels described earlier, the agora was a conscious act of giving form and pride of place to functions that concerned every citizen in the city, not a privileged few. Thus it was the core of the entire design of the city and a democratic instrument giving unity to the physical and social structure.

In archaic Greek cities the agora was casually sited, amorphous, and irregular in shape—no more than an enlarged end of a street. Activities in the space varied from season to season and city to city, but usually the agora began as a large, informal meeting ground for spontaneous encounters and other events more adventitious than ceremonial and formal. This function of interchange continued throughout history as the agora took on new faces. Forum, campo, plaza, grande place, and downtown are lineal descendants.

Unhappily, the ability of the agora to perform its function became limited by two factors. First, sizable numbers of the population were disenfranchised. These second-class citizens could not freely use the agora and thus continue to contribute effectively to daily life or momentous decisions. Secondly, as population grew, the agora no longer could comfortably accommodate face-to-face contact. In Greek society the breakdown of immediacy in communications with no technical substitutes available had dire consequences. But for a century at least (say B.C. 500 to 400) a perfecting, though not perfect, balance had been reached. The center had a distinguished architecture and the capacity to accommodate public participation in constructive dialogue on all aspects of a common culture.

It was no accident that in the best architectural examples of the agora, man becomes the measure of the space, and man's concept of that space, symbolizes his views on the natural environment. In a fundamental and important scholarly contribution, C. Doxiadis reports that those who entered the site of a designed Greek city "immediately perceived its plan and were at once led to their objective." The visitor "saw every form clearly and as a whole, all of them in front of him; yet he was free to move about. His path led to no building, it was free; it was directed towards nature, and the whole site was designed according to natural laws."

The "natural" here represents points of view about space, finite or infinite. In the Doric compositions the line of travel serves as an axis, dividing the building sites so that the axis is clear and the eye of the pedestrian can be carried beyond the site into infinity. Often the axis is arranged to the cardinal points of the sun, i.e. east or west. In the Ionic period the reverse is largely true: Space, as experienced along the line of movement, is enclosed and limited by buildings.

What do these design experiences mean for our own environment today? Perhaps only that the value we give is the value we receive. Again, the quality of an environment is a mirror image of the culture.

The importance of the Greek public spaces was not lost on those who ruled Rome. Throughout Roman history the expansion of the Empire was marked by large public works, especially forums, where prodigal expenditures of wealth and labor were made at the expense of traffic, sanitation, and housing. The Imperial forums command attention because of their technology, but their contribution to contemporary concepts of environmental design is of Greek origin, i.e. the idealization of a central place.

CIVITAS

Throughout history the center endures, its shape, content, and location adjusting to rise and decline of cultures, empires, and civilizations. The coexistence of permanence and change has captivated urban geographers, sociologists, aestheticians—anyone concerned with the morphology of city form. Numerous attempts have been made to classify the forms of the center, perhaps largely in the hope of identifying the strategic element or critical mass that creates an epitome or criterion suitable for emulation. Yet this approach to design—systematic stock-taking or the science and art of typology—is less than three hundred years old.

The act of classification eliminates extraneous material and, through analogy and homology, leads to the reasons why things come into being. Traditionally, centers have been classified by similarities in configuration (patterns in space) or historic sequence (patterns in time). Neither system is entirely appropriate except to the extent that it illuminates the values held by the classifier and the cultural temperament of his period and place.

In extracting aesthetic significance for our time from early centers, the problems of classification include simultaneity (similar forms appearing at the same time with no apparent connection); physical remains too gross to permit differentiation between the purely utilitarian and the purely aesthetic purpose (if such existed); and changing boundaries in classification that come about as current taste, invention, and insight demand periodic reassessments of all that occurred before.

Nevertheless, the Medieval and Renaissance centers may be viewed today as an unbroken series evolving from promorphic to neomorphic forms. G. Kubler defines promorphic as physical solutions that are "technically simple, energetically inexpensive, expressively clear." Neomorphs are "costly, difficult, intricate, recondite."

The Medieval centers can be discussed as a promorphic group because they share these four characteristics: simplicity in design, a Western European culture, preindustrial technologies, and similarities in the reasons why the centers took the shape and content that they did. Withdrawal of Roman power to the Italian peninsula about A.D. 500 was followed by five centuries of a largely agricultural existence in which forests were cleared, the Christian religion was propagated, and strongholds of sacred and secular power were situated at the intersections of major travel routes. To these nascent centers there accrued certain privileges in self-government, so that by the end of the twelfth century these urban places had become politically and physically distinctive developments. Universally the *civitas*, the Medieval town, was a center of administration with fortifications, a legal constitution, an inviolable charter of rights, and a population primarily engaged in industry and trade. From these crossroad prototypes, urbanization spread in a phenomenal growth of cities to such an extent that most European cities today can date their founding from 1200 to 1400 A.D.

Mont St. Michel *Source: Harvard Graduate School of Design Library*

Aigues Mortes. *Photo: French Government Tourist Office*

Carcassonne *Photo: French Government Tourist Office*

The impact of site, situation, and natural conditions on the design of these early towns has already been discussed. An examination of promorphic designs shows at least three basic arrangements, each reflecting a method for compactly grouping houses, streets, and centers; and geographers have classified these resemblances into families of street-plan types: rib street patterns, parallel street patterns, and spindle or elliptical street patterns. These patterns were occasionally juxtaposed on the site in response to a plan, but more often than not the location and extension of the street patterns were adventitious, following site decisions rather than generating urban form. The origin, function, and shape of the nucleus, or center, was the predominant influence on the promorphic urban plan.

Air views of relatively old European towns show three distinct forms of commercial center: the triangular market place formed at the convergence of two or more main roads, the rectangular market fitting into a rectilinear street pattern, and the market formed from a greatly enlarged street produced by pulling down a block of buildings for a public open space. Later, as growth of the market could not be contained in the old center, it moved to another part of town or evolved into a series of specialized markets spread throughout the city. In the planned new towns of the late Middle Ages, founded largely to undertake commercial enterprises, accommodation for the market place and related buildings received first priority, after defense, in the town design.

While cathedral and market place dominated the design structure, diocese and parish were the fundamental political divisions, scaling down the larger urban place into familiar units. In some towns there were as many as one parish church for every one hundred families. Furthermore, members of craft guilds were occasionally organized by districts and neighborhoods, their size and scale depending on the overall size of the settlement. The resulting decentralization provided in the larger towns a heightened sense of individuality among people who shared face-to-face contact through parish life. Festive costumes, local saints, special holidays, and district architecture gave uniqueness to the parish. At the same time the populace shared the excitement of participation in the circle of activities that belonged to the whole city and which took place in the central-market-and-cathedral place.

Unlike ancient Asiatic cities and classical Greek and Roman settlements, the Medieval cities had a special quality of life and activity in both the dominant center and the subcenters, giving identity to both the parts and the whole in both formal and informal physical settings. Remnants of the resulting design structure can be sensed in Florence, Sienna, and Venice; but the best example extant is perhaps Bruges.

Church squares, local market places, large private gardens, and generous distances between houses and the canals and streets suggest communal respect and interest among all who lived in Bruges. A sense of order, spacing, and internal planning prevails, but not at the price of monotony. Consistency in texture, color, and scale, and an awareness (intuitive perhaps) of the effect of sun and shadow on buildings and open spaces make for a memorable urban scene from the town wall to the main center, and a richness of detail climaxed by the surprising scale and grandeur of the Bourg and Grande Place, the civic center and market place, respectively.

At the height of its prosperity, Bruges was a textile center of Europe, with thirty-five thousand homes and 200,000 people;

Early Plan Of Bruges *Photo: Belgium National Tourist Office*

but the silting of a water connection to the North Sea ended maritime trading, and the city's economy dramatically declined. Thousands of homes were vacated, and the population dropped by two thirds. Demolitions and reconstruction have been carried on since, but much of the Medieval flavor remains.

For all their similarities, there are at the same time important differences in form among the Medieval centers, the ways in which they grew, and the general effect they had on the overall design structure of the urban place in which they were situated. Further, the German models differ from those developed in the Low Countries as much as the Italian Gothic cities differ from the English. Site, historical antecedents, forms of government, availability of technical and aesthetic skills, and other factors play a role in differentiations.

Many early Medieval French towns have their roots in the defensive settlements, the *castra*, laid out by the Romans. An evolution from the Greek gridiron plan, the *castra* had two additional physical features: The walls gave a strongly defined rectangular boundary around the town, and the enclosed settlement was divided into four parts by two intersecting axes, the *cardo* and *decamanus*, which cut through the walls at their midpoints. The intersection of the axes was enlarged for the marketplace or forum.

In the Roman colonies town planning and development was carried on for two purposes: the resettlement of population and the consolidation of military power. The Roman method of fighting required freedom of maneuver and a highly organized order of battle. The four-square camp served both needs. The pitching of the legionnaire's camp and the construction of an increasingly permanent defensive position became standardized in detail, almost a drill; and the practical advantages of the military layout made it easy to adopt as a model for urban growth from tent city to stone city. As natural defensive sites came to be used, the plan was adapted to the terrain. Simplicity in surveying and in allocating plots of land to settlers reinforced the practical and authoritarian design.

Gradually, the rectilinear shape of Roman towns in France gave way to the oval, eccentric, and curving plan associated with Medieval cities. For defense the cathedrals and monastery buildings were constructed close to the walls of the *castra* and gradually new streets and homes focused on this religious center. In time the settlement showed several nuclei, reflecting (as mentioned earlier) the growth of various economic and political powers. These nuclei include: the *cité* clustered around the cathedral, the *bourg* or merchants' colony, the *vicus* or manorial settlement supporting the monastery, and *château* or seat of the secular lord. Eventually, all were enclosed by a wall, which was later extended as new *faubourgs*, or neighborhoods, were developed outside the original lines of defense. Of this type, Bourges is a good example.

A second type of French Medieval town and city is marked by an ecclesiastical monument (church, cathedral, abbey, or monastery) surrounded by curvilinear blocks and, in turn, an oval-shaped wall. Such towns and cities were established later than the Roman *castra*, largely as fortified places for early Christian orders. Extension from the nucleus varied according to relationship to trading routes as well as to site conditions. The most dramatic example of the latter influence is Mont-Saint-Michel, whose growth is aborted by steep slopes and ocean tides. It thus remains a well-preserved example of the period.

A cell-like pattern of urban growth can be observed in a third kind of Medieval French plan: the castle town. Here the center consisted originally of the *donjon* or residence of the lord, *motte* or soldiers' quarters, and *basse-cour* or outer defenses. This area domiciled the servants and dependents of the lord and soldiers. In time of danger local residents and their animals were admitted, as well as refugees in war. Merchants and artisans were attracted to form a *bourg*, and to serve their needs, churches and markets were opened. This nucleus then had some value, and accordingly defensive palisades and walls were placed around it. Fair representation of this type of development can be seen in children's book drawings of knights in armor riding through a busy market place with crenellated towers and turrets, bailey, moat, and drawbridge just ahead.

French geographers' distinguish between *villes créées* and *villes spontanées*, those that are planned and those that just grow. To the former group belong the *bastides* or fortress cities designed in a formal manner between 1150 and 1350 so as to attract settlers, hold their loyalty, and develop sources of income for the ruling class.

F. R. Hiorns, among others, claims that the *bastides* represent a reform movement in Medieval town planning. He suggests that they follow classical precedent "insofar as it was understood." Whatever the cause, the towns differed from many previous settlements in the sense that proscribed laws and regulations controlled many of the design features in advance of construction. Charter rights given to the community occasionally included the regulation of size and depth of house lots, the requirement to extend upper floors to form an arcade at the street level, and controls on the selection of building materials. Mon-tauban, Aigues-Mortes, and Carcassonne are excellent surviving examples of these formally designed cities within fortifications on hilltop and plain.

Montauban is rated as a superb example of Medieval urban design. The site lies high on a peninsula overlooking the River Tarn. The fortifications were given trapezoidal shape, conforming to the natural land form. Streets and blocks of uniform size were skillfully set out with the central market parallel to the walls, all situated in advance of development.

Malaria, poor access to major trading routes, and limitations on cultivable land led to the abandonment of Aigues-Mortes shortly after it was completed in 1270. As a result the major features seen today are much as they were seven centuries ago when the town served as the embarkation port for the Crusades led by Louis IX. The town is laid out in a grid-like pattern, forming a parallelogram about twice as long as it is wide and emphasizing the design's *castra* lineage. Nine town gates give entry to the main streets, which lead to the center, to the west of the main axis. The street patterns break the force of prevailing winds coming in off the sea. Well-ordered geometrically, Aigues-Mortes presents a bleak picture today, probably because the streets and spaces have not been softened by centuries of hard use and those incremental changes that occur when many people modify the physical environment over time to suit their moods and needs.

Proportion scale view, and vista of street and center, commonly associated with urban design in the Renaissance, were not entirely missing from the Medieval scene; but it is the picturesque qualities of these ancient places that stand out, as the morphology of Medieval cities well illustrates. To what extent were these qualities de-signed? The house of critics seems split on the issue. P. Lavedan suggests that the aesthetics were casual by-products and that natural conditions of site and practical matters of defense, commerce, government, and habitation designed the towns. In contrast, L. Mumford sees the same outcome as the result of intelligent and sensitive accommodation of old and new with no one style or aesthetic dominating but a gradual accretion and blending of all elements together in "effortless spontaneity and artless unconsciousness."

Nowhere in Europe can this latter effect be seen to better advantage today, in its simplest sinuosity, than in Carcassonne. Nineteen centuries of construction, at least, comprise the town: Roman mosaics, streets, castle, cathedral, church, shops, houses, walls, towers, prison, and barracks all melded together. Wall surfaces differ slightly in texture and detail. Fenestration, vaults, and towers take varying shapes and forms, yet seem cut from a single piece of stone.

The Medieval scene yields three cues for contemporary environmental design: demonstration of the importance of centers and subcenters as nuclei for rational urban growth and social vitality; demonstration that accommodation of constant change can be carried out without loss of visual unity; and demonstration of the idea that the hierarchical ordering of parts into a whole is a common feature of design structure.

The Visual Exuberance Of a Renaissance Courtyard *Photo: Richard P. Dober*

ATTEMPTS AT A STRONG ORDER

Great examples of environmental design are not the product of isolated events but rather a manifestation of a continuous process of change. Whenever the details of change reinforce one another sufficiently to indicate a new direction or repeat and enlarge themselves as a single theme, one can identify a style. The Renaissance and its afterglow, the Baroque, are illustrations. Their era produced a strong aesthetic, so strong that many of our everyday concepts of visual order and disorder and the vocabulary of civic design date from this period.

The Renaissance utilization of linear perspective marks the point of departure from Medieval concepts of space. In painting, sculpture, and architecture, the static and flat representations of objects and motion were replaced with the notion that a one-station viewpoint of volumes, shapes, and relationships—their arrangement as in the vision of a single man—was scientifically correct and thus aesthetically true.

This aesthetic attitude carried over into the other arts. In Medieval theater, convention allowed all the stage sets to be placed next to each other and the actors to remain on stage when not taking part in the action. The audience noticed neither the silent players nor the unused sets until the action moved to them. The Renaissance aesthetic found this ludicrous and unjustifiable. "For in the new conception of art," writes A. Hauser, "the spectator wants to be able to take in the whole range of the stage with a single glance, just as he grasps the whole space of a painting organized on the principles of a single glance."

If all the community is a stage, how did the Renaissance aesthetic deal with this space? To begin with, community development in Renaissance Italy had its own peculiar impetus. In general, community development is dependent on opportunities presented by social and economic change. It cannot live apart from those forces, as might painting and sculpture; and during the Middle Ages those forces often entailed founding and expansion of urban centers. Relatively few new urban settlements were founded, however, in Western and Central Europe between 1500 and 1800. The general size of population remained constant, particularly the ratio between urban and rural populations. Thus the inspiration and substance of Renaissance community development did not lie in new and shifting population centers. Rather, cultural changes account for new aesthetic influences on the design of the Renaissance environment, and the incentive for physical change in a relatively short period of time can be explained largely by political upheaval and advances in the state of warfare. In addition, these cultural influences and political and military factors were closely intertwined in their development.

J. Burckhardt notes that in Italy the Renaissance city-state was itself a work of art—"the fruit of reflection and careful adaptation." For two hundred years the struggle between despotism and republicanism engendered opportunities for designers to demonstrate both technical and aesthetic skills in the building of fortifications and improving the layout of the space they enclosed. The regular protecting walls, such as those that encircled Aigues-Mortes, were broken and pushed forward at selected places by bastions from which flanking fire could be directed at the attacker. Like arrowheads, the bastions extended the military rectangle into a polygon with many facets. Extended in further elaboration, the boundaries became the star-shaped *città ideale* of classical art. Additional refinements were made as successive designers found in its simplicity an easy way to combine in city planning both art and science with overtones of astrology. Finally, the genuflection towards antiquity which graced the Renaissance arts with the respectability that comes from antecedents was ever present. A classical preoccupation with abstract notions of beauty is clearly evident in the Renaissance infatuation with complete form, for true beauty to the Renaissance designer was that to which nothing can be added or taken away.

G. Scott succinctly summarizes how the practical necessity of designing cities for defense was turned into an academic art through ". . . the revival of scholarship, the invention of printing, and the discovery of Vitruvius, the Roman architect and engineer. Scholarship set up the ideal of an exact and textural subservience to the antique; Vitruvius provided the code, the printing disseminated it." Of the many versions of Vitruvian principles, Alberti is given credit for having made the first Renaissance translation; but the most interesting one is Cesare Ciserano's. Published in 1512, it incorporated new material and illustrations that connected the ideals of ancient Greece and Rome to more recent Medieval town planning principles. Further, Ciserano's version of Vitruvius gave instruction in the management and technology of city building from the viewpoint of social gains to be derived from an ordered, hygienic, and utilitarian layout. At the same time Ciserano may be considered a minor force that turned the concept of the ideal away from a preconceived standard to flexible models, such as those proposed by Scamozzi.

In yet another version of Vitruvius, Scamozzi introduced some original features to the ideal scheme, such as water canals and the artful distribution of open spaces, so as to give life and interest to the various parts of the city. The attractiveness of his native Venice was not lost upon him nor on Francesco di Giorgio, the Siennese who warped the flat Vitruvian plans to fit hillside sites by using corkscrew roads. Scamozzi later designed the Venetian frontier town of Palma Nova, still extant south of Udine. Palma Nova is a fascinating monument to a major design influence and one of the few Utopias ever built.

With few exceptions, the *città ideale* remained a paper diagram, for as S. Giedeon notes, "the idea of a town as an entity in which the interactions of thousands of separate existences can be coordinated was foreign to the temper of that era." The major Renaissance contributions to environmental design lay elsewhere, and certainly one was the handling of the architectural details of public spaces—especially the design of walls, floors, and the sequence of interior and exterior spaces brought together through the processional qualities of monumental staircases and formal connecting streets. Not the least of the overriding objectives behind these designs was the positioning of a central element (space or building) in the design and the grouping of buildings to effect a painterly perspective. In accomplishing these measures, the town centers of Medieval times, similar in comparison, were transformed with such an outburst of artistic versatility that no public surface was free for long from paint, carving, and encrustations celebrating secular and sacred events. Simultaneously, the designer broke away from pure craftsmanship and constraints of guild to achieve an independence of stature comparable to that of poet and scholar. This emancipation led to freedom to explore new media, to desert one commission in favor of another, and to compete both as patron and artist. The transitory quality of novelty

as a force in design was not unlike the values of our own time and in strong contrast to the Medieval attitude which, as A. Hauser explains, "recognized no independent value in intellectual originality and spontaneity, recommended the imitation of the masters, and considered plagiarism permissible . . . and was at the most, superficially touched, but in no sense dominated by the idea of intellectual competition."

The Renaissance is thus marked by exuberance and conflict rather than order and self-control, by the acceptance of an ancient ideal rather than the copying of recent, authoritative models or the search for a new ideal, the enjoyment of novelty notwithstanding. The best aspects of the Renaissance physical environment were already present in the Medieval town, the better blossomed in the Baroque.

Di Giorgio, that modest Siennese painter and interpreter of Vitruvius, articulated another important Renaissance principle of consequence to our time: The designer has only to establish "the main lines of the plan and leave it to life itself to make adjustments where necessary." The vitality of the principle is expressed best in Pope Sixtus V's plan for Rome; the carrying out of the scheme itself is an exemplary study of human enterprise, optimism, energy, and good will.

Sixtus V was of peasant stock, a mendicant friar of the Franciscan order, zealous, erudite, articulate, and a master organizer. Appointed cardinal at forty-eight, he came to Rome and spent the next fourteen years in relative obscurity, having been deliberately cut off from Church affairs by Gregory XIII.

The years were not uneventful, however. Engaging the architect Domenico Fontana, Sixtus V developed a great estate, Villa Montalto, on Esquiline Hill, which he found deserted and waterless and left as a land-

Palma Nova, 1967 A remarkable, historic town whose configuration closely parallels the Vitruvian ideal. *Photo: Aerofilms Limited (Copyright reserved)*

mark in garden design. Unfortunately, Villa Montalto has been lost to history, and the site is now occupied by the main railroad station in Rome.

These fourteen years were a testing period for the future Sixtus V and Fontana. Both master planner and architect shared similar attitudes about organization and detail and concerned themselves with technique as much as with effect. The ideas thus germinated were harvested with dispatch on Sixtus V's election as Pope, though with much ruthlessness, understandable in the context of external problems. England was lost as a Catholic country, Spain defeated as a sea power in the Channel, Germany divided, and France leaning to Protestantism. In this light any domestic achievements would have been admirable.

Sixtus V became Pope, however, in the first wave of affirmation and liberalism that marked the Catholic Restoration and the beginning of the Baroque. The late sixteenth century seemed tired of war without resolution: the internecine conflicts between competing city-states, wherein successful intrigue was climaxed by the securing of the Papal throne for a winning nobleman, usually aged and thus of short Papal tenure and little significance. In addition, compared with other city-states—Venice, Genoa, Florence, and Sienna—Rome was a second-class city. Politically and aesthetically, power and achievement appeared to lie elsewhere. It was a time and place of opportunity for a man of vigor and large designs.

Sixtus V made great strides in his five years as Pope: One forceful and practical step after another led from public safety to physical and economic health and city design. Gangsterism was eliminated and Rome made safe for visitors. The papal treasury was increased twentyfold, and from this new wealth public works were financed. A

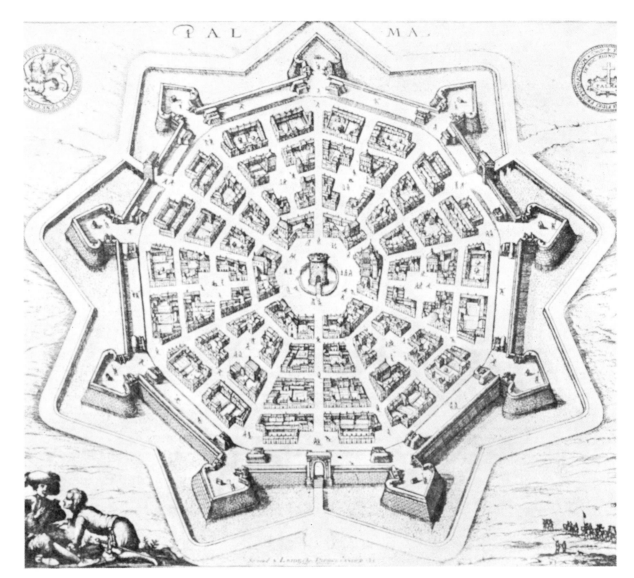

Plan Of Palma Nova C. 1593 *Source: Harvard Graduate School of Design Library*

MACHINARVM TRACTORIARVM AD ONERVM ELEVATIONES·CVM
QVIBVS MOVETVR CELERITER ET EXPEDITE·FIGVRÆ

A QVAEDVCTIONVM DIVERSIMODA AFIGVRATIO VT AGRESTIV FRVC/
TVVM AFFLVENTEM VBERTATEM: NON MINVS TVTIONEM MOENI
VM SVPLEMENTVM Q3 NECESSARIVM IN CIVITATIBVS HABEAT

Cesaraino's Vitruvius Late Renaissance versions of Vitruvius not only showed the rules of design but also recognized the ways and means of actually building cities. *Source: Houghton Library*

start was made on draining the Pontine marshes, a source of malaria and constraint on urban growth. Within eighteen months new water was piped to the five highest hills in Rome over eight miles of aqueducts and through eight miles of underground conduits. On the Strada Pia Sixtus V constructed a public square and large basin for rinsing wool, an encouragement to new industry. (Later it was redesigned as the Fontana di Trevi.) Public washing places with sheltering enclosures were put up in other parts of the city, improving public health as well as giving protection against inclement weather.

Finding a city of beggars and unemployed, Sixtus V built poorhouses. When his public works were not enough to occupy all the idle and willing hands, he promoted a scheme for restoring the traditional Roman manufacture of textiles. Mulberry trees were planted everywhere. By papal decree, an extraordinary plan was drawn by Fontana for converting the Colosseum into workshops and living quarters for wool-spinners. The cupola on St. Peter's, designed by Michelangelo and left unfinished by two popes, was completed, Giedeon reports, by eight hundred workmen "on the job, day and night, weekends and holidays" for twenty-two months. Though much of Medieval Rome was left extant, a new set of streets was cut through the city to tie together the fragments of public spaces and monuments left by Sixtus V's predecessors. It is this act of linking centers that unified his plan for Rome and left the most enduring impress on the city's design structure.

While the practical foundation was economics, the impulse for the grand design was ecclesiastical: to sustain Rome as a pilgrimage center. Seven important churches and shrines were connected by streets so that all could be visited in a single day. It was intended, wrote Fontana, that the lines be carried straight through, "without concern for either the hills or valley," but not insensitively, for the lines were graded to gentle slopes, with "charming sites, revealing in several places which they pass, the lowest portions of the city, with various and diverse perspectives; so that aside from devotion, they also nourish with their charm the senses of the body."

In this objective one sees for the first time an intentional blending of function and aesthetics along lines of circulation connecting centers on a city-wide scale. Further, the lines are not the special precincts of tyrant and conqueror but places for general public participation. In its size and scale and length, the design of the streets anticipated the requirements of the new kinds of transport which were to replace the horse and sedan chair—the carriage and coach.

The impulse to connect was also used to open, enlarge, and mark the critical places in the plan of Rome. The best inhabitable areas were made accessible and tied to the city by new and broad streets. The Medieval and Renaissance open spaces, largely backstreet market places, were given new prominance. Moving ancient landmarks with technical genius, writes Giedeon, Sixtus V was "like a man with a divining rod . . . placing his obelisks at points where, during the coming centuries, the most marvelous square would develop." Following a heightened instinct for design opportunities, Sixtus V moved the imposing stone work of the Egyptian sun god from Nero's circus to St. Peter's Cathedral—centering it so well that Bernini had no trouble a century later placing his colonnades around the obelisk, enclosing the space in front of the church to complete Sixtus' plan.

Thus the pontificate of Sixtus V stands as a magnificent moment of change. As the Church accommodated itself to the losses of the Reformation, important immediate measures were taken to ameliorate suffering in this world instead of postponing relief until the next; and the coldness of the Inquisition was replaced by a delight in the pleasures and joys of secular life. Under Sixtus V's direction, in five short years, Rome was transformed. "Everything seems to be new, edifices, streets, squares, fountains, aqueducts, obelisks," wrote a contemporary traveler returning to the city after the death of the Pope. And though the completion of the large scheme during Sixtus V's tenure was at best fragmentary and critics consider the work dull and flavorless, artistically mediocre, in specific architectural terms, the self-fulfilling qualities of the design structure were powerful enough to draw together several generations of better designers who followed. In a thoughtfully documented presentation, E. Bacon points out the strength of Sixtus V's plan. It not only brought order to a city without form but at the same time set into motion design opportunities for buildings and open spaces yet to come. By establishing a circulation system which directed movement from center to center and a set of monuments which decided the terminal points, the plan fixed design elements which could not be moved and to which all succeeding ideas had to accommodate themselves.

From the Baroque onwards the center as an element in design structure only repeats in variant forms the intention of Greek, Medieval, and Renaissance examples: the center serves as a point of gravity; as a place for intensive contact, communication and exchange; as an instrument for symbolizing public purpose through its special aesthetic effects; and as an origin, terminus, or juncture for lines of movement.

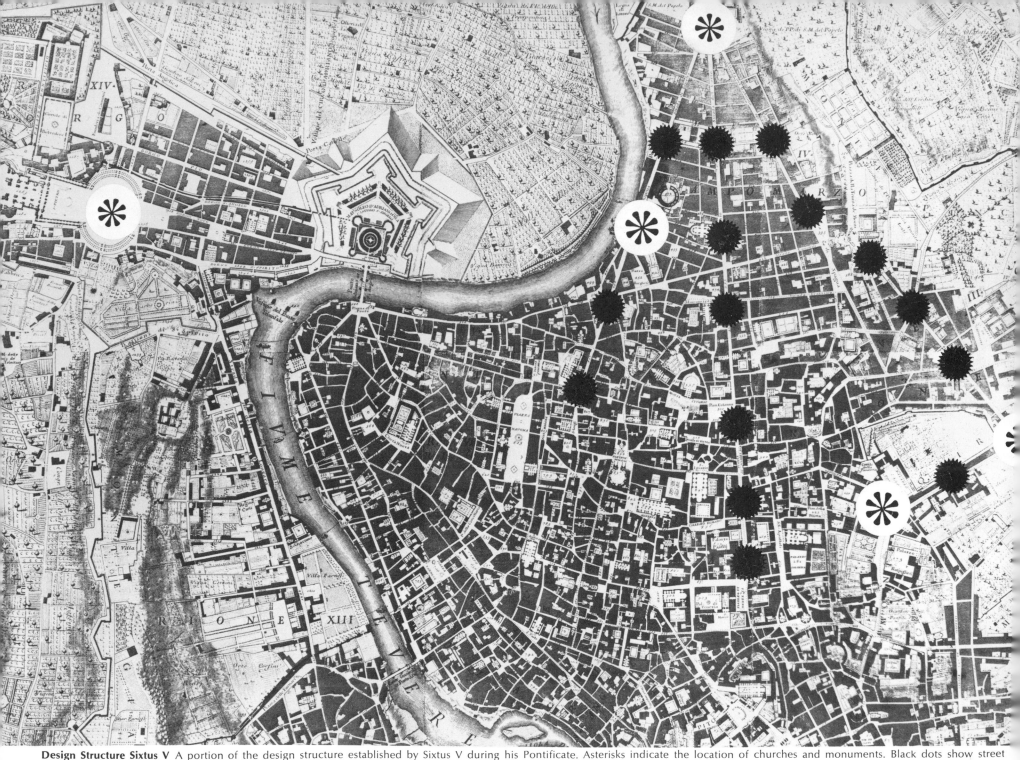

Design Structure Sixtus V A portion of the design structure established by Sixtus V during his Pontificate. Asterisks indicate the location of churches and monuments. Black dots show street construction. These are the beginnings of the Baroque city design. For a splendid discussion of Sixtus V's work, see *The Design Of Cities*, Edmund N. Bacon, New York, 1967. *Source: Harvard Graduate School of Design Library*

BAROQUE ATTITUDES

Since it still influences present habits of conceptualizing design structure, further comments on the Baroque are in order—especially its utility in the large-scale designs that accompanied the rise of absolutism and then their diminution during the subsequent periods of industrial urbanization.

The Baroque may be considered the first contemporary aesthetics: a set of attitudes that doesn't exhaust its resources when crossing national boundaries or those of the sciences and the arts. The word baroque was originally used to describe the extravagant, capricious, and bizarre tendencies which appeared in seventeenth-century painting and architecture as a way of rejecting formulae and classical approaches to design. Baroque now means a fusing of opposite tendencies into a unified whole and it connotes a tolerance for ambiguity, a preference for open-end systems rather than closed-end, for dynamic designs over static forms, for the interdependence of observer and observed, for the obliteration of frontiers and fixed contours, for the replacement of the absolute by the relative, for an accommodation of the accidental, the improvised, and the ephemeral within an intentional unity.

There are overtones here of the late Medieval universal outlook, but where the latter stems from divine immanence and transcendentalism, the Baroque order is built on discoveries in the natural sciences. Natural law makes no exceptions, the Baroque philosophers believed: the universe is a continuous system organized on principles as infinite as they are uniform.

This impelling vision had no territorial limits in either geography or intellectual discipline. In Italian architecture it allowed Borromini to shatter all precedents in his departure from classical principles as seen in the undulating wall surfaces of San Carlo alle Quattro Fontane. At the same time he would take pains to study ancient buildings and even incorporate fragments from the past when restoring and enlarging older edifices. In France the Baroque outlook affected Descartes and mathematical concepts; in Germany, Bach and music; in Holland, the Dutch landscape painters.

There is a truism that community design expresses only what society will permit. There are good reasons to feel that the Baroque attitudes in painting and architecture were never fully transferred to large-scale design. Projects cited by historians as Baroque examples of town planning are not aesthetically satisfactory; indeed their dependency on absolutes and subservience to formal unity seems antithetical to the spirit of the Baroque.

This may be explained by suggesting that the requisites of political survival deflected an aesthetic impulse of marked dynamism and violent contrasts into a new form of classicism. In the Piazza del Popolo and Piazza San Pietro, monumental open spaces were created to accommodate great crowds drawn to acknowledge secular and sacred powers. A cultural interchange then occurred between waning and rising political forces, Italy and France. For reasons of prestige, however, France insensitively adopted the superficial aspects of Baroque town planning and in so doing aborted further refinement, just as later other countries would fashionably adopt the French models and St. Petersburg (Leningrad), for example, would become: "more French than Paris itself."

With shift of political power from Rome to Paris and the evolution there of monarchal absolutism, the conditions for *les grands plans* as French Baroque of the seventeenth century was known, fully emerged. Social and cultural relationships were frozen. A new royal attitude was apparent: the French kings saw themselves as setting a world style. And they did. An official art appeared, controlled by a remarkably centralized government. The exuberance and variety of individual arts were suppressed in the interest of amalgamating everything into a monumental unity. Painting, sculpture, interior decoration, architecture, individual buildings, gardens, and roads, were all assembled into a single design, and integrated art, all encompassing.

At the same time, however, government ended the private relationship between artist and public. The Renaissance notion of the individual—with special aspirations and viewpoints—was replaced by an Academy that was to educate, administer, and direct all artistic production to a single purpose: the glorification of the king and state. The system dominated manufacture. Studios and workshops turned out products "of impeccable taste and technical perfection," all cast, sculpted, turned, moulded, and painted to a uniform quality and a monotonous iconography. Though flowering in France the Baroque maintained obeisance to its place of origin. Among the highest achievements in the fine arts in Paris was the Rome prize.

The château and surrounding garden, street, and public square were the dominant civic design elements of seventeenth-century France, all commanded by the king and his troops. Of these, Versailles—"the little house of the greatest king on earth"—was the most imposing. The hunting lodge is constructed and reconstructed under successive rulers. Under Le Vaux, Le Brun, and Le Nôtre, the palace becomes an architectural masterpiece. According to Lavedan, "from whatever angle we consider it, either from the entrance, with its ever narrowing courts, or from the garden, with the immense central recess in the façade, there is movement throughout. Opulence, perspective, the desire to astonish, all remain very close to the Italian spirit."

On Mansard's appointment as royal architect, contrast, movement, and fantasy were moderated in favor of a strong horizontal expression. The new wings of the Palais de Versailles enlarged the façade to 550 yards. In front, the Place d'Armes and boulevard were constructed and extended to Paris in a straight line many miles away. The straight line and the square represent "the will of a number of men exercising their reason, which disposed them in this way."

In Paris, paralleling the development of the châteaux, especially from the time of Henry IV onward, the royal spaces were meant to serve as places for setting the statue of the sovereign, a symbolic staking out of majestic rule. Architecturally, they approached perfection, as in the Place des Vosges. This square, five hundred feet on each side, is lined with houses of all one type with open arcades at the ground level forming a continuous covered walk. A single type of dormers and slate roofs unifies the design. Its imagery was imposed in Rennes, Bordeaux, Reims, and Nancy.

In Nancy Louis XV's son-in-law welded together two previously separated sectors of the town and three independent squares into a civic space that commemorated the conquest of this regional capital of Lorraine. Here more than in any other existing French city the often cited impeccable taste may be seen as it was. The monumental open space is composed of building façades, ground plane, statuary, fountains, colonnade, grillwork, and landscape. All are subordinated as individual works of art.

This discipline is not accidental or adventitious. It rises from implied values as exportable from France as the designs themselves. The ideal of design integrity can be seen in such disparate works as the civic square of Helsinki and John Nash's Cumberland Terrace—the latter snobbery a group of individual houses sharing a common plaster coating of columns, porticos, and pediments giving the air of a great mansion overlooking a great park. The order is imposed as easily by the demands for fashion made by private speculation as it is by government prescription.

The tidiness of the eighteenth- and early nineteenth-century designs, their purity and unity, are attractive contrasts to the visual anarchy of recent urbanization. However, the order may disappear when the controls of Absolutism are removed, as it did in the Royal Squares of Paris after the Revolution, or when political support for the design declines as in L'Enfant's plan for Washington, D.C.

L'Enfant's design was conceived as a set of principal buildings and squares

. . . on the most advantageous grounds, commanding the most extensive prospects. Lines or Avenues of direct communication have been devised, to connect the separate and most distant objects with the principal, and to preserve through the whole a reciprocity of sight at the same time.

L'Enfant's own description of the plan is sprinkled with Baroque design concepts and absolutes. In the various squares "Statues, Columns, Obelisks would be erected to the memory of the States;" at another key intersection an "historic column would be raised . . . from whose station . . . all distances of places throughout the country could be calculated." The "President's palace would exhibit a sumptuous aspect" and

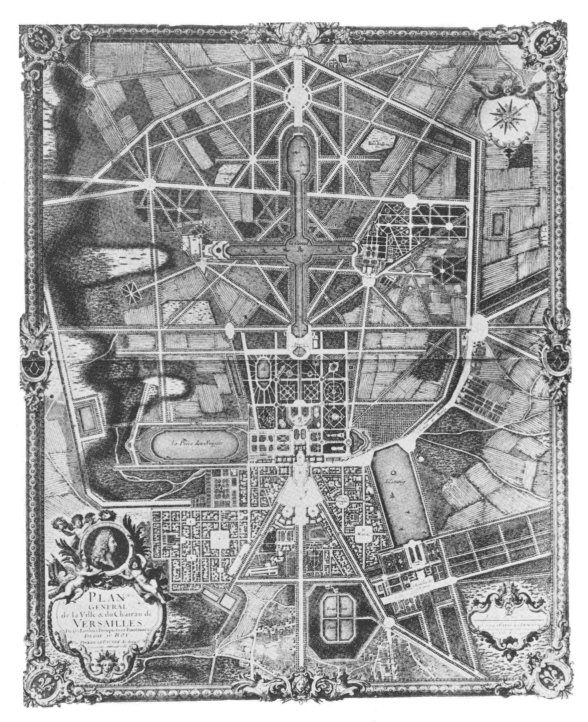

Plan Of Versailles, 1717 *Source: Harvard Graduate School of Design Library*

. . . claim the suffrage of crowds and visitors. Every House within the City will stand square on the Streets, and every lot, even those on the divergent avenues will run square with their fronts.

These were the controlling elements L'Enfant used: monument, square, avenue, and uniform building line. They failed not so much because they were beyond the capacity of technology to construct but because there were no means to bring the scheme to early completion. There were neither reasons nor funds nor population to fill the stage.

Further, the carrying out of the plan had not been given as much thought as its design. Unlike Sixtus V before him and Georges Haussmann later, L'Enfant was not skilled enough to manage the execution of a grand design. He was at constant odds with the developers and landowners, apparently fearing speculators who had, in L'Enfant's view, "organized to engross the most of the [land] sale and master the whole business." Against George Washington's advice, he intended to raze houses which were constructed in the District prior to his plan and thus remove impingements on his design.

"Having the beauty and regularity of your plan in view, you pursue it as if every thing and person were obliged to yield to it," wrote Washington. L'Efant would not yield, and so Jefferson, writing as Washington's agent, notified the Commissioners overseeing the development of the District,

It having been found impracticable to employ Major L'Enfant in that degree of subordination which was lawful and proper, he has been notified that his services were at an end.

For seventy years thereafter the plan of Washington was the butt of jokes: the wilderness city, the city of magnificent distances, the village monumental, streets without houses and houses without streets.

Gradually, however, infilling took place, and parts of the city took the shape L'Enfant intended. Reporting to a special Senate Committee on the improvement and development of Washington, The District Park Commissioner's Consultants (1902) concluded in their report:

The original plan of the city of Washington, having stood the test of a century, has met universal approval. The departures from that plan are to be regretted, and whenever possible, remedied.

Not coincidently, all four commissioners were leaders in the City Beautiful movement, itself an attempt, in the spirit of L'Enfant, to bring visual order to the city through boulevards and civic centers, though with too little concern or feeling for the urbanization that surrounded them, as well documented in John Reps's *The Making of Urban America*.

L'Enfant's design camouflages a Washington that existed only in his dreams, a design structure honored when convenient, an obstacle thrown up against any genuine attempts to find a design for Washington appropriate to the time. Like the great American novel, a new plan for Washington may be beyond our conceptual capacity until another kind of Renaissance is reached.

In the Baroque influences one sees the full swing of a pendulum from designs which anticipate and accommodate change to those that carry the seeds of their own diminishment. In the Baroque one also sees emphasis on dynamic designs, in which movement through space generates an architecture of its own, clear indication that the presence and placement of circulation elements can create design structure at the largest scale of human habitation.

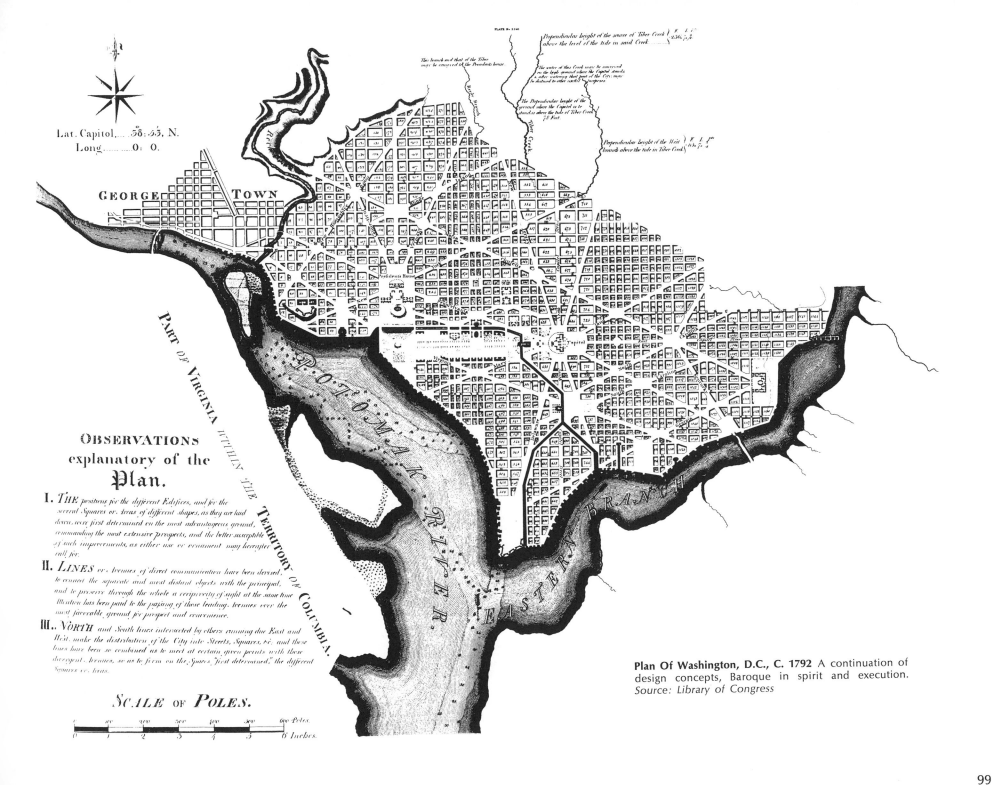

Lat. Capitol.....58:53, N.
Long.........0: 0.

GEORGE TOWN

PART OF VIRGINIA WITHIN THE TERRITORY OF COLUMBIA.

POTOMAK RIVER

EASTERN BRANCH

OBSERVATIONS
explanatory of the
Plan.

I. THE positions for the different Edifices, and for the several Squares or Areas of different shapes, as they are laid down, were first determined on the most advantageous ground, commanding the most extensive prospects, and the better susceptible of such improvements, as either use or ornament may hereafter call for.

II. LINES or Avenues of direct communication have been devised, to connect the separate and most distant objects with the principal, and to preserve through the whole a reciprocity of sight at the same time Attention has been paid to the passing of those leading Avenues over the most favorable ground for prospect and convenience.

III. NORTH and South lines intersected by others running due East and West, make the distribution of the City into Streets, Squares, &c. and these lines have been so combined as to meet at certain given points with those divergent Avenues, so as to form on the Spaces "first determined," the different Squares or Areas.

SCALE OF POLES.

Plan Of Washington, D.C., C. 1792 A continuation of design concepts, Baroque in spirit and execution.
Source: Library of Congress

99

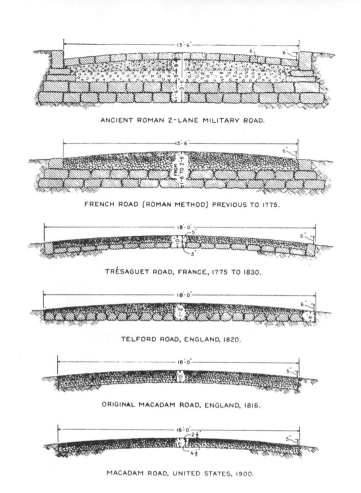

ANCIENT ROMAN 2-LANE MILITARY ROAD.

FRENCH ROAD (ROMAN METHOD) PREVIOUS TO 1775.

TRÉSAGUET ROAD, FRANCE, 1775 TO 1830.

TELFORD ROAD, ENGLAND, 1820.

ORIGINAL MACADAM ROAD, ENGLAND, 1816.

MACADAM ROAD, UNITED STATES, 1900.

Twenty Centuries Of Road Profiles Source: Harvard Graduate School of Design Library

Early American Post Station Well into the eighteenth century, American roads and transport facilities had not yet reached the level of Rome's achievements. *Source: Library of Congress*

Proposal For Reuse Of Early American Canals The remnants of the nineteenth-century canal system are well located to serve the recreation needs of nearby urban population. *Planners and designers: Dober, Paddock, Upton and Associates, Inc., and Richard C. Stauffer, Architect*

CIRCULATION

Clearly a center, as defined earlier, can establish design structure, whether public place, monument, single building, or point of origin, a center gives form and animation to the physical environment. It appears, however, that a center cannot endure for long independent of the context around it, nor independent of the circulation networks which connect it to other parts of a larger design context.

Circulation is the act of passing from place to place. Design structure can be established by using lines of circulation to give boundaries to environmental areas, to make them accessible, to tie them together. Circulation is an essential form of communication without which symbiosis cannot be achieved within any system of human habitation. In very important ways the kinds of circulation available establish the size and scale of the environment and in some instances its quality.

Historic and contemporary community designs can be grouped by circulation characteristics, for example, type of locomotion: foot, water, animal, and machine. Each of these methods of movement has time, distance, and functional limitations and capabilities. Each is associated with specialized channels of movement. Each produces distinctive design forms.

The most primitive forms of human existence and settlement were fixed by the limits of circulation. While nomadic tribes could follow the supply of game and fish from one district to another, the *location* of a permanent settlement was dependent on the amount of travel that could be economically carried on between the sources of food supply and the permanent camp. The *size* of the camp was determined by the density of food within reasonable travel distances. Substitute job opportunities for food supply, and the same relationship exists in modern life. The place of work and the place of residence is fixed by the limits of commuting. The size of the residential community is fixed by the number of job opportunities available.

The introduction of cultivation in ancient cultures had the effect of increasing the density of food supply, and accordingly fewer people were needed for production and transport of food over long distances. The surplus time is believed to have allowed an increase in population and the emergence of specialized occupations. As described earlier, it was thus possible for urbanization to take place. Beyond this initial stage, however, further urbanization is again dependent on the kinds of circulation available for the transport of fuel, building materials, artifacts, and people.

Studying relative modes of primitive transport, C. Clark has estimated that the cost of transporting by water one ton of grain one kilometer in India today is the same as it was in the Ancient world: about one kilogram of grain. Human portage is four to ten times as expensive as water transport, and pack-animal transport is at least twice as costly as water—even with a good road network. Accordingly, the first cities come into being, survive, and grow in importance because of their proximity to terrain which yields the highest amounts of grain and simultaneously has navigable waterways which will allow the cheapest form of transport—circumstances that were present in the Indus and Nile Valleys.

Since the dawn of cities, the invention and enlargement of means of circulation and transportation have been accompanied by new forms of civilization and new patterns of urbanization.

Rome was the first successful civilization not tied to water routes—the first paved road empire. The Roman road network has been considered by historians as one of the Romans' greatest achievements. The terminal points were the boundaries of the Empire, their major juncture the all-powerful center, the Forum of Rome. Over fifty-three thousand miles were constructed. (The U.S. Interstate Highway System has forty-one thousand planned, twenty thousand completed.)

The Roman looked upon his roads as instruments for subjugating, walling, fortifying, and controlling the world, though function was not always the builder's sole concern. The work of C. Gracchus (the greatest Roman engineer, says Plutarch) is instructive and illuminating. "His main care was the construction of roads, in which he paid regard to their useful and ornamental side."

Roads were driven across the landscape directly, without regard to natural obstacles; and design was uniform on main routes throughout the Empire. Two parallel trenches were dug, ten to fifteen feet apart, and the material in between removed. A triple foundation was laid: first a layer of smaller stones, then one mixed with mortar and firmly tamped, then a bed of concrete. The covering pavement was cut in rectangular or polygonal shape from material indigenous to the area.

The straight line appeared in the landscape as a continuous wall with a constant height throughout. Hollows were filled and ravines bridged to maintain a consistent elevation for military reasons. Thus the road was like a platform. A view could be had of the immediate surroundings, which reduced in hostile territory the chances of surprise attack. Because of the directness and ease of movement the roads afforded, legionnaires could assemble and travel with great dispatch. Along these routes to conquest, administrators, commercial travelers, students, and teachers would follow, taking Roman culture outwards and other influences back.

The network was an extraordinary movement system and so remained for centuries, a national network not even closely approximated in England or the United States until well into the eighteenth century. C. B. Troedsson describes how Sir Robert Peel was notified in Rome of his selection as Prime Minister. "Sparing neither expense or himself," he traveled by fastest transport to London. The trip took fourteen days, exactly the same time of travel expended by Roman administrators over a similar stretch of road seventeen hundred years earlier.

Benefits both military and cultural induced the institution of the *post*, the first world-wide, regulated transportation system. A hierarchy of buildings and activities was set at regular points along the road network. Every six miles there was a change station, which by law had to have at least forty horses, mules, and donkeys ever ready. Every fifteen miles there were posting places where teams, drivers, and vehicles as well as single animals could be exchanged. Employees at the post included doctors, blacksmiths, conductors, and mechanics.

By the fourth century A.D. the entire system was run by a central office, and for everything that the nineteenth-century train or steamship would later offer, Rome had a counterpart: fast and slow carriages; first-, second-, and third-class tickets; stopover privileges; sleeping cars; ordinary and express freight; mail service; overnight accommodations; meal service; guide service; travel tickets good for one to five years; and occasionally special travel privileges granted for the lifetime of the emperor. The system was comprehensive as well. As the empire grew beyond the continental limits, sea travel was included. The Mediterranean became "their sea," carrying bulk cargo of grain from Africa, olive oil from Spain, tin from Great Britain, and works of art from Greece.

The Roman roads permitted several centuries of city-building, carrying the beginnings of contemporary urbanization to France, England, Spain, and the lands across the Rhine and Danube. Along these roads Christianity spread, giving the new-found cities a reason for being when the center of the empire itself fell. As the empire collapsed, however, portions of the road were obliterated to impede the progress of invading armies. Other segments fell into complete disuse with the shrinking of cities, except as boundary markers for dioceses, parishes, counties, and estates. But eventually even the ruts proved to be important. The wheels of British stagecoaches were sized to fit them, and when the stages were put on iron tracks, the distances between the rails were set. Thus the gauges of many a continental railroad are Roman in origin.

Until well into the sixteenth century, improvements in land transport and circulation did not go beyond such inventions as the horseshoe, wagon swivel, horse bit, and compass, and the substitution of the stagecoach for the wheeled carriage. Access to water transport was therefore a condition for urban growth. Eighteenth-century empire builders, such as the French treasurer Colbert, attempted to develop a national economy by tying together roads and canals, although the tax policy for carrying out the scheme in France eventually led to the downfall of the regime. Throughout Europe, lack of road transportation, meant that inland cities were relatively isolated, carrying on a separate existence, producing for themselves little more than the necessities of life and, with few exceptions, remaining unable to use surplus goods or develop specializations.

For similar reasons, the initial settlement of the American east coast was concentrated in seaport cities. These grew rapidly and prosperously. Boston, New York, and Philadelphia served as regional centers, supplying the rural hinterland and exporting their surplus. But they and towns like Salem, Massachusetts; Newport, Rhode Island; and New London, Connecticut, thrived to a large extent only because they were tied to maritime enterprises.

Ingenious business practices in these new communities allowed capital to accumulate and the beneficial aspects of civilization to spread, especially through the device of triangular trade routes. As every American history course teaches, flour from Philadelphia and New York, domesticated animals from Connecticut, and fish and lumber from New England were shipped to the West Indies. The cargoes were exchanged there for sugar, indigo, and coffee which were carried to England. The Caribbean raw goods would in turn be exchanged for articles of manufacture and then taken back to the East Coast.

As can be expected, designs of all early American coastal cities reflect the importance of the port. Traffic from the inland towns aimed at that one destination, which set much of the existing street patterns in the early American cities. Significantly, the financial centers of several historic communities, such as Wall Street in New York and State Street in Boston, have never moved far from their original site on the threshold to the sea. The best early American architecture is within walking distance of the original waterfront. Furthermore, many such waterfront districts now await creative reclamation, being superb sites for city life. They are natural places for quality environment, and the historic artifacts remaining offer unique opportunities for a civic design that has continuity with the past and a flavor distinctive to the particular city.

The extension of coastal civilization inland was motivated by religious beliefs, mercantile instincts, and political desire to stake out claims to wilderness tracts. Initially, these second-stage settlements were founded along the banks of the larger rivers which flowed to towns adjacent to coastal indentations, thus being suitable as smaller ports between the major cities. The network of cities was fashioned by water.

This was not to be wondered at. As in Great Britain, travel by land was primarily along toll roads, built by private trusts. Progress was slow, costs were high, and the means of travel limited to pack mule, saddle horse, and stagecoach. The latter at best could take six to eight passengers. In 1802 a one-hundred-mile trip from New York to Philadelphia, took one and a half times longer than a similar trip on an old Roman road from Pompeii to Naples. Furthermore the American road was filled with dust in summer and mud in spring and autumn. Land travel was neither attractive, comfortable, nor convenient.

Although Congress authorized public road construction into the Ohio Valley in 1802 and additional construction in 1811 and 1837, the significant transport until the advent of the railroad was by water. Rivers and lakes were used to open the continent and canals dug where nature was improvident. The canals allowed business acumen and initiative "to move from wharf to waterfall," the latter giving an inexpensive form of power for early manufacture. Canals gave access to new markets and stimulus to city development. The most spectacular was the Erie Canal, constructed between 1817 and 1825. It linked Buffalo, on Lake Erie, with the Hudson River at Albany by way of the Mohawk Valley. A second waterway was constructed northward to Lake Champlain. The canal proved to be the major factor in New York's rise to economic supremacy in the Northeast.

The relationship of commerce and prosperity to water site was not unnoticed in other sections of the country. Canal construction gave impetus to speculation in city development as well as to other business enterprises—"town jobbing", Reps calls it—carried on by visionaries more adept "with the pen and the drawing instruments, than with the apparatus of the mill." Lands adjacent to stream, river, and canal were bought and platted in hopes of booming profits, just as later the coming of the railroad puffed land values and new town development beyond all reason and need. Most of these schemes, early and late, went bust; and the town sites that did survive developed at a pace much slower than avarice and greed had planned. They still suffer from land butchery, spending enormous sums to erase the damage done by speculation. Poor street design, environmental conditions such as building in flood plains, and the general inadequacies of "backwoods baroque" planning continue to impede urban development in these towns and cities a century and more after their founding.

By the time of the Civil War, six thousand miles of canals had been cut, of which one thousand are still in use today. The presence of these old rights-of-way offers an unusual opportunity for a new kind of national park. Restored and used for recreational purposes, they could serve again as an economic resource to communities bypassed and in relative decline because of the ascendancy of other modes of transport. As described on page 100, the National Canal Park could be a living museum, an environmental contribution to both urban and rural areas, and a resource within reach of the population group in the United States having the least outdoor recreational land.

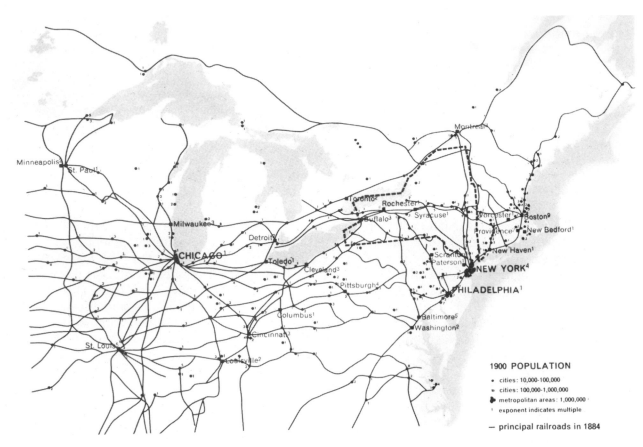

1900 POPULATION

- · cities: 10,000-100,000
- ● cities: 100,000-1,000,000
- ▲ metropolitan areas: 1,000,000
- ¹ exponent indicates multiple

— principal railroads in 1884

The Growth Of Cities And Rail Transport As the railroads expanded westward, new cities were founded and existing cities grew in size and wealth. *Source: Change, Challenge, Response. Office of Regional Development, Albany, New York, 1964*

City Density One of the by-products of intensive urbanization: congestion. *Source: Harper's Magazine, 1884*

BY RAIL POSSESSED

The nineteenth century was an era of immense urbanization and city expansion largely because of relatively cheap transportation. The steamboat and the railroad enabled large numbers of people to settle vast areas of the world. Via ocean, river, and land transport, virgin country was occupied and industrialized; and small cities grew into large metropolises.

In previous ages major cities came into being because of their religious, military, or political importance. From the introduction of the railroad onward, commercial and industrial activities became the dominant reasons for the birth and growth of cities. Rail transport overcame all natural barriers to concentration. It brought to inland areas economic opportunity heretofore restricted to sea, river, and canal ports.

At the time of the Civil War, eight of nine cities in the United States over 100,000 in population, were thriving seaports. The ninth, Brooklyn, lay at the edge of New York harbor. Between Chicago and San Francisco there was no urban place larger than Dubuque, Iowa, with thirteen thousand residents. By 1920, the peak of urbanization, twenty-five cities had more than 200,000 people, among them Chicago, St. Louis, Denver, Minneapolis, St. Paul, Seattle, Los Angeles, Atlanta, Birmingham, and Kansas City —cities that didn't exist or were small villages seventy years earlier. In the same period the number of urban places leaped from a few hundred to over two thousand. The wide geographical distribution and population concentration characterizing this urbanization was only made possible by the railroad and its urban counterpart, the trolley car.

In environmental design, the nineteenth century was a frantic one. An early English railroad, built by Stephenson and opened in 1830, was greeted with dire warnings as to its impact on safety and health. It was predicted that within its environs women would give premature birth, hens would not lay, and the countryside would be set afire. For a short time the train had to be preceded by a man on foot carrying a warning flag announcing that the "infernal machine" was coming.

The strongest immediate impact in England was on the agricultural economy. Rail transport allowed regional specialization. Livestock and dairy farming could be concentrated in the wetter and hillier western counties, and grain production concentrated in the drier east. Further, the railroad permitted shipping of cheap, industrially produced goods which competed with local handicrafts. The self-sufficient village was broken up and a large population made available for urban employment. With a head start in industrial techniques, surplus capital, and surplus labor, England was able to manufacture and export industrial goods in exchange for food supply, furthering the migration from farm to city, again allowing industrial expansion, and bigger cities.

In America the railroads first tied together the Eastern coastal cities and by 1850 were extended southward and westward to the edge of the great American plains, giving great benefit to the territories they connected. During the Civil War, rights-of-way were continually threatened by one army or the other; for the movement of goods by track had marked advantages, though foot, horse, and water were still the soldier's transport.

With Appomattox and peace, the same era that saw communication to Europe facilitated by the transatlantic cable saw protective tariffs and a turning inward in economic development. The maritime trade, already weakened by embargoes during the Civil War, declined. It became evident that the best opportunity for commercial profit lay in internal expansion and the opening of the West.

For this development the railroad became the chief instrument, with Congressional encouragement through land-grant rights and laissez-faire financial practices. Investing $10 billion in capital equipment and rights-of-way, private enterprise laced the continent together with over 160,000 miles of track in thirty years. Up to World War I, railroad construction and operation were the most important single aspect of the national economy. The railroads controlled one-sixth of the nation's wealth, and their capital was ten times larger than that of all banks and trust companies combined.

By 1870 coast-to-coast travel was reduced to six days from the one hundred days by combined rail and prairie schooner, dependent on the weather and the mood of the Indians, of a decade earlier. Of more importance than travel time was the cheapness of transport. Beyond the immediate rail line the country was still using horse and wagon. All the hard-surfaced roads laid end to end would not reach from New York to Boston. To every city the railroad meant the difference between economic growth and stagnation. It was the all-important factor in city development, "the life-giving line." Its symbolic importance can be seen in early city designs: the Union Station occupied a central position, and in the older cities its architecture was probably the best of its time.

Thus the impact of the railroad was not only great on a national level but on local life as well. The rail lines fixed the position of land uses and in turn social values—there being a right and wrong side to every track. Wherever the railroad stopped, a community would spring up; and commercial and manufacturing activities would first be concentrated near the station. Cities grew to greatest size when several railroads came to a common juncture, such as in Kansas City and Chicago. To these central places gravitated commerce and exchange and the main streets and roads. Densities rose to 100,000 people to the square mile in the vicinity of the major terminals.

From the eighteenth-century onwards only the richest people could afford to live more than three to four miles from work. The average man was tied to his place of employment by the length of a comfortable walk. Accordingly, dense blue-collar districts grew up around the manufacturing works, and white-collar districts developed close to downtown.

In New York City, except for those who could commute by steam train or ferry, the journey from work to the more attractive outskirts was too far for a daily trip. The densest housing clustered around the city's main industries: clothing, tobacco, retailing. In the entire American experience, Manhattan perhaps had the poorest housing of all for the working class, for even the meanest miner's shack or company town had some contact with nature, and the overcrowding was not community-wide. By 1875 New York had thirty thousand tenements holding more than one million people. Though substantial effort was made to eliminate the worst conditions, there were still close to 100,000 inhabited rooms with no windows at the century's end. Large parts of the city were filled with uniform buildings, four to five stories high, unrelieved by any nearby open space and crowded well beyond their designed capacity.

The misery of urbanization conjures up any number of images—some stereotyped exaggerations, others real-to-life. But certainly in the Satanic picture gallery, two documents of chaotic design and its foul atmo-sphere hang side by side: The English mill town and the East Side slum. Neither does justice to the horrible conditions in the environment where, in Jacob Riis's expression, "the other half lived."

As many cities grew up at river sites, the location of bridges and ferries continued to influence as they always had the location of lines of circulation and the adjacent land uses but, in terms of the technology of the railroad age. Central city and across-the-river suburb became tied into a continuous urban area by newly invented wrought iron and steel cantilevered bridges, carrying horse-drawn, electric powered and steam traction cars.

The internal design of the city was as affected by cheap mass transport along fixed rail lines as the overall city design was shaped by the railroad. In London and other regional capitals of the western world, the density of population continually rose from 1800 to the introduction of the horse tram and then steadily declined with the use of the underground steam train (later electrified) and the electric tram with the resulting convenience of residential dispersal.

Cable cars, first tried in San Francisco in 1872, offered a way to reduce horse-drawn traffic. This was a coming need, for congestion in all central business districts was fierce, with express cars in New York City being pulled by as many as twelve horses. But the cost of cable operation was high and spurred inventors to search for a cheaper form of mechanical traction. Thomas A. Edison successfully developed an electric motor, and one of his associates found a way to electrify the horse car. Richmond, Virginia, opened the first trolley line in 1887, and twenty-five additional cities commenced similar operations within a year.

Though railroads were moving fifty million commuters annually in the ten largest cities, within a decade of their invention trolley companies had twice as many coaches in operation.

The tram, horse-drawn and electric, increased the volume of business downtown, encouraging entrepeneurs to replace one- and two-story brick and wooden structures with masonry blocks three- to five-stories high. Cast-iron fronts and other industrialized building components began to give a uniform appearance to commercial centers all over the country. The visitor could quickly tell whether everything was up-to-date in Kansas City or anywhere else—an aspect of urban cosmetics still commonly felt. Land values rose, and the cheapest downtown housing was replaced with small factories and warehouses, now more than ever convenient for a larger labor pool.

Not only housing demolition but also new housing development were stimulated by mass transit. The first horsecar line in Rochester (1863) was built and operated by real estate speculators. B. McKelvey notes that though the transportation was unremunerative, the housing paid well, diverting city growth from its earlier east-west direction and eventually fostering the construction of hundreds of modest homes on the city's outskirts. Boston's horsecars similarly opened new sites for inexpensive housing. Cheap transport seemingly allowed an inexpensive environment for everyone.

The spread of mass transit saw the revolution of mass-housing styles. Philadelphia and Baltimore housed their rising population by constructing miles of row housing—small individual homes fronting little streets, with tiny back yards and few neighborhood amenities. It was not entirely an architecture of the poor but rather a city archetype, for even the wealthy citizens lived cheek to cheek. Baltimore's Park Avenue still has several examples of excellent period row-house architecture saved from urban blight by an intelligent renewal program and handsomely restored for city life by imaginative architects. Until the increase in auto traffic poisoned the air and the noise became intolerable, the porch-fronted row houses of North Broad Street in Philadelphia were an equally comfortable and pleasant urban environment.

Other cities increased density by changing the housing type. Boston introduced the four story three-decker: a balloon-frame building of brick or wood with porches on the three upper stories in front or rear. Chicago built repetitive blocks of monotonous architecture combining the row house dignity of Philadelphia with the vertical concentration of Boston.

As the old centers became less attractive places to live, the railroad and trolley opened up more and more new land for development at the city outskirts and beyond. For the urban masses, weekend recreation was improved as amusement parks and picnic areas were constructed at the end of the lines. Interurban travel was made more comfortable and convenient. The ubiquitous electric trolley connected most of the major cities on the East Coast together. Those who could afford to do so moved to the suburbs, which sprang up along the transport lines that radiated outwards from the larger cities. This spoke-like design was reinforced when hard-paved roads were introduced, thus firming the skeleton which became the early twentieth-century city.

Downtown At The Turn Of The Century The technology of centralization improved the efficiency of the commercial life of the city, but not the appearance of the environment. *Photo: Library of Congress*

Burnham And Root's Montauk Block *Source: Harvard Graduate School of Design Library*

The Architecture Of Finance, New Orleans, Louisiana, C. 1895 *Reprinted from: Architectural Record, 1898*

TECHNOLOGIES OF CENTRALIZATION AND DESIGN EFFECT

The decades from 1880 to 1910 were pivotal yet paradoxical years. The national population doubled; the frontier vanished. While the farm lands of the Midwest were being filled, most of the new population crowded into the cities; and the new technologies inexorably and decisively turned the country, at the very time of great rural expansion, into an increasingly urban and industrial nation. While the railroad induced centralization, it was also beginning to bring an opposite effect. The many lines that came together in the center of Chicago, creating a strong central area, also allowed stockyards, lumber mills, foundries, and machine shops to be scattered along the same spoke-like rights-of-way. Around these employment zones cheap housing would spring up and in time churches and schools and small stores—the beginnings of a metropolitan area. Again, in Detroit by 1900 factories and warehouses had moved from waterfront locations along Lake Michigan to railroad-oriented sites, while the trolley lines that reinforced downtown also pulled commerce out, like taffy, along the routes of heaviest use. Caught between the two, in-and-out forces of development, cellular enclaves of residence, new and old, were boxed in by the transportation system with no functional, much less design, structure. But as fast as the city was extended outward by iron rail, other technologies were encouring centralization: electric lighting, the elevator, telephone, and all the marvels of architecture that went into skyscrapers, high-rise apartments, and hotels.

The history of these environmental changes could be written in terms of energy sources alone. By 1910 ninety per cent of the power in the United States was produced by soft coal—much of the coal moved by railroad, the trackage again affecting the location of industry and in turn community, and the coal produced much of the grime.

The Corliss engine displayed in the Philadelphia Centennial Exhibition (1876) showed how cheap electricity could be produced by steam engines. Electricity was welcomed as a clean power that reinforced centralization. As mentioned, its first influence was on forms of traction: trolley car and later subway. Electricity permitted a twenty-four-hour urban environment. With Thomas Edison's invention of an economical incandescent bulb (1878) exterior and interior environments could be lit to significantly higher levels of illumination cheaper than with gas or oil, irrespective of time and season.

With the application of electricity to elevators and escalators (1890), greater densities of people could be moved vertically at speeds faster than with the hydraulic lifts of the previous decades, and taller buildings followed. As a power source for the telegraph, and more importantly the telephone, electricity sustained concentration of business activities.

Invented in 1876, the telephone might be considered the first step towards twentieth-century automation. There was one telephone per twelve people by 1910 and now one telephone per three people in the United States. Telephone-using countries are urbanized countries, and the ranking of nations by per-capita income parallels the ranking of nations by number of telephones per person.

Exchange of information is a requisite for commercial survival. There is some evidence to suggest that the rapidity of flow of communications is a factor in business success. Messengers on foot and mail service are slow means of communication next to the telephone which offers instant exchange of ideas, advice, data, conjecture—all the nec-

Rockefeller Center The first, the largest, and perhaps still the best designed central-city office complex. *Photo: Thomas Airviews, courtesy of Rockefeller Center, Inc.*

essary and superfluous or not so superfluous discussion that accompanies a transaction. Because it can give access internally as quickly as externally, the telephone allows concentrations of people to be in easy contact inside a building, inside a central business district, inside a city, inside a metropolis, inside a region. Thus the telephone could glue together all the other conditions for centralizing business and commercial districts: a large pool of diversified, competent, specialized labor; professional people and their services; and people with technical skills—all having easy opportunity for contact and exchange.

The architecture of commerce which followed centralization and intensification of business in the central city was divided by two influences: the rational designs of engineering and the fads and styles of a society that looked to Europe to set its fashions.

The housing of this density of activities constitutes two main architectural trends: the Italianate architecture of commerce and the skyscraper, both of which had English roots.

In England, where specialized activities such as banking and insurance required fire-resistant structures and that element of prestige that comes from the employment of an architect of established reputation, the early commercial buildings with their heavy masonry walls and Italianate detailing looked like Renaissance palaces and indeed were fair copies of the Italian models. Soon, however, several external conditions distorted the pseudo-Renaissance designs. The requirements for internal light (some three decades before Edison's economical light bulb) led to enlarged windows, placed close together, a solution aided by removal of the English use tax on windows (1851). Further, the value of urban land suggested four- and five-story buildings in both American and English cities rather than the prototype style of three stories or less. Land values and construction costs rather than aesthetics eventually stimulated technical innovations and architectural changes in commercial structures, but for reasons of prestige, those who could do so resisted tall buildings.

The implied continuity from Florentine banker to industrial banker was deliberately symbolized and maintained by client and designer. The superficial style was not thrown off until recent years when depositors discovered that an imposing façade was no guarantee of sound financial practices, and the banks found out that profits were directly related to how many people they could attract to use their services. For a century, though, a Renaissance image marked the important places of financial exchange, both in large city and country town, in the central business district and in any neighborhood moneyed enough to support a bank—one example of how implied virtues influence the designed environment and the role of architecture therein. For a long time only the strongest willed designers such as Louis Sullivan could find a way to toss away the accepted mold, as he did in his Merchant's National Bank in Winona, Minnesota.

With the exception of banks and some insurance firms, most commercial office space was influenced more by urban location and competitive real estate practices than by aesthetic considerations. The size and mixture of space was determined by lot size and entrepreneurial skill. Cast-iron façades could be bought by catalog number and assembled by skilled builders, designs which are highly valued today as examples of nineteenth-century manufacture and period artifacts. Scaled to uniform dimensions by technical limitations, they fitted together without pretension to form a unified design for city blocks.

Until the advent of the elevator, commercial buildings were rarely more than five floors. No matter how valuable the site or prestigious the firm, few clients would walk up more than four flights. The Bunker Hill Monument near Boston had a steam-hoisted passenger lift in 1844; but the first elevator in a commercial building was installed, after Elisha G. Otis's design, in a downtown New York City store in 1857. The introduction of an elevator solution helped win the Equitable Building design competition for the architects Gilman, Kimball, and Post (1871). The construction of the building had three immediate effects. First, relatively small existing buildings were topped with one-story mansards and then had elevators installed to reach the top floors. Second, all new commercial buildings were designed with elevators in mind, and third, the height of the average structure was doubled. The skylines of the big cities, once marked by church steeples, now were dominated by a new building type: the skyscraper.

The evolution of the skyscraper, the only significant innovation to come out of formal American architecture of the nineteenth-century shows the pressures of function and commercial prestige alternating in shaping what was to become the largest single architectural element in the designed environment. Rightfully or not, the skyscraper, the by-product of concentrated lines of circulation and urban density, is the symbol of the late nineteenth- and early twentieth-century city and encompasses in a single form all the technologies of centralization: the iron and then steel I beam, electric power, and the telephone.

In Chicago the accidents of history propitiously combined to result in the most significant expressions of this new building form. The city had been destroyed by the Great Fire of 1871. Chicago's strategic location, however, encouraged rapid reconstruction and continued growth. The economic depression that followed the Panic of 1873, which dampened the building industry elsewhere, was not felt as much in Chicago; and it thus drew architects and engineers from all over the country. The opportunities were immense. As Henry-Russell Hitchcock notes, there were "no established traditions, no real professional leaders, and ignorance of all architectural styles past and present." In this milieu William Le Baron Jenney was prompted to introduce the first fire-retarding skyscraper construction (steel framework and masonry cladding); Burnham and Root to discover spread foundations (which kept tall and heavy buildings from sinking into the Chicago mud); and Dankamar Adler, Louis Sullivan, and Frank Lloyd Wright to arrive at designs free from academic traditions and details.

The tradition of function determining form was carried by Chicago's leading architects to other cities, such as St. Louis and Buffalo, where Adler's Wainwright Building and Sullivan's Guarantee Building were erected. The relatively simple lines of the Chicago School sharply contrasted with the New York skyscrapers, where each succeeding building not only tried to rise higher than its neighbor but also sought a visual advantage through stylistic distortions as grotesque as they were distinctive—a notable example being Cass Gilbert's interpretation of the Late Gothic in his Woolworth Tower (1913).

This competitive urge, commercially based, still continues. Because values accrue with high densities, our present system of land ownership, city development, and building technology uses the skyscraper as the most convenient way to increase densities, through building and rebuilding, and thus profit from rising land values. The million-square-foot building is no longer an exception, nor is the fifty-story height, whether tower or slab; but despite continuing construction unprecedented in volume, each new addition to the skyline seeks to be unique and dominant.

The same attitude now affects the urban space around the base of the building. The design strength of linear streets is now being weakened as it is chopped up by small plazas at the foot of each new skyscraper. Tiny, tinny, faddish, disconnected, functionally irrelevant, and of no aesthetic consequence, many are no more than visual burps that accompany an undigestible design philosophy—all this despite the felicitous example of Rockefeller Center. Here a visual unity was established that is distinguished when viewed from a distance and memorable when experienced close up. An otherwise modest plaza has become vital because of the building around it. Massing and materials complement each other—a design concept from which the latest additions to the center have unfortunately departed.

In the larger environmental perspective the design of the American skyscraper has not been solved. Elevator speeds have been increased. There have been refinements in mechanical controls. New forms of cladding have been introduced. But it now takes forty tons of building materials to shelter each employee. One well-praised office building must run its air conditioning systems in subfreezing winter temperatures because solar radiation occurs through a poorly conceived fenestration pattern. The configurations of office floors in general commercial buildings are derived from a formula based on the distance from windows to the center of the building rather than an understanding of how human beings work. Typical city blocks are collections of unrelated skyscrapers—lost opportunities for coherent civic designs. The cityscape is the worse for all these events.

New Forces, New Design Wichita, Kansas, central area. The growth of the town was based on river and rail access. Concentration was permitted because of the technologies of centralization. Now the highway system sets the dimensions and scale of city development. *Source: Wichita City Planning Department*

THE NEW FORCE

The iron rails opened the country and filled the nineteenth and early twentieth-century American cities through unbridled private interests which created as much conflict as vitality. The spectacular features of technological advances, hardly altruistic, managed to cover over blight, inefficiencies, inequities, disease, and graft. In response, good government marched up the hill and down again, at best reducing social tensions but hardly introducing permanent urban reforms. The monopolies of economic power were difficult to pierce. The tides of immigration continually brought language and cultural differences that made it difficult to muster a persuasive and universally acceptable case. The City Beautiful Movement helped gain large parks, impressive civic centers, and streets embellished with lamp standards and trees; but in the interior blocks behind the glitter, the urban environment remained the way it was: inhuman, unrewarding, and disordered.

However lugubrious this account of environmental decay, the very same cities experiencing it were laboratories of progress, testing grounds for an urban democratic society yet to come. The traditional values of community life were not entirely lost. Through formal associations and informal contacts, new sets of relationships were being forged among immigrants and native born in politics and public welfare. Public education created bridges and ladders for social and economic advancement, often within two generations. Both the arts and the sciences thrived. The exacerbations of urban life did lead to some thoughtful measures whose full effects have not yet been diminished: public health, social-service agencies, recreation, museums, orchestras, and libraries, free education.

The social and economic evolution did produce unprecedented opportunity and an ever-increasing state of material well-being for most of the nation. That so little of this produced or took place in a designed environment is a historic fact. Whether some comprehensive and purposeful order would eventually have emerged will never be known, for at the time some improvement was being offered through social reform and city planning, a new technical achievement pervasive in influence upset all previous images, assumptions, and predictions and shaped the new environment: the automobile.

Expressway Scale: By-Product Of The Age Of The Automobile A portion of the downtown expressway system, Cincinnati, Ohio.
Photo: Department of Public Works, State of Ohio

ACCOMMODATING THE MOTOR VEHICLE

From the viewpoint of safety, general comfort, and aesthetic effect on both large-scale and small-scale designs, the motor vehicle today is the major influence on the environment.

In 1967 there were more than 100 million licensed drivers in the United States and almost as many licensed vehicles. The automobile generated almost one fifth of the gross national product. Automobile drivers traveled 935 billion miles and had 13.6 million accidents, resulting in fifty-three thousand dead and 1.5 million people permanently disabled. The premiums alone on automobile-casualty insurance came to 9.2 billion dollars, twice the total Federal support for elementary, secondary, and higher education. The predominant number of injuries and fatalities occurred in built-up areas and along slow-speed roads. There has been no urban area unaffected by the automobile's power to damage by impact, fumes, noise, and ugliness.

As the number of high-speed safety-engineered expressways and roadways increases, the congestion and danger on local thoroughfares seem to increase. The ease of driving on fast-speed roads may affect the psychology of driving on lower speed streets. In any case, drivers invade and infiltrate local streets and neighborhoods, impatient and frustrated, seeking a way to avoid congestion and delay along the older routes of travel. The automobile has been criticized as being unsafe at any *speed*. A more fundamental problem is that it has made unsafe any *place*.

As mentioned earlier, vehicular fumes are a major cause of atmospheric pollution. The problem is more than discomfort. An impressive body of circumstantial evidence indicates that high cancer rates and cardio-respiratory diseases as well as constant minor irritations such as rasping coughs, smarting eyes, nausea, and irritability can be traced to the unpleasant by-products of the internal combustion engine. Heavily traveled automobile routes are atmospheric sewers. Locating them in habitable areas is as barbaric a custom as handling effluents in open ditches.

An equal nuisance is noise: engine roaring, rumbling, and backfiring, squealing brakes, honking horns, shifting gears, the uneven hum of tires on road surfaces, door slamming, and the squeaking, penetrating clatter of loose chassis and groaning wheels. While no conclusive evidence points to ill health, the noise of traffic is prejudicial to the general enjoyment of urban life, destructive to sleep and the small pleasantries of conversation and social intercourse, and an obstacle to the efficient conduct of any business or activity which abuts a heavily traveled street.

The visual intrusion is of equal consequence, and the steady encroachment of the automobile into the urbanscape is unending. In the briefest of glances the eyes cannot avoid an encounter with gross and fine visual exasperations. In motion or at rest the automobile is the dominant element in the city visual; and while its presence is most keenly felt in the crowded core areas, the replacement of landscape and architectural amenities outside the central city with a surface of automobiles and pavements for automobiles seems a universal condition of life. Servicing, maintenance, and parking areas give the older city fabric a moth-eaten appearance from the sidewalk as well as the air. On the ground the automobile stains the surface of roads and walks, catches and holds dirt and debris, reflects light and heat.

Not the least of visual problems is the

chaotic clutter of signs, lights, and direction signals necessary to control traffic movement and the advertisements and divertisements created to catch the attention of driver and rider—each competing for attention with compelling insistence. As the messages get confused and lost, the visual noise rises and the sensitive eye tunes out the pervasive ugliness: damaged street surfaces, broken street furniture, battered signs, derelict car dumps, dreary and formless parking lots, and highway improvements (how debased the word) out of scale with the environment they are designed to serve.

The design conflicts between vehicle and environment may be considered a matter of technical lag, i.e. the invention of the automobile was not accompanied by appropriate inventions for moving and controlling traffic. Much of our urban environment was never intended to serve the motor vehicle. The accidental and deliberate putting together of buildings, open spaces, and streets to form an urban area was never done in anticipation that each family would have its own means of private locomotion, mechanically contrived for easy acquisition and use, capable of both slow and fast speeds and short- and long-distance travel. The fact that historic boulevards such as the Champs Elysées in Paris or Constitution Avenue in Washington, D.C., are suitable for moving heavy volumes of traffic today is sheer happenstance. Certainly, the blighting effect that such traffic has had on areas adjacent to it is proof enough that no rational mind was devising a suitable prototype of late twentieth-century engineering technology when such boulevards were designed.

"Is this trip necessary—in an automobile?" The utility of the motor vehicle could be questioned on the grounds that cumulating ill effects at an accelerating rate are inimical to human life. Outside of some deep-rooted philosophical line of reasoning, such as Mumford's view of technology and civilization, there are no clear-cut answers if one assumes that environmental problems are not related to the machine itself but rather to the use of the machine.

The motor vehicle has distinct advantages. It provides mobile services such as fire fighting and police protection and to outlying areas, health clinics and culture, through bookmobiles and artmobiles. Further, the motor vehicle is an economic form of transport for manufactured items, food, certain low-weight and high-value products, personal goods and objects and regional mail delivery. It can carry large numbers of people along flexible but designed routes of transport (such as bus lines) or a few people at odd times and along many arbitrary travel lines (such as in a private automobile).

Personal mobility is a large factor in automobile ownership and will surely increase. Recent studies indicate that automobile owners take four times as many trips as nonautomobile owners. In fact the predominant number of vehicle trips each year are for private pleasure and individual convenience. Since the amount of leisure time available at the end of this century is predicted to be three times as much as it was at the beginning, further increase in traffic above and beyond population increase can be expected. The new leisure is expected to free large numbers of people not only from fixed places of work and residence, but also probably from a fixed calendar for leisure-time pursuits. The pattern of winter work and summer vacation, already changed for some income groups, is likely to disappear in favor of several free-from-work periods during the course of the year. The population will be younger, because of education more heterogeneous in its outlook, occupation and interests, distinctly mobile, and either on its way to the excesses that marked the decline of Rome, or moving with energy and exuberance to the ever higher plateau of civilization that marks a progressive and productive society. In either case, the population will drive, not because the automobile is a status symbol, but because it is the accepted and conventional mode of personal transport.

Personal mobility may also increase because of changing patterns in employment opportunities. Personal and professional services are replacing industrial and related jobs as the predominant means of earning a living. The location of the new employment centers will not be as dependent on power and transportation systems as manufacturing and goods distribution have been in the past. Accordingly, there may be greater environmental compatibility between places of work, recreation, and residence. Instead of distinctive land-use areas and a resultant strongly defined work-to-home commuting pattern (supporting some form of mass transport as well as the individual automobile), a variegated land-use pattern may emerge. People may be free from the requisites of traveling along historic fixed lines in their journey to work. A kind of scrambled pattern of movement may occur with many exchanges of people between many communities that can each support a full range of urban activities— in contrast with the present pattern of bedroom community separate from place-of-work community.

Within the realm of technical feasibility, it is possible to imagine an alternate form of individual locomotion. Experimental models of miniature jet propulsion engines, strapped to the human back, have been manufactured. Antigravity machines have become a favorite interest of basement inventors. Such devices have difficulty in operating in inclement weather. The navigational hazards are possibly greater than for ground vehicles because of free access to air space. Strict regulation would seem necessary in this respect, and unless controls were to be exercised the ultimate environmental barrier might be breached by air view and the individual's privacy threatened. Finally, personal and public liability insurance costs alone may rule out the introduction of such machines.

Technical innovations, it would seem, are more likely to be aimed at improving the individual ground vehicle rather than inventing a new one not tied to the road. Recent safety measures are the first steps in this direction, as well as substantial research being carried out by automobile manufacturers and public agencies.

Of special import to the environment is the search for an alternative power source to the fossil-fuel engine. Battery-generated automobiles are on the horizon, quieter and perhaps more efficient than combustion engines. This line of research may lead to a family of vehicles for use in different kinds of environmental zones: smaller, relatively slow-speed, electrically powered vehicles for short trips and large and faster vehicles for long-distance travel. The latter may be linked together in train-like fashion and controlled by computers. The management of this traffic may have to be handled as a public utility. This in turn may lead to new modes of licensing the use of public space for circulation and ownership rights to private vehicles.

The concept of individual transport as a public utility (like water, instantly available) is further described in later sections on micromovement systems and modern transportation technology. The point made here is that private and public transportation must be conceived as a total movement system. Similarly, any discussion of private vehicles in the designed environment cannot be carried on without reference to technological advances in mass transport. Hovercraft, monorail, pneumatic tubes, conveyor belts, and moving sidewalks are all part of the solution to the conflict between man and motor. The manner in which high-volume transport enters the urban environment, the connections between modes of travel, and the implications this has for design forms at the termini are significant issues in environmental design.

But after all these things are said, the compelling advantages of a relatively small, independent; self-powered, highly maneuverable ground-vehicle still have such practical and psychological appeal to individuals that designs for the future cannot logically be premised on the decline of the automobile or equivalent machine. And while alternative forms of movement are likely to emerge, including an extension of present mass transportation technology, the central question remains: How does the automobile influence design structure?

As stated categorically above, little of our present environment has been designed for accommodating the automobile or truck, yet one can see and trace the effects that the motor vehicle has had on the pattern of recent urbanization.

Much of the present conflict between motor vehicle and environment is rooted in the historic design structure of urban places. As a result, the first order of relief has been to decentralize the denser urban areas. Though this movement has not been primarily brought about by the automobile, the motor vehicle has fostered it initially, for two reasons. First, economic demands have caused parking and lines of circulation (such as improved highways) to replace less profitable land uses. The initial cost of improvements can be shown to be economically beneficial through the resulting reduction of travel time or increase of occupancy and value of land due to provision of access and parking. (The flaw in this line of reasoning is that it lacks recognition of longer-term consequences and environmental deficiencies which often are not entered into the equation of benefits and cost analysis.) Secondly, the attractiveness of individual mobility is only brought to its highest form (under present technology) when intensive urban activities requiring large amounts of parking (for example in a shopping center or industrial park) are relocated to less developed sites, typically on the periphery of the central city. This naturally gives further impetus for decentralization.

Lately an intermediate pattern of urbanization is developing between low-density suburbs and high density urban centers. Clusters of medium-density housing and associated commercial and professional services are being located near the intersection of the major expressways and the mass transport interchanges, thus having the advantages of highway access and urban density.

Accommodation of the automobile is not, however, the prime cause of urban decentralization and suburban growth. Equally important, for instance, are such technical innovations as a system of cheap, long-term, personal home mortgages. Furthermore, it is important to recall that the city itself was not a flawless environment before the advent of the automobile. For political, social, and economic reasons, it showed little capacity for meeting the requisites of new technological invention, urbanization, or human aspirations. The vast urban-to-suburban migrations, interregional movements from harsh climates to the gentler natural environments along the seacoasts, and the phenomenon of ghost towns and ghost cities may have been made possible by the motor vehicles: but they were not the stimulus. Given the historic conditions of inadequate public services, a diminished and demeaning physical environment, and the latent attractiveness of living and working on a new site free of urban blight, the depopulation of the nineteenth-century urban place might have happened through the twentieth century. Decentralization would then have been symptomatic of a total new concept of American life—low-density settlements as dramatic and pervasive in their own way as the nineteenth-century city was in the age of the iron rail. Such a trend is not occurring and will not occur for the following important reasons:

Some degree of compactness and proximity between large groups of people is necessary to support a rich and varied life. Concentrations of people encourage face-to-face contact and make possible a diversity of interests, services, and opportunities. With the advent of computer technology, decision-making has not been decentralized, though computers could be a force for scattering commerce, and government over the metropolis. Finance, business, government, medical services, education, and culture have not all followed the outward movement of new residences and shopping centers to the suburbs. Proximity of these specialized activities to each other has proven to be beneficial and desirable. They have, in fact, through private and public renewal, enlarged and improved the districts in which they are sited. They in turn attract and hold such activities as restaurants, specialty shops, and supporting services. To an important extent the presence and interaction of these activities have given cause for high-density housing to develop nearby. In a real way the last twenty years have been a period in which the core of the city has been pulled apart and put together again in a pleasant and functional manner. High density has not proven to be inimical to amenity in these new designs.

Further, there are economic limitations to an all-motor, low-density urbanization pattern. The cost of constructing, maintaining, and operating the infrastructure would be substantial. In addition, dependence on a single mode of movement (the automobile) would be impractical and the travel necessary to encompass a suitable range of services intolerable and self-defeating, if spread over a vast area.

Ecologically, low-density development is defended on the grounds that it may not induce irreversible changes in the metropolitan landscape and biota and that higher densities will. But less than three percent of the national land stock is urbanized, say fifty million acres; and only half that much again may be changed ecologically by the end of the century. Further, some degree of degradation may be tolerable during the change from raw land to urbanized land while one viable ecological system replaces another. Higher densities in existing urban areas may in fact help insure conservation of natural resources.

Finally, there is the sense of contrast, the chord of novelty, which biological man must hear for intellectual and physical well-being. There are beneficial effects to be accrued by having the best of all possible worlds—high-density urban places on one hand and countryside on the other. While the United States could accommodate five times its present population at a density no higher than pastoral Europe the major concentrations of people are likely to be contained in a relatively few metropolitan areas. A monotonous pattern of low-density development could lead to a visual landscape contrary to human needs.

Yet a magnificent landscape beyond the metropolitan border is little everyday comfort to the urbanite. However much leisure time he has to get there, city environs will be the commoner visual experience. Large, close-in, distinctive, and contrasting landscapes in the city are called for, not a lower density landscape everywhere.

Thus the automobile as a force on design structure has accompanied, furthered, and to some extent thwarted and been thwarted by other major forces of human life pushing toward concentration and dispersal of urbanization. But unquestionably, the automobile will continue to influence the larger environment. The cluttered landscape, junkyard vistas, and pollution are all well-known problems. While politics and the science of environmental improvements are now being directed to solving these issues, the central question is: What constitutes an appropriate mix of natural and man-made environments when the automobile (or some other form of individual transport) is as common an environmental object as a house or a tree. One answer lies in the arrangement of lines of circulation and open space and various densities of urban areas so as to create a strong community design. Essentially, this has become a matter of adjusting the automobile and other forms of movement to regional-scale designs and a matter of the regard and respect given to the pedestrian's environment.

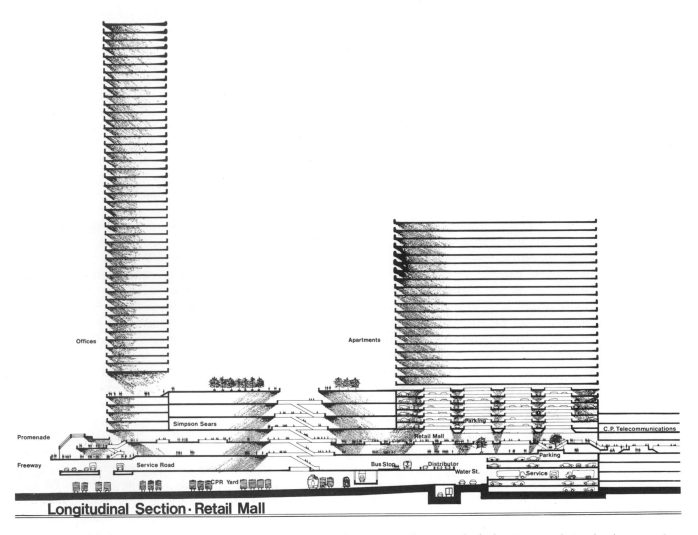

Longitudinal Section · Retail Mall

Offices — Apartments — Simpson Sears — Parking — C.P. Telecommunications — Retail Mall — Promenade — Freeway — Service Road — Bus Stop — Distributor — Water St. — Parking — Service — CPR Yard

Environmental Zones Proposed treatment of circulation systems for a twenty-three-acre high-density central-city development. Separate and distinct channels are set aside for various movement systems leaving sizable portions of urban space exclusively for human habitation. *Source: A Plan For Project 200, Vancouver, B.C. Designs: Dober, Paddock, Upton and Associates, Inc.*

TECHNICAL REQUIREMENTS

On a technical level the problem of accommodating the motor vehicle in the environment and the criteria for judging any solution can be briefly stated. First, the inherited street and road system is unsatisfactory for automobile traffic because its design is simply not suitable for motor-vehicle movement. There are few clear and lengthy rights-of-way. The flow of traffic is interrupted at frequent intervals by vehicles turning into or across the lines of movement. Secondly, while the volume of traffic is often close to the capacity of the channel carrying it, most of the existing streets and roads also provide points of access to the buildings along the right-of-way. These traffic channels thus have to provide room for standing and parking in addition to movement and consequently are heavily congested. Thirdly, local traffic is mixed with through traffic. The resulting "static" impedes the flow, and this is further interrupted because of constant acceleration and deceleration. Fourthly, by-passes and other diversionary measures intended to clear up points of congestion and mixed traffic themselves draw additional traffic. Fifthly, the amount of public investment in forms of movement other than the automobile is insufficient. There is an overdependence on a single mode of transport and few options exercised for using high-density transportation. The environmental effects of all this obsolescence have already been touched on: the dangers, nuisances, and aesthetic detriments.

The literature on traffic engineering, traffic planning, and traffic design solutions is extensive. Research, carried on with sizable funding from both government and private groups, is increasing. Therefore, significant improvements can be expected from these approaches. What can environmental design contribute? The environmental problems have been described and briefly recapitulated; the criteria for a solution can be as simply stated and a solution offered.

Any technical solution should be judged by how well it serves as an economic and safe allocation of urban land for channels of movement which provide optimum accessibility to other land uses by a reasonable number of vehicles in such a way that the environment is aesthetically and functionally satisfactory. An elaboration of these criteria is helpful as a way of grasping some fundamental issues.

"Economic" here implies a rational relationship between costs and benefits. The driver and automobile owner should pay a fair charge for the use of urban land, both for traveling and parking. Also the cost of accommodating the automobile should be reckoned in terms of its impact on other forms of circulation and transport. Accordingly, any calculation of costs and benefits should be done within the framework of a comprehensive circulation system.

So far, a mathematical model for making these calculations is not available, and hence the case for what is *economic* is intuitive. Much of the work being done on cost and benefit analyses uses past trends as a base for projecting future solutions. So far, little recognition is given to the possible improvement that may follow if some new design or technique is introduced so as to deliberately direct investments to new, different goals.

Public investments, determined on a cost and benefit basis, are also complicated by the fact that anticipated profitability alone is not a sufficient guide for making public choices. If this had been true in the past, we would have no radios, commercial airlines, or synthetic fibers.

"Safe" covers all dimensions of health, from the elimination of region-wide air pollution to the minimization of injury and annoyance. Safety objectives include full separation of pedestrians and vehicles, an elimination of dangerous speeds, no major conflicts between channels of flow and the points where they join together, suitable designs for stopping places where drivers become pedestrians and vice-versa, separation of through traffic channels from local traffic channels.

"Allocation" implies systematic management and control of an important resource, in which the multipurpose aspects of the allocation are considered. The environmental dimensions and the engineering dimensions of accommodating the automobile may require separate accounting systems to determine allocations. The former may be improved by research in the behavioral sciences and by pragmatic, intuitive judgments derived from the design professions. The latter has to include improved methods of measuring the secondary and long-range consequences of any decisions based on an initial evaluation of what is economic and efficient.

For convenience, "urban land" here means any land in any place defined by the Bureau of Census as an urban area or a metropolitan area.

"Channels of movement" are the designated rights-of-way for vehicular travel. These include streets, roads, highways, tunnels, overpasses, and so forth.

"Optimum" is a state of balance between what is desired and what is economic.

"Accessibility" is the making possible of movement from one place to another with directness and dispatch, at safe speeds, with some comfort, and with convenience. On arrival, the driver should be able to stop and park without restriction close to his final destination. Designs should provide for ease of movement, turning, and maneuvering for service and emergency vehicles, and of course for vehicles in the parking areas.

"Other land uses" covers buildings, open spaces, transportation lines, any and all special or mixed areas of activity or ground in an urban area.

"Reasonable numbers" is related to the economic capability of the channels of movement and a judgment (perhaps arbitrary) on the environmental impact those numbers have on the land uses adjacent.

"Vehicles" are any self-propelled forms of locomotion and, within the context of this section, mostly internal combustion engines.

"Environment" is used in a broad and general sense to mean the space used by man.

"Aesthetically" connotes beauty, and "function," utility. "Satisfactory" is a quality of performance that is more than adequate and is satisfying.

From the viewpoint of accommodating the motor vehicle, beauty and utility would reflect these features: The network systems approaching the environmental and service zones would be clearly designed, the points of entrance and departure well marked, the travel routes visually interesting but not demanding. Buildings, open spaces, roads, and streets would be in scale with each other. When the driver became pedestrian, the environment would not be dominated by moving or parked vehicles nor cluttered with street furniture associated with the management and control of traffic. Through traffic would be kept out of the relatively dense urban environments, and there would be no proximity of pedestrians and buildings to noisy traffic, whether light, medium, or heavy in volume.

The only "solution" that approaches the ideal criteria above is the containment of vehicular traffic in channels and areas designed solely for their own use. The resulting design structure, as shown in the illustrations for Project 200, then can consist of three parts: environmental zones set aside for human activity, access zones for parking and points of service related to the motor vehicle, and corridors set aside for vehicular movement. The relationship between the corridors and the zones is strictly a service relationship. Traffic and corridors are not ends in themselves but a supporting element for the environmental zones. The idea is neither new nor startling and is fundamental to many current conceptualizations of environmental design. Pursued to its logical conclusion, an urban place would consist of a series of zones laced together by a network of corridors or channels. The cellular structure would be designed in a hierarchical fashion, i.e. the size, number, and location of corridors would vary in accordance with the kind of activity that is adjacent.

This zonal concept serves as a point of departure for equating the density and quality of environmental development with the capacity of the network that will serve it. Obviously, when the density is valuable enough in its own right and the capacity of the accompanying network is constrained by topographic conditions, existing urbanization patterns, environmental considerations, or costs, the automobile network may be replaced by some other form of circulation, such as mass transport.

The application of the idea on new sites is simple, its introduction into existing areas difficult. Colin D. Buchanan, the British traffic engineer, believes that three variables are involved in making these circulation-design decisions: the standard of environment desired, the level of accessibility required, and the cost that can be borne to carry out the scheme. Stated as Buchanan's Law, this means the establishment of environmental standards should automatically determine the requirements for accessibility, but the latter can be improved according to the amount of money that can be spent.

The idea that each environmental zone has a definable traffic requirement is a commonsense observation. Factories, schools, and houses are constructed with preconceived notions of how many people and activities will be contained therein. Though some elasticity in use can occur, economic, social, physical, and environmental breakdowns take place when capacity has been passed. Defining the acceptable levels of accommodation for the automobile by environmental zones is similarly a matter of determining the desirable, or at least tolerable, mix when bringing the environmental and service zones and corridors into proximity with one another.

The conceptualization of the environmental design, as total design in which the zones and corridors cannot be considered independently of each other, not only overcomes the problem of disunity and dysfunction in accommodating the automobile but also encourages a search for new architectural and engineering solutions which will combine movement and nonmovement designs. Traffic then becomes treated as a utility, with its own special rights-of-way and functional requirements, integrated into a larger design structure.

There are absolute limits on the use of the automobile, if environmental conditions are respected. The weight of evidence is that man and motor are incompatible when flung together in large numbers, only when their spheres of activity are separated will improvements in the environment occur. At best, present approaches are palliative. The magnitude of what has to be done is as appalling as it is challenging. But the issues are not complicated, nor the choices unclear. New designs must be advanced, as Buchanan has written, "without confusion over purpose, without timidity over means, and above all without delay."

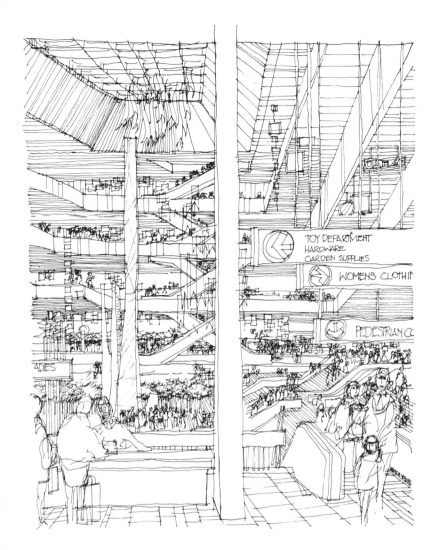

By-Products Of Scale The elimination of environmental deficiencies related to the automobile depends on having projects large enough to finance the necessary technical solutions for environmental zoning. A test of the efficacy of large-scale development is whether exterior and interior public spaces can be created free of vehicular circulation. Design studies for Project 200 indicate that environmental zoning can achieve this objective. *Designs and drawings: Dober, Paddock, Upton and Associates, inc.*

Streets And Roads Giving Design Structure Grid, crescent, circle, and curve combining to give form to Kapuskasing, Ontario. *Source: Fox Valley Regional Planning Commission*

PATTERNS AND NETWORKS

In accommodating the automobile in environmental design, what might vehicular and pedestrian channels for movement be like in form and function? First of all, these channels, formally designated for movement, circulation, and transportation and ranging in scale from expressway to footpath, can be viewed in design terms as lines. Lines help establish design structure by forming recognizable skeleton-like networks, by connecting activity areas together, by serving as boundaries and edges, and by contrast in function and appearance. Some general observations will be made about such lines in terms of vehicular channels first. Later, after describing plans to accommodate these channels, the environmental qualities of pedestian networks will be explored.

There are four basic designs for movement networks: the radial, the ring, the grid, and the linear. Like a spider's web, the radial has a strong center, with lines moving outward. Where the flow of movement has a common origin, destination, or intersection and volumes arriving at the center can be controlled, the radial network gives the most practical line of travel. It works well in gathering and distributing traffic to and from the peripheries.

To by-pass the center of the radial system, a ring road may be introduced. The resulting wheel-like design still favors movement to the center but also has the effect of giving circumferential accessibility across the radials. This pattern has been used extensively by the Interstate Highway System designers to surround and cut through metropolitan areas. For example, multilane expressways now girdle Kansas City, Indianapolis, Atlanta, and Columbus. Diagonals from the outer ring connect and pierce the central business districts. As compelling as

they are dramatic, these superhighways have set the regional design skeleton for decades to come. At construction costs of $25 million a mile in urban areas, these structures are unquestionably permanent features in the environment.

Radial designs are most useful at the microscale when laying out circulation systems for campus plans, hospital complexes, and superblocks. Parking and vehicular service can be brought in close to the building groups without cutting through the prime environmental zone in the center, by cutting off and looping together the spokes of the "wheel" some distance back from the center (see adjacent diagram).

Rectangular grids are probably the most common network pattern. They can be fitted into the topography and adjusted in size and frequency to accommodate varying traffic demands without losing their strong and easily grasped geometrical design. Because of historic associations with speculative real-estate development and many poorly executed plans, this type of grid is often rejected outright, in favor of curvilinear designs, without regard to adjustments that can be made in the gridiron to relieve visual monotony, differentiate between modes and volumes of traffic, and control traffic patterns by disconnecting certain links and thus reducing the hazards of through traffic.

Buchanan suggests that in those instances where a uniform spread of development can be predicted and controlled, a hexagonal pattern may be superimposed as a skeleton for enclosing and servicing land uses. The three-way intersections are economical and are fourteen times safer than four-way intersections.

Robert Le Ricolais's work at the University of Pennsylvania's Institute of Architectural Research suggests new concepts in

network design of genuine import for environmental design. He notes that network designers tend to visualize solutions in historic terms, largely the anthropomorphically and astronomically based orthogonal geometry of early Western civilization. The application of modern topological theory, however, suggests a design quite different from the gridiron, namely a mosaic of hexagons subdivided into triangles, which Le Ricolais calls a trihex.

In a ten-block area analyzed for accessibility, Le Rocolais found his trihex grid yielded about twenty percent fewer intersections than the orthogonal grid. Using an elegant but simple mathematical model, Le Ricolais compares the trihex and orthogonal systems in terms of moving pedestrians and motor vehicles when both are separated from each other. The qualitative analysis indicates that a trihex vehicular grid at one level, coupled with a trigrid for pedestrians at another, produces the most ordered and efficient system of differentiated movement. Automobiles would travel in straight lines, free flowing, with few traffic lights and signals and a minimum number of intersections. Channel capacity would rise but chances for accidents would be reduced. Pedestrians would move across vehicular lines along triangular routes which are the shortest paths for travel. They would have no contact with motor vehicles.

The pattern makes it possible to differentiate between small and large areas and public and private areas and also affords valuable architectural opportunities. The trihex, for example, can embrace a large land-use area with optimum accessibility at the perimeter. The space enclosed could be set aside for a shopping center, park, or complex of large public-oriented buildings. The triangulation afforded inside the trihex allows easy subdivisions of the space for private

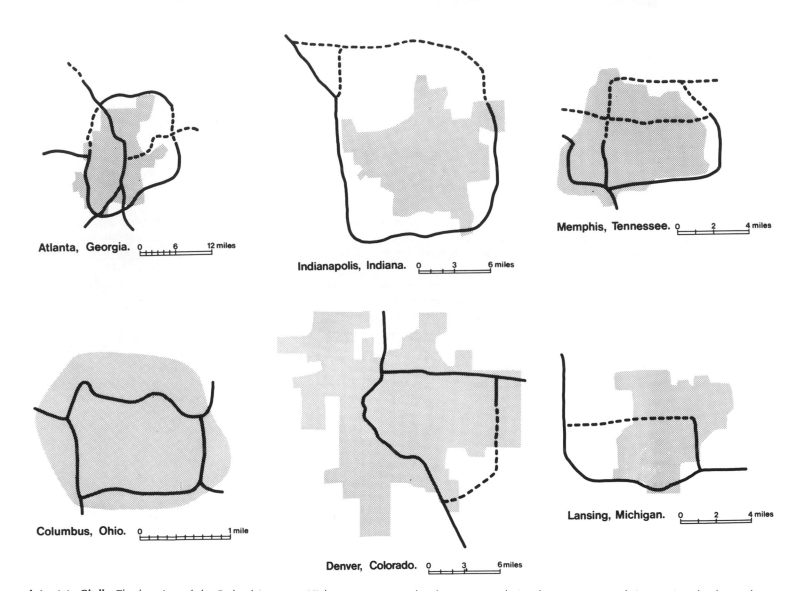

Atlanta, Georgia. 0 — 6 — 12 miles

Indianapolis, Indiana. 0 — 3 — 6 miles

Memphis, Tennessee. 0 — 2 — 4 miles

Columbus, Ohio. 0 — 1 mile

Denver, Colorado. 0 — 3 — 6 miles

Lansing, Michigan. 0 — 2 — 4 miles

Interstate Girdle The location of the Federal Interstate Highway system may be the strongest design force now at work in creating the form of many American cities. Constructed segments of the road are shown by a dark line; segments under design are shown by a broken line. *Drawing: Dober, Paddock, Upton and Associates, Inc.*

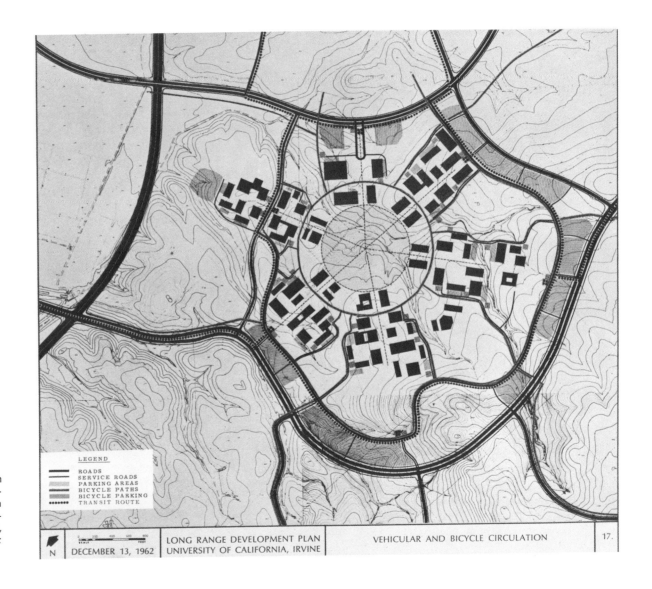

Circulation System, University Of California, Irvine An ingeniously worked out vehicular and bicycle circulation system allowing access to the central campus in both the first and final stages of development. The design creates pleasant and safe areas for pedestrians, while giving an overall form to the campus. *Source: William L. Pereira, Architect*

LEGEND
ROADS
SERVICE ROADS
PARKING AREAS
BICYCLE PATHS
BICYCLE PARKING
TRANSIT ROUTE

N

DECEMBER 13, 1962

LONG RANGE DEVELOPMENT PLAN
UNIVERSITY OF CALIFORNIA, IRVINE

VEHICULAR AND BICYCLE CIRCULATION

17.

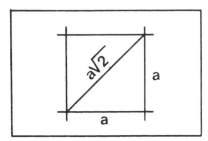

ORTHOGONAL

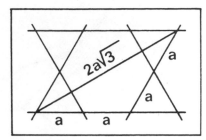

TRIHEX

The trihex grid was examined as an urban street system. These factors became evident.

1. *One- or two-way system:* The trihex grid is usable for both systems; but the analysis of turning movement shows that the one-way system that makes use of the 60° intersection allows closer spacing of the grid's streets.

2. *Scale of the grid:* The minimum distance between intersections or crossings should be 750 feet. Two streets of the same direction will then have about 2,500 feet between them.

3. *Relation to topography and implementation:* The trihex grid can be adapted to topography without destroying the system. The system can be made to work in an incomplete form either in adaptation to the topography or to the development process, i.e., staging of construction over an extended period of time.

Patterns The use of regular geometric configurations in urban planning is based on the desire for a rational layout of cities or other forms of settlements. The geometrical patterns used today can be classified in very general terms as radial, rectangular or gridiron, spine and rib, and the "loose" circulation layout. In reality there is very often a mixture of these systems. (See p. 124 for a good example of mixed patterns.)

The triangular or hexagonal grids appear often in planning theory, but there are few examples actually constructed that use more than parts of these configurations; yet the use of hexagonal grids continues to hold the attention of theorists. Walter Cristaller and August Lösch applied hexagonal patterns in their theories for the determination of the location of the optimum places for economic activities. Cristaller developed several plans for regions on this base. Arthur C. Comey applied a hexagonal grid as a base for regional development. (See Ill. 105). Recently, Colin D. Buchanan in England proposed a hexagonal grid as one alternative circulation system for a renewal area in London.

The trihex grid as a configuration is a combination of triangles and hexagons. It eliminates some of the problems which both triangular and hexagonal grids have, if they are used separately. It does not have three directions crossing at every intersection as the triangular configuration has, but only two; and it allows straight movement, which is not possible in the hexagonal grid.

Using simple mathematical formulas, Professor R. Le Ricolais has compared an orthogonal and trihex grid, both with the same-length unit for squares, triangles, and hexagons, and reached these conclusions.

1. *Street length:* The savings in street length are 13.5% for the trihex grid.

2. *Number of intersections:* The trihex has 13.5% fewer intersections. (In grids with unit-length partitions, street length and number of intersections are related in direct proportion to each other.)

3. *Length of way compared to linear distance:* The maximum extra way in the orthogonal grid is 41%; in the trihex grid, 15.%; the difference is 25.4%.

4. *Average length of way:* Compared with the orthogonal grid (100%) the trihex grid's average length of way is 12.5% shorter. (The average direct connection is 21.5% shorter.)

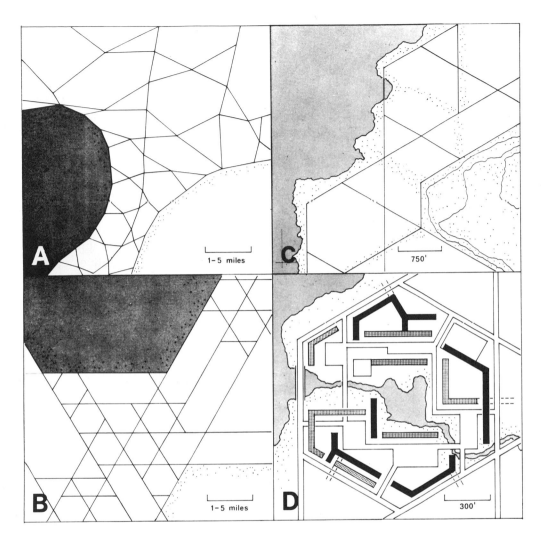

Possible Application In attempts to superimpose the grid in development areas, its rigid geometry very often shows no respect for site or situation. The orthogonal grid, for example, in most cases is not able to follow strong topographical elements.

Diagram A shows that the trihex grid can follow these elements without breaking up their design character. This application is especially useful on a large scale.

Diagram B shows the application of the grid in relation to everyday construction practices. It does not have to be built at once but can, after the original layout, be developed at different points with parts of different sizes.

The use of the trihex partition is obviously an advantage in large-scale urban development. But applied at the very small scale, the grid conflicts with orthogonal and vertical building systems. One resolution to this problem is to use the present building system inside a relatively wide-spaced trihex grid.

Diagrams C and D illustrate the case for a trihex grid with parallel streets 0.25 miles apart. The hexagon contains about thirty-two acres. Densities from ten to 150 persons per acre would then result in three hundred to forty-five hundred persons living in such an area. Diagram D. suggests a high-density area.

Le Ricolais believes the trihex grid, if not totally applicable today, nevertheless contains elements which are superior to the present principles of street layout. The fact that it provides more and more efficient connections in large-scale development is of importance for the reasons that urban areas are becoming increasingly larger, and the demands for efficient circulation are rising. *Source:* R. LeRicolais

129

uses such as housing. In both instances the geometry of the movement system may be mirrored in the geometry of the buildings with significant cost reductions in constructions. Triangular floor framing can bring about a thirty-percent saving over orthogonal systems. Further, Le Ricolais notes, by truncation a triangular building lot can be turned into a hexagonal plan. The number of necessary supports and columns per unit area of structure would be reduced, in wall areas for instance, by seven percent.

The linear circulation system consists of a single line, or pair of lines, carrying movement from point to point. Lines may be tied together to form a loop, oval, or circle. The main line may act as a spine, with minor lines branching off. The linear system is typically used for railroad and water rights-of-way and pedestrian access through superblocks. For vehicular movement it is largely a secondary system in most urban areas.

Obviously from the foregoing, circulation networks occupy a significant part of the urban environment. Statistically, Harland Bartholomew's review of land uses in American cities in the early 1950's indicated that about twenty-eight percent of the land was devoted to roads, streets, and other rights-of-way. A 1967 study of the central business district in Springfield, Illinois, showed that half the surface was used for circulation and parking lots. A sampling of current land-use studies in other cities suggests that the percentage of land used in downtown areas for the motor vehicle may be increasing at a faster rate than population growth and urbanization. Environmental deficiencies brought about by this pervasive situation have been described earlier. What can be done?

In general, of course, comprehensive transportation planning can prevent further extension of automobile use trends and con-temporary designs can mitigate against unreasonable encroachment into the environment occupied by man. Specifically more efficient network design can help. As can be seen from the discussion of four basic designs for networks, the opportunities here lie not so much in the specific network patterns (radial versus gridiron, for instance) as they do in setting up a hierarchy of lines and channels when establishing an overall circulation solution and then choosing a pattern or patterns to suit it.

A hierarchical network may be likened to a tree. The size and number of elements and their location and connection to each other are rational and reasonable. A tree has a trunk, limbs, branches, and twigs. Planning manuals are filled with nomenclature and typical cross sections of rights-of-way that are analogous: expressways, collector roads, and local streets. A hierarchical network is a common-sense one, historic, and still useful for collecting, moving, and distributing vehicles. But to work well it must allow differentiation in speed, volume, and direction. The size of the lines and how they link together are designed accordingly.

Friction-free network design, implicit in a hierarchical solution, is an objective in itself—relentlessly pursued by engineers as traffic volumes rise and accident statistics become grimmer. The answers have been wider lanes, separation of traffic moving in opposite directions, ingenious ramp and interchange designs, computerized signals, and expressways. The latest of these, the forty-four-thousand-mile Interstate Highway System, is impeccably engineered, reasonably safe, and not unattractive. The remaining 500,000 miles of urban streets and roads are not as well designed and do not tie together very successfully into a unified whole.

Discontinuous systems may be explained away as historic accident, just as the impact of traffic on adjacent land uses may be excused on the grounds that it wasn't planned, it happened. The unfortunate fact, however, is that not much is being done to solve either problem. There is little consistency in public policy. Expressways have been built, but not equivalent improvements in the adjacent road networks. High-speed highways get better; the maintenance of low-speed roads declines. Few freeway location decisions have been made with regard to the microenvironment through which the highways pass.

The resolution of these problems is simple enough, almost reiterative: continuity in design, rational connections between the parts of the network, segregation of traffic by type and volume, and reduction of the impact of movement systems by placing them on the periphery of environmental zones or subordinating them as they pass through. Here again, the environmental problem is not so much what to do, but how to get it done.

DESIGN FOR MOTION

A line may be straight or discursive, rise or fall; but whatever its physical characteristics, a line possesses the ability to induce predictable psychological and physiological reactions in a person, who passes along it. His movement is generally purposeful, and will follow the path of least resistance, along the easiest grades, seeking the greatest degree of comfort and protection. His motion along the line can be influenced by signs and signals, consciously or unconsciously given and received; and obviously, speeds will be limited not only by the form of locomotion (foot, bike, cab, etc.) but also by the form and capacities of the channel. Origin and destination, progression, the

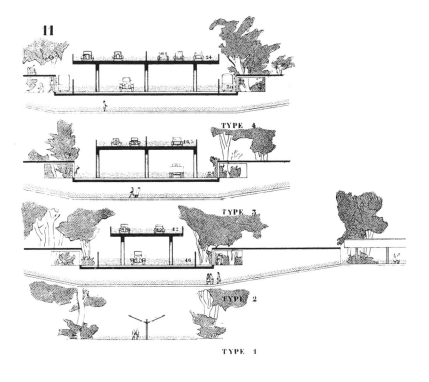

11

TYPE 4

TYPE 3

TYPE 2

TYPE 1

0   10   20

Rational Road Systems In the 1920's designers such as Le Corbusier recognized that the character and volume of traffic would affect the architectural and engineering aspects of environmental design. They proposed establishing a rational design order, a concept still useful in solving today's problems. *Source: Harvard Graduate School of Design Library*

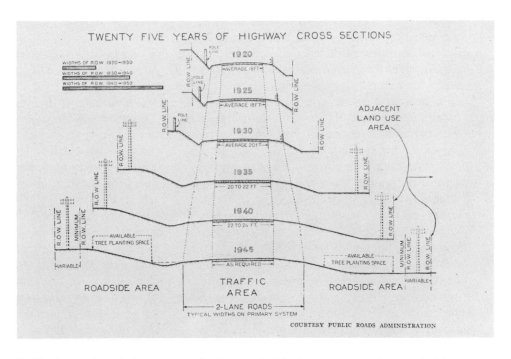

Traffic Impact Largely for reasons of safety, typical highway cross sections tripled in width in twenty-five years as the volume of traffic and the speed of the automobile increased. *Reprinted courtesy of: Bureau of Public Works*

view from the road, the quality of the journey from place to place—all these can be anticipated in planning. The circulation network can be designed with these causes and effects in mind, taking into account the technical requirements for construction and human physiological responses, as well as more general considerations of ease, safety, and aesthetics. This is design for motion.

Technical standards for vehicular rights-of-way are fairly uniform throughout the United States, growing as they do out of common necessities. The alignments (vertical and horizontal position of the channel in space) of streets, roads, and highways vary according to the traffic volumes and traffic types they are intended to serve and the velocity of movement. The latter is contingent on the number and frequency of exits and entrances into the flow of traffic, the width of the lanes in the channel, surface conditions during various kinds of weather, and illumination and sightlines (the distance required to see and react to typical road conditions).

Vehicular rights-of-way are usually designed for speed limits well below the technical capacity of the automobile and truck. For safety the space between vehicles should increase with speed in order to allow enough reaction time for turning movements, sudden stops, and other shifts in momentum. Thus the high speeds reduce the capacity of the channel by increasing the distance between cars. For two-lane roads the optimum speed is around thirty miles per hour. For multilane and divided highways, speeds of forty to fifty miles per hour give optimum results. To get the highest and safest design capacity, some urban expressways have minimum and maximum speed limits.

Physiological responses affect road design. The eye can react to no more than twelve separate images a second when ob-

ject and viewer are in a relatively fixed position. Highway engineers calculate that it takes about three-quarters of a second to change focus from speedometer to some point on the road ahead. In that brief instant the observer traveling forty miles an hour has moved forward sixty-six feet in space. Close objects have passed and the medium distant objectives have changed in location and scale while the far horizon seems to remain fixed in space. Looking straight ahead, the eye can encompass a visual field of approximately 180 degrees on a horizontal plain, sixty degrees downward, and seventy-six degrees upwards. Pivoting the head will increase the range of vision but not the field. Levels of illumination determine the detail seen, other conditions being constant. Inherent in the human body, these physiological constraints are substantial and have an extraordinary impact on how objects in the visual field are arranged and how paths of movement can be designed in reaction to them.

A classic study (1937) by J. R. Hamilton and Louise L. Thurstone identified the basic response of human physiology when driving in automobiles. The study is fundamental to environmental design because of what it tells about human limitations imposed on designs for channels or vehicles.

For instance, fast driving requires concentration. As one moves through space at increasing speed, the number and frequency of visual objects encountered will multiply. On the other hand, the faster the rate of movement, the greater the need for attention on the path ahead. The pedestrian's visual field may be enriched by increasing detail; the opposite is true for the driver from the point of view of safety.

As speeds rise, the point of concentration ahead advances: from six hundred to two thousand feet when accelerating from

twenty-five miles per hour to sixty-five miles per hour. Foreground details begin to disappear, and the perception of space is seriously impaired. Objectives cannot be seen clearly any closer than eighty feet when traveling forty miles an hour. At sixty miles per hour foreground perception is negligible. Only large and simple shapes, forms, textures, and silhouettes can be readily seen. Detailed objects that come into visual range are blurred, easily misinterpreted, likely to produce tension and engender danger. The perception and comprehension of speed and distance may then be affected, with further possibilities for loss of control and accidents. Further, the cone of vision shrinks. The horizontal angle of peripheral vision drops from one hundred degrees to less than forty degrees. Objects placed horizontally, parallel to the road, are difficult to see, those placed perpendicular to the road, are more prominent. Accordingly, irrelevant elements do not belong in the visual field of high-speed roads, peripheral objects should be designed for least intrusion. Only the important features should be located along the axis of vision and positioned for maximum visibility.

The combination of an ever-diminishing side vision and an increasing need for focusing on a distant horizon produces a tunnel-like visual demand. To reduce the hypnotic effects of this continuing, straight-ahead vision, a curvilinear road layout may be introduced in high-speed locations so as to change the position of distant frontal features. Subtly and with sophistication, highways can thus use an urban panorama or country landscape as a focus for high-speed driving, eliminating monotony and increasing enjoyment and safety.

Route 85 from Albuquerque to Santa Fe, New Mexico, runs straight as an arrow through an arid, flat landscape. At mid-

Futurama The 1939 World's Fair (General Motors Exhibition) projected the delights of high-speed automobile travel. Exhibition designers suggested road location principles which are only now coming into being. *Photo: Richard Garrison*

London Bridge Is Going Up A Ponte Vecchio solution. The proposal calls for using air rights over the Thames River crossing as the site for hotels and concert halls. *Design concept: G. A. Jellicoe*

point the road curves to pass a small filling station. The story is told locally that the owner refused the State's offer to purchase his property for road construction, and rather than wait for condemnation proceedings, the State built the highway around him. This single building brings a strong landmark into the visual field of the driver, relieving an otherwise boring right-of-way.

Especially on a clear night, the approach to Boulder, Colorado, via the expressway from Denver is an excellent example of good design for motion and the delights of driving a half hour at sixty miles per hour. The silhouette of the Rockies is first seen, then a twinkle of lights as the city's dimensions grow horizontally while the mountains become larger. Then the road rises and falls, and the city seems to move gently from side to side. Low-density suburban housing begins to define the sides of the channel ahead. Details come into focus, and major buildings of the University announce the edge of the city.

Along Route 6 on Cape Cod, Shootflying Hill plays a similar role: a horizon feature strong enough to attract and hold the driver's attention as he moves towards it but interesting enough in silhouette to change in dimensions as the right-of-way is modulated vertically in space.

As he is freer to view the environmental world, the automobile rider's perception is different from that of the driver; but through the windshield the general effects of tunnel vision are the same. In rapid transit and railroad travel, when the passenger moves at high speeds, the visual range is limited to picture-window panoramas. When passing through dense urban areas or landscaped rights-of-way, very little of the environs can be seen. Accordingly, engineers working on the proposed express railroad from Boston to Washington believe that the train may as

well be placed underground along selected parts of the right-of-way. The airplane introduces for the passenger a new scale in viewing points and extremes of viewing speeds. Some observers speculate that the design of the metropolis may be laid out with the air traveler in mind, a readily perceived three-dimensional picture of immense scale and detail.

As noted, safety and pleasure also determine road alignments. While the shortest distance between two points on a flat surface is a straight line, the easiest, most pleasant, and natural way to move is in a gently curving line. The spiral movement, for instance, is a universal, natural phenomenon —in the double helix of DNA, in galaxies, in the motions of solids through liquids, in the path taken by a blindfolded person. Illuminating research by Boris Puskarev illustrates the advantages of the spiral in highway design.

The essential characteristics of a vehicle and its motion require that all turning movements be made in some form of a curve. The radii differ according to the length of the vehicle and the speed. Conceptually, the perfect highway alignment would be a continuous curve allowing constant but gradual changes in direction. In practice, the horizontal alignment is composed of straight lines, circular arcs, and other transitional curves.

Straight lines are simple to lay out, direct, and predictable, but at high speeds monotonous, fatiguing, and lead to unsafe driving. The arc is similarly easy to lay out and a logical way to direct turning movements. It commands the driver's attention and gives excellent optical guidance, as its horizon may be seen slightly to one side rather than directly ahead.

When straight line and small circular arc are combined, the juncture can be dis-

comforting, visually and viscerally. At high speeds a sharp change will require sudden braking. A spiral transition curve can smooth out the juncture. A combination of circular and spiral curves and tangential straight lines can be orchestrated so that the continuity of the roads flows together. The degree and frequency of curvature can give clues of impending change, for example as an interurban highway changes into an urban expressway.

Vertical alignments are of equal importance. The grade, or inclination of the line, must be controlled for ease in movement. Typically, the rise or fall in elevation should be no more than three feet every one hundred feet for trucks and five feet to eight feet for automobiles. Other engineering constraints on elevations include safe sightlines when moving through crest curves at hilltops and sag curves at valley bottoms and requirements for drainage—all these being technical standards necessary for safe roads. The visual qualities of the vertical road alignment come from graceful grades that undulate over the landscape and especially the manner in which the horizons are opened and closed as motion is carried up and down grades.

Continuous, free-flowing, cordinated vertical and horizontal alignments have to be carefully planned from the beginning. The sculptural designs require a unity of plan and profile. Fortunately, computer-assisted graphics can now be used to study various alignments and their relative cost and benefits. On the highest speed roads at least, aesthetic satisfaction can be matched with engineering requirements for superior designs.

On the slower speed networks, street planning for those in motion has been a neglected art, with few real changes in almost fifty years. The exception has been plans sponsored by astute real-estate developers who have learned that imaginative street design can increase sales and upgrade the general appearance of their communities at little extra expense. The story here is not so much one of innovation as of common-sense application of well-known site-design principles. Typically, in the past the overall subdivision design was laid out by an engineer interested in obtaining the maximum number of lots along the minimum right-of-way. As a result, perimeter streets ran parallel to heavily trafficked roads, and adjacent houses suffered accordingly. In addition the best natural sites were cut off from the general public. Further there was no sense of threshold when arriving at the community and little distinction between the interior roads. When adjacent communities came up with something better, real-estate dealers were quick to copy to avoid losing a market advantage.

Privacy, safety, and individuality have become current objectives in subdivision design. A hierarchy of local streets is imposed on the site and the internal network tied to the nearby through streets at as few places as possible. The interior network may be a large loop road, or several loops, giving access to short streets, smaller loops, or cul-de-sacs. Often the automobile network is crossed over or under by a separate pedestrian path system which connects to community facilities and public spaces. The best of these street designs make good use of existing site features, especially topographic changes and trees, and keep through automobile traffic away from the subneighborhoods and districts, yet make it easy for visitors, service representatives, and others to find their way around.

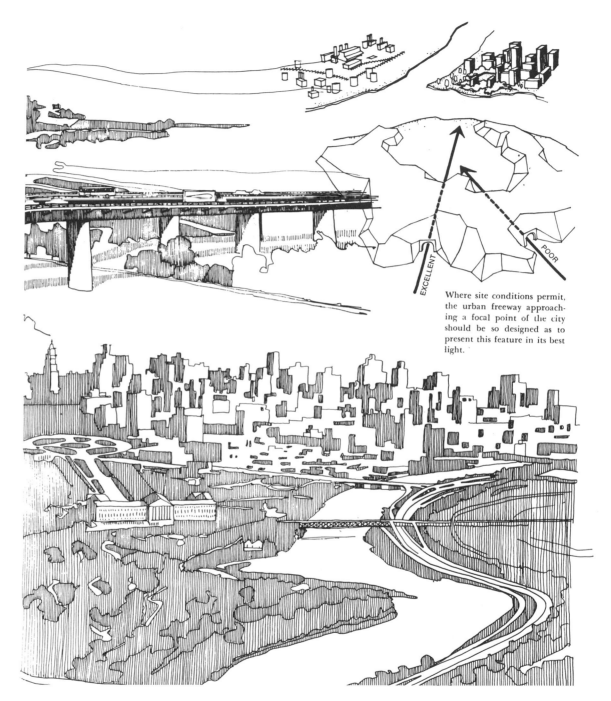

Where site conditions permit, the urban freeway approaching a focal point of the city should be so designed as to present this feature in its best light.

Design In Motion A simple principle of urban design: When options are available, design for visual effect. *Reprinted from: Freeway In The City, Washington, D.C., 1967*

LINEAL CITIES THEORY

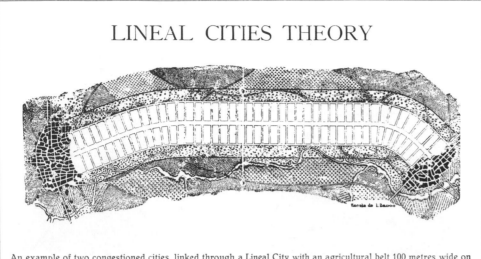

An example of two congestioned cities linked through a Lineal City with an agricultural belt 100 metres wide on each side, joining two old agglomerated cities.

LINEAL CITIES THEORY

Plan of a Lineal City fragment with the main street 40 metres wide, the transverses 20 and the posteriors 10 meters in width, all with trees and shrubberies. The division the blocks of houses into plots at different sizes is arranged in order that all dwelling-houses should be surrounded with orchards and gardens. The Lineal City is bound en both sides with agricultural zone and woods.

Soria Y Matta The basic concepts of Soria y Matta's proposals can be gleaned from these two diagrams. *Source: Harvard Graduate School of Design Library*

136

THE LINE AND CELL IN COMMUNITY DESIGN

As a premonition of things to come, Arturo Soria y Matta's scheme (1880) for developing the suburbs of Madrid is a significant recent historic example of transport and circulation networks providing design structure. Original and provocative, Soria's work has been overlooked by those who see architecture only as a type of building, rather than an environment, and severely chastised by those who devalue his ideas of community design because of its supposedly superficial social implications.

Soria's concept can best be appreciated within the context of its period. It began as a practical alternative to the continuing congestion of the central city and the capricious and arbitrary extension of city boundaries. The 1880's were the peak of unbridled city development. The lure of the city led to the shame of the city. In varying degrees throughout all industrializing countries, sporadic building, despoliation, and barbaric housing and working conditions overlapped each other until vast central districts were, in Lamb's words, one continuous "miserable region of damp, dilapidation, and decay." Worse still: To those slums which were created as people overused old housing, instant misery was added in the form of insubstantial new shelter thrown up cheaply for quick profit. Many of these developments were built on sites opened up by horsecars and tramways.

At the time Soria first published his ideas (1882), the counteraction to the ills of the industrial city had only just begun. As described earlier, benevolent manufacturers such as Salt, Cadbury, and Lever had constructed model industrial towns in England; but the Utopian Edward Bellamy had not yet looked backwards. Ebenezer Howard was yet to commit to paper his doctrinaire vision of Garden Cities of Tomorrow, and parliamentarians around the world had not cleared their throats to address themselves to the problems of the poor and the diseases of urban life. In his *Linear City—New Archichecture for Cities*, Soria set out his principal ideas for organizing urbanization and designing the environment, some of which may be as applicable to our time as they could have been to Soria's.

Soria believed that transportation was the fundamental form-maker for all kinds of urban agglomerations: transportation determined the natural distances between places of everyday importance, especially the journey from work to home. He stressed the separation of work and residence as well as the separation of new communities from old. To those who knew the pollution, accidents, grime, and filth brought on by close proximity of housing to industry in the nineteenth-century city, this idea of separation must have had great appeal.

Of the technologies of transport known to him the electric tram appealed most to Soria's engineering experience. Accordingly, in his mind the most beneficial design solution was a town plan that adapted itself to the linear geometry of the railway. The principal railway and parallel street in his plan were like a spinal column, an axis, which fixed the transverse streets and points of intersections.

Soria suggested that the main spine should be at least 180 feet wide and the transverse streets at least sixty feet. In addition the spine and connecting roads would serve as rights-of-way for water, sewers, and electricity. It was this vertebrae-like design that made the linear city different, in Soria's view, from the classical city designs and the modern garden city.

The geometry of the street system helped create a cellular design structure, established in part by the form of the building lots. The frontage in Soria's view would range from three hundred to three thousand feet. Rectangular and trapezoidal lot shapes were preferred, because they were more orderly and systematic than the tortuous shapes of older cities and also cheaper and more convenient to service with modern utilities.

At intersections of the principal streets, Soria set aside large and small public spaces for fountains and public monuments. All the streets and squares were abundantly planted, the number, type, and size of trees helping to distinguish the major civic spaces from the minor.

Housing in the Linear City was a harbinger of the Great American Dream: "To each family a house, for each house a garden." Further, Soria felt that it was an error to believe that construction of row housing or apartments was cheaper than the building of a single house. Higher costs of single houses were counterbalanced by protection against fire, the assurance of privacy, the reduction of vandalism and noise, and the gain in light and air.

In principle, all the buildings in Linear City, public or private, rich or modest, were to occupy no more than a fifth of their sites. Each house would have its own bath. Varying according to the location of the lot in relation to the main streets, setback lines were established to give aesthetic order to the street and allow its future widening.

Soria wrote that this new mode of urban development cannot go from thought to execution and real life, "becoming crystalized in visible and tangible creations," without adaption to the surrounding environment created by geography and history. In this respect he noted that the value of land fell, along concentric lines, as one moved from the center of the old cities to the suburbs. He anticipated that, in the same way, values along the proposed lines of development would decrease in direct relationship to the distance from the new main line. Accordingly, he proposed that existing cities and his proposed linear communities be united by special "avenues of juncture". In this way there would be a vast triangular network of urbanization in each country. The triangle would be established by lines extending from the old centers, along some of which industry and agriculture might be concentrated. The central space enclosed would be preserved for recreation and keep one urban place distinctive from another.

Thus the design was conceived as a regional plan. It would adapt itself to existing urbanization by linking together old towns with new. The debasement of agricultural land by industry escaping from the central city would be controlled by rational development, so that both public and private interests would gain in the increasing value of land. The deployment of manufacture and food production would thus support public needs.

Soria's philosophy was criticized by his contemporaries for its radical decentralization and ribbon-like continuation of the contemporary urban pattern, especially so by those who supported the Garden City ideas of Ebenezer Howard. Soria's defense was lucid and comprehensive. The Linear City was a clear-cut, practical and efficient design, he wrote, "constant at all times and perfect for the whole surface of the earth." It allowed easiest access to both the pleasures of urban life and the advantages of country living.

Earlier in this book, the design of Garden Cities was described. They were conceived as satellite communities of relatively fixed dimensions girdled by open space. The typical plan contains a nucleus, or town cen-

ter, and various cell-like districts that cluster around it. The linear concept, on the other hand, is recognizable by a linear network and the organization of activities into cells along the spine. Typically, the highest density is closest to the spine. The line itself has no fixed limits.

It is of some consequence to note that the polarization of two different approaches to community design came about, in response to a similar set of problems. Both Soria and Howard directed their vision to the improvement of man's state. Their two strands of the same cord have too long been held apart as separate, perhaps because Soria lacks a good biographer and Howard's principles are continually applied in full or partial form at least once a year by a public or private developer.

Not that Soria's achievements are insubstantial. Believing that social problems could not be left solely to the uncertainties of private philanthropy, Soria backed his convictions by founding Compañía Madrileña de Urbanización (CMU), which was incorporated (1894) to perform six functions in Madrid: acquire and sell land to prospective builders; design and construct houses for a variety of income groups; provide electric, water, and sewer services; and operate the tramway system. Individually, none of these activities broke precedent for private corporations. As noted earlier, this was the way much of nineteenth-century America was built by profit-seeking companies. Soria's achievement lies in the comprehensiveness of his approach and design concept at a regional scale; and despite economic catastrophes, CMU was still operating in the late 1950's.

The early years of CMU were filled with obstacles not unlike those faced by present-day private developers in new-town development. Local government was disinterested;

the company lacked the legal rights to exercise eminent domain; the site was bare and raw; and power, water, and sewer lines were far away. However, the tramway operations were immediately successful. Transport income was sufficient to enable CMU to change from animal to steam to electric traction in the first eight years of the company's existence. The profits later subsidized the buying of land and the construction of four miles of total linear community development about six kilometers from the center of Madrid, and continual operations of the tramway until the Spanish Civil War and its demise in favor of motor buses stilled further construction.

Today though Soria's linear community remains in a significant way aborted, a remarkable aspect of the whole endeavor is how close the actual construction comes to the specifications set out in his manifestos. The district has a degree of visual order that is in sharp contrast with the hodgepodge urbanization that one passes through to get there. The main avenue is gently curving, about fifty yards wide, and attractively planted with pines. The right-of-way for the tramway runs down the center, and other traffic passes along the roadways on either side. The vertebrae effect is carried out by perpendicular streets about one hundred yards apart. Transverse roads are also planted and have the generous width that Soria envisioned. About three thousand homes were constructed, all sited in conformance with CMU regulations: uniform setback lines, ground coverage, and height lines. Today the buildings have little merit, but the townscape is delightful, with gardens and street trees dominating the scene.

Unfortunately for the history of environmental design, Ciudad Lineal, as it is locally called, never became a self-sufficient town or even a district. It contained no provision

Figures illustrating the work carried out by the Compañía Madrileña de Urbanización, founder of the Lineal City.

10.500.000 square meters of land purchased.

5.200 lineal meters of the first track of the Lineal City nearly completely developed.

17.000 lineal meters of the 2nd and 3rd tracks in way of developing.

8.000 meters of roads, including 4.000 paved or macadamised in the main road of the Lineal City.

12.000 meters of transverse streets.

100.000 trees planted, watered and living in the main and transverse roads of the Lineal City.

120 kilometers of water elevating and distributing mains for the service of a population of more than 30.000.

52 kilometers of railways including the tramlines of «Cuatro Caminos-Chamartin-Ciudad Lineal - Ventas», «Cuatro Caminos - Fuencarral», «Cuatro Caminos-Dehesa de la Villa», «Ventas-Canillejas», «Pacífico-Vallecas-Canteras» and the railroad «Cuatro Caminos-Colmenar Viejo».

100 kilometers of railways applied for and under study.

An electricity plant and 150 kilometers of electric mains for light and power, distributed to the towns and suburbs of Barajas, Canillas, Canillejas, Cuarenta Fanegas, Chamartín de la Rosa, Ciudad Lineal, Fuenlabrada, Hortaleza, Humanes, Madrid Moderno, Moraleja, Móstoles, Pueblo Nuevo, San Fernando, Tetuán, Ventas del Espíritu Santo, Vicálvaro, Villaverde y Villaviciosa de Odón.

1.800 dwellings on the Company's land, including cottages for the well-to-do, middle and working classes, all of which, even the humblest, in the best conditions as to light, sunshine, water and air, without danger of fire or epidemics.

5 preferred business: Land, Water, Railways, Buildings and Electricity supply, and **5 auxiliary business,** viz.: Tile-yard, Store-houses, Printing press, Groves and Amusements' park, with hundreds of employees and workingmen.

An humble cottage garden in the Lineal City

Accomplishment In a report to the supporters of the Lineal City idea, Soria y Matta demonstrated the efficacy of his idea, if not its widespread application. *Source: Harvard Graduate School of Design Library*

for communal services, shopping, or leisure. For a time the area did hold a good mixture of income groups and social classes—artisans, storekeepers, skilled workers, as well as professionals, writers, and artists. A survey of occupants in 1931 indicated that thirty percent of the residents were middle and upper class; the remaining were "either in humble employ or so described their occupation as to conceal its true status." In the age of the automobile and the suburbanization of Madrid, however, there are physical signs that the fortunes of those living along Calle de Arturo Soria are now quickly declining. What remains is a shadow image of a strong but incomplete design concept.

How much influence Soria's work had on later linear and cellular developments is not entirely clear. George R. Collins suggests that through publication, word-of-mouth, and personal contact, knowledge of the concept has spread throughout the world. Unquestionably, Soria's ideas are well-reflected in a number of idealized linear-community designs that chronologically followed his own. Whether these were inspired by Soria's work or constitute examples of an idea whose time has come or are simply coincidental cannot be documented. But the thread of family resemblances among a group of schemes is worth noting and following as illumination of the usefulness of Soria's techniques of design structure.

Edgar Chambless' *Roadtown* (1910) is intriguing as an integrated design. He used a subsurface monorail to establish a linear spine over which enclosed space and pedestrian walkways are superimposed in one continuous row. At designated points the row-housing is broken off for factories, shops, and community buildings. Gardens and farmlands lie on either side. The scheme aims at bringing all the aspects of daily life—work, exchange, leisure—into one area. There is an agrarian bias in the setting and a dedication

to urban technology, especially in the proposals that accommodate the design for communal services, such as centralized cooking. This comprehensive kind of approach to physical and social environment, in which the direction and shape of the form is set by the lines of circulation, continues to attract designers, as shown in recent linear schemes for reconstruction of Bedford-Stuyvesant and the crossing of the Thames (see page 133).

Arthur Comey's (1923) regional attack on sprawl asumed that urbanization should not be contained in self-limiting enclaves such as Garden Cities but should be guided to "natural" limits. Since city growth follows lines of communications, gives value to the land adjacent, and determines use, he believed that the intersections of regional highways and railroads should be used for concentrating commerce. From these centers, ribbons of industry, housing, and community facilities would then extend along the lines of transport. Agricultural and other open spaces would separate the ribbons from each other. The star-shaped patterns would be connected to each other by a triangular network, having advantages of optimizing movement and access. A system of differentiated travel routes was proposed, the size and location of each reflecting speed and environmental considerations.

Though the design is shaped by a grid-iron pattern of superhighways, Richard Neutra's *Rush City* (1932) has overtones of Soria's interests in a hierarchical circulation and open-space system defining the character of housing groups. Neutra's idealized city consisted of a central business area connected to numerous districts via a high-speed access road. The districts were composed of bands of land uses, each bounded by a major thoroughfare. The residential bands were intended to be homogeneous in social classes and income—this being further reflected physically by house type and amount of

open space allotted to each band. Several bands inside each district were set aside for light industry and recreation, thus reducing travel and commuting from district to district.

Thomas A. Reiner, in his prospect of ideal communities, emphasizes that schemes like Neutra's and Comey's have basic shortcomings: The plans lack grounding in social and economic realities or are one-sided in viewpoint. Furthermore, Reiner felt the conclusions reached in design form do not always flow from the assumptions; the designs presented are only a few alternatives among several possible.

The advantages these utopian schemes offer are, however, multiple: constructive criticism of existing solutions, detailed descriptions of needed research, and polemic illustrations of an improved environment fascinating enough to stir men's minds to action. As Reiner notes, they "are an indispensable link leading to public understanding, acceptance, and enthusiasm." The contagion of these utopias, however, simplified and incomplete, is well represented in Le Corbusier's linear designs and the work of his many colleagues, notably José Luis Sert.

Le Corbusier continually pushed rationality and technical order beyond the borders of implementation, but always in a way that clarified basic principles and objectives. In the *City of Tomorrow* (1924) he presciently anticipated the new scales of city design: highways broader than rivers, skyscrapers with populations the size of large towns—all organized into superblocks served by underground mass transit and the motor vehicle.

In his postwar (1945) work, Le Corbusier placed these concentrated urban centers into a regional setting that idealized national urbanization. He thought three styles of life could be served by three types of community: agriculture, carried on in communal villages; industry, clustered and attached to linear green-towns; and activities requiring concentrated populations in face-to-face contact placed in high-density centers at regional junctures.

As in other early utopias, Le Corbusier attempted to order the environment through the use of hierarchies of roads, open spaces, and housing types. He disliked sprawl and emphasized having the lines of movement fix the design. Further, he gave particular concern to the scales of motion and the separation of pedestrian from vehicular traffic.

Le Corbusier's architectural forms, massing, and relationship of built-up areas to open space have supplied innumerable designers with starting points for inspired architecture and site design, a factor which by itself makes his work more important than Soria's.

For José Luis Sert hierarchical relationships are the major organizing elements in the community and site planning. The design stems from a desire to produce optimum opportunities for contact and communication among people, activities, and environments. In community design this is achieved by controlling the size of the environmental unit in accordance with the kind of activity being accommodated therein. Each unit is self-sustaining as far as possible. The intensity and variety of activity within the unit increases with the size of the population using it.

Site planning is affected by the volume of traffic that will move through the unit and whether this traffic is on foot or in vehicles. Conflicts between pedestrian and vehicles, either in motion or at rest, are minimized. The generating points of pedestrian and vehicular flow between activity areas are strongly articulated and, along with a respect for the natural site conditions, are woven into a basic diagram. The containment of activities by environmental units and the directness of circulation tend to create strong

linear patterns, but Sert will blur these in response to programmatic considerations. Circulation elements and open space are positioned to break down large units into constituent parts.

Hierarchical approaches to design structure could lead to doctrinaire, pedantic, and stereotyped solutions. In actual practice, however, a hierarchical approach is not a formula applied to a problem but a way of examining the existing situation and categorizing solutions. Sert's plan for Havana, Cuba (1956), is a good example.

Civic nuclei such as the government center and university precinct in the Havana scheme are organized as self-sufficient units and serve as accent points in the overall design pattern. A provisional network of lines of circulation is established using a differentiated street system, i.e. the widths and design profiles of the streets relate to the type of traffic carried and its speed. The network is adjusted to the existing city, keeping in mind the natural boundaries of smaller districts, sites of historic and aesthetic merit, and the environmental impact the network will have on people living in the neighborhoods adjacent. Further distinction is given to the overall design by using linear parks and open spaces in a way that emphasizes design structure. The resulting plan is both ordered and varied, with vehicular circulation channeled and well-related to the neighborhood and districts served.

The MARS (Modern Architectural Research) Plan (London, 1942), another linear design, called for a primary east-to-west corridor more or less paralleling the Thames and existing railroad lines. Within this zone, industry, commerce, and administration would be concentrated. Historic London would be a distinct area within the network.

Residential zones about one mile wide and separated from each other by two miles of countryside would be established north

and south of the corridor perpendicular to it. The effect would be a herringbone, or spine-and-rib design. The outer ends of the ribs would be used for concentration of heavy industry, each subcenter being connected by a belt railroad line forming a loop around all the rib ends. Thus the loop would intersect and intercept national highways and railroads at the perimeter of the London metropolis. Though never adopted, the MARS Plan has influenced later linear designs for small and large communities.

The linear concept presents a reasonable approach to handling problems at the microscale, especially when the development has to be staged over a period of time. The concept plan for Hook New Town shows to good advantage its virtues. The central area grows like the core in an apple, not the stone in a plum. This linear technique allows major lines of movement to be laid down in advance of construction, as well as areas of specified use and locations of special buildings without commitment to a specific architectural solution at the time the long-range concept is prepared. The plan for Runcorn is another recent example of linearity and cellular structure well-used on a new site.

Less attractive than any of the above schemes are the simplifications of linear design ideas used by the Soviets in the First Year Plan (1928-32). The designs were aimed at systematic breaking down of social distinctions between rural and urban proletariat. The Plan called for a rigid set of parallel strips of industry, railroad, greenbelt, highway, residence, and countryside. The residential strip would contain public buildings and services. Rural and urban workers would thus be forced into a common environment. It is this kind of doctrinairism and rigidity which can blur and blemish the image of linear designs in the eyes of critics and clients. This suggests that concepts of design structure succeed or fail only in application.

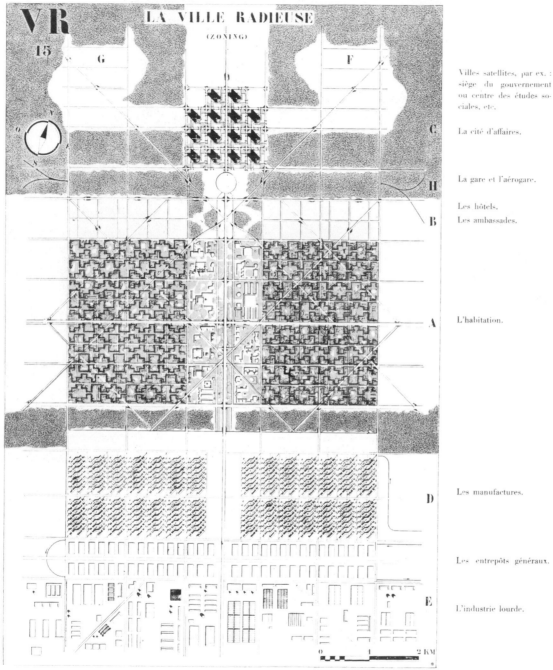

La cité d'affaires.

La gare et l'aérogare.

Les hôtels.
Les ambassades.

L'habitation.

Les manufactures.

Les entrepôts généraux.

L'industrie lourde.

Villes satellites, par ex. : siège du gouvernement ou centre des études sociales, etc.

Le Corbusier's La Ville Radieuse

VARIATIONS ON A THEME Linearity and cellular development can be seen as universal themes in a number of visionary plans for twentieth-century cities.

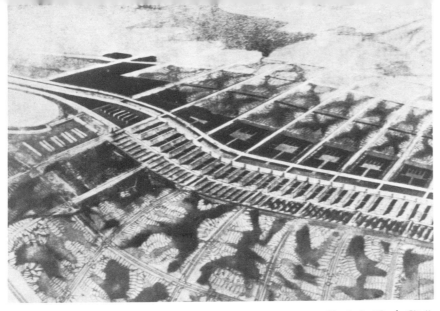

Neutra's "Rush City"

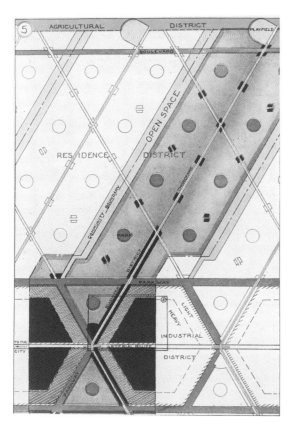

Comey's Regional City Design

Plan for Havana (1955), Weiner, Sert and Schulz, Planners. *Source: Harvard Graduate School of Design Library*

143

Proposed Linear Development Through Bedford-Stuyvesant Neighborhood, Brooklyn, New York *Source: McMillan, Griffis, Mileto, Architects*

North

open space

Hook New Town Plan of pedestrian system. Line and cell, containment of the town by a green belt, and a recognition of the design requirements for accommodating vehicular circulation gave incentive for a fresh look at large-scale design throughout England. Though the plan was never carried out, many of the principles suggested in the scheme were used in later projects such as Cumbernauld New Town, the University of Essex, and the University of Bath. *Source: Greater London Council*

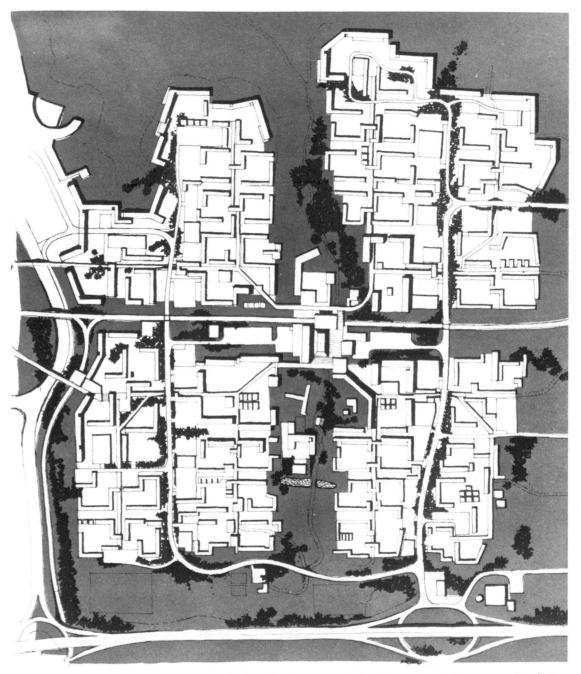

Town Center, Runcorn New Town A further development of the line and cell in community design.
Source: Runcorn Town Plan, 1966

145

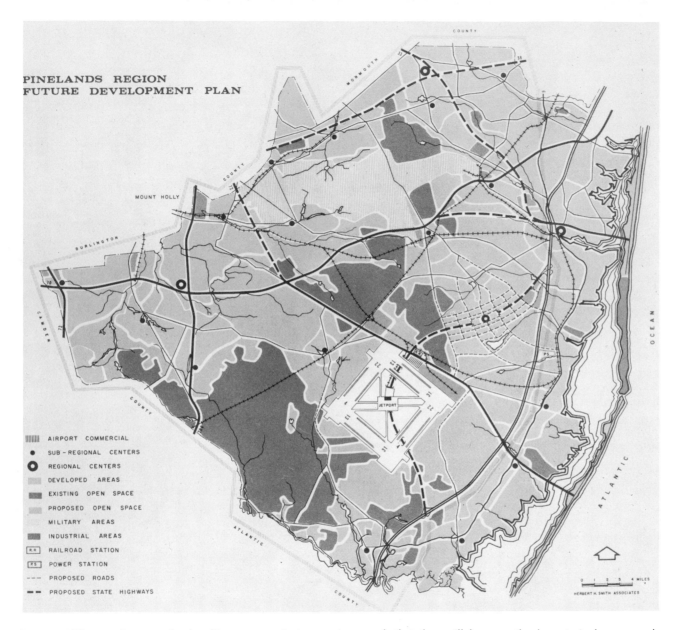

MOUNT HOLLY

OCEAN

ATLANTIC

ATLANTIC

JETPORT

|||||||| AIRPORT COMMERCIAL

● SUB-REGIONAL CENTERS

◉ REGIONAL CENTERS

DEVELOPED AREAS

EXISTING OPEN SPACE

PROPOSED OPEN SPACE

MILITARY AREAS

INDUSTRIAL AREAS

R.R. RAILROAD STATION

P.S. POWER STATION

– – – PROPOSED ROADS

▬ ▬ PROPOSED STATE HIGHWAYS

0 1 2 3 4 MILES

HERBERT H. SMITH ASSOCIATES

Superport The requirements for handling supersonic transport are such that they will become the largest single man-made element in the environment. The scale of design can be seen in the sixty-four-square-mile jetport proposed for southern New Jersey. *Source: The Pinelands Regional Development Plan, 1967. Herbert H. Smith, Associates, Planners*

INFLUENCES OF MODERN TRANSPORT TECHNOLOGY ON DESIGN

Land and water transport have played influential roles in transforming places of human habitation and mastering frontiers by extending the range, speed, volume, and degree of comfort of travel. Modern transport technology continues to advance in these respects but with some important differences from former objectives. There is a growing recognition that the new transport technology will encourage new and improved settlement patterns and bring about environmental controls of higher order and greater sophistication than in the past—if for no other reason than to reduce the nuisances created by the introduction of new machines on land, on water, and in the air.

An approximation of this future may be immediately sensed by examining existing transportation systems and vehicles, both those in operation and research prototypes. Further, the early impact of these devices on the environment may be deduced, since the span between the introduction of improved technologies and their widespread diffusion is less than a decade and a half—this being the average time for applying technological change since World War II. The rapidity with which beneficial technology can be put to work is encouraging but does present immediate dilemmas in planning, especially those decisions which affect the location of transport facilities and the readjustment of land uses around them. Environmental designers face the paradox of having beneficial inventions available and at the same time few methods for felicitously placing these objects in habitable areas at a pace that keeps up with their discovery. Air-space technology is a striking example of this kind of lag.

For air transportation and for space research and technology respectively, the pro-posed 1969 Federal Budget anticipates an outlay of $1.2 billion and $4.5 billion. These figures total twice the amount of the Federal outlay for all programs in the Department of Housing and Urban Development and do not include expenditures made by the Defense Department, whose total investment in air technology to date has been estimated to be roughly $500 billion.

This enormous overall expenditure has had immense consequences for the design and production of aircraft, spacecraft, and missiles, and for aviation in general. The increase in civil aviation alone in the last twenty years proceeds from these fgures: The number of airports and aircraft has increased about fifty percent, the number of revenue miles and total air miles by one hundred fifty percent, the number of passengers by four hundred fifty percent, and the freight tons flown by six hundred forty percent. There are now more miles of paved Federal airport runways than there are of Federal highways.

Research and development are proceeding on superspeed and supersize aircraft with the likelihood that the latter type will be introduced in large numbers before the former. Enlarged versions of craft now flying may accommodate up to three hundred passengers. The next step would be supercraft carrying one thousand passengers or one hundred fifty tons of cargo. A prototype model for military use is already under design. Civilian versions, traveling at subsonic speeds and capable of using existing airports, could operate at costs comparable to intercity bus travel on six-hundred- to twelve-hundred-mile journeys.

The first effect of supersize aircraft would be to allow an expansion of air travel without a parallel expansion in the number of aircraft flying, but from the point of view of design, the on-the-ground impact would be the most significant result and pose great difficulty. Just the volume of traffic generated would require expansion of terminal facilities and vast changes in nearby land uses and circulation networks. In addition, improvements would have to be made in handling baggage, cargo, and personal transport to and from the airport. Few East Coast airports have sites capable of meeting this degree of expansion, and thus significant changes in metropolitan land uses would be required. One solution forecasted is the development of huge airports at the peripheries of metropolitan areas and high-speed transit to reach central-city destinations.

Such superports, however, are enigmas to land planners. The probable size of the superports, fifty square miles, and the general scarcity of unencumbered landholdings of that size in metropolitan areas constitute serious problems. In the New York-New Jersey air corridor, substantial natural preserves, unusual ecological areas, are in danger of being confiscated for superports because other open lands do not seem to be available. On another level there are few precedents for establishing performance standards for land-use controls in relation to superports and no readily available government agency well-enough equipped through laws, personnel, or jurisdiction to carry on the necessary advanced planning.

Furthermore, supersonic craft will also require enlarged airports, not so much because of any resulting increase in numbers of vehicles but because of frequency of flights and special safety requirements. Speeds ranging from two thousand to fifteen thousand miles per hour are predicted, though the latter may be engendered more by a missile-type vehicle than by aircraft.

The acute environmental problem especially associated with jet-aircraft operations, however, is noise. The evidence to date is that human habitation is injured by subsonic and supersonic vibrations and noise. Daily activities affected include instrumentation associated with industrial and research work, film and sound recordings, outdoor cultural performances, and, in the immediate vicinity of flight paths, general living. Tests cases have been won in court by house owners whose properties have been damaged by noisy flight approaches. In other instances, airport authorities have been successfully sued for trespassing and committing nuisances (noise) in the air space over private homes. Special districts have been set up through zoning around new airports in the United States in order to restrict development that might be affected by air noise, but this approach has not been retroactively applied to existing facilities. In Boston the local authorities are willing to purchase at fair market price any home in the direct flight path of incoming airplanes. British authorities will pay one-half the cost of insulating parts of the house against the noise of nearby aircraft operations.

These measures alleviate but don't remove the major source of noise, current engine operation; and for superspeed craft the picture is gloomier. Aerodynamic studies to date hold little promise of eliminating the supersonic boom that takes place when acceleration occurs beyond the speed of sound. Without noise-suppression, terrorizing and other psychic affects as well as widespread physical damage may accompany the introduction of supersonic aircraft and missiles. For reasons of environmental control, such craft may be restricted in their use and flight paths and perhaps be designated solely for trans-oceanic travel. Scientists speculate also that the continuation of supersonic flights in themselves may alter weather patterns in ways inimical to human habitation.

At the other end of the size and speed scale, helicopters and related vehicles have proven to be invaluable forms of transport for certain kinds of police and fire work, emergency transport, and for passing over traffic-congested urban areas. Helicopters are popular for connecting metropolitan airports together and bringing air travelers from the main terminals to downtown and suburban destinations faster than ground transportation. Currently, however, scheduled commercial flights by helicopter must be subsidized, for the costs for individual ownership and operation are high, terminal facilities limited, and noise levels intolerable, as with jet craft.

A number of experimental studies have been carried on to discover whether helicopters might be workable substitutes for fixed-wing aircraft, since the former have the advantage of vertical or short-run take-off and landing. Everyday use of vertical-lift aircraft would be increased if simplified designs and volume production were undertaken, safety measures introduced, and acoustical controls improved, in the belief of Melville C. Branch, an astute observer of the relationship of transportation to city development.

A family-type air vehicle probably would engender new designs for neighborhoods and cities. That family life would be advanced by another form of personal mobility is conjectural, but universal accessibility to air space would have greater impact on urban development than any mechanical invention to date. The opportunity for beneficial use of personal air vehicles would seem to lie more in low-density communities than built-up, metropolitan complexes. National planning and air-space controls would be required, with new communities possibly filling the interstices of the continent. Personal travel for recreation, leisure, and work for that portion of the population that chose such a living style would be measured in hundreds rather than tens of miles. From the point of view of expense, technically a personal air vehicle might be manufactured at a cost comparable to a new automobile, but the enormous public expenditures that would be needed to accommodate the vehicle seem out of scale with its utility, especially in light of higher priorities for government funding of education, health, and community services.

As well as the air, the oceans, seas, lakes, and rivers are part of the last frontier, territory unabused by permanent human occupation. The science of oceanography promises early use of water areas for farming, mineral extraction, recreation, desalination, and possibly small, permanent research communities investigating ways and means of living in natural environments usually hostile to man.

The extension of shore-line activities into shallow waters may be the first step into the new frontier. Historically, the edge of land and water has been indiscriminately exploited rather than well-planned. As there are limits to this critical space, there should be early consideration of environment-design standards and development objectives for shore-line areas, perhaps even a national environmental plan for all our coastlines, public and private.

In the immediate future, improved water transport may have greater impact, beneficial and otherwise to human habitation, than aircraft travel. While extensive underwater transport is still problematical, ocean engineers are working on subsurface sea trains. One- and two-man underwater scooters and small submarines are available, but the craft cannot go as deep as equivalent air machines can rise high in the air and are accordingly limited in range.

Surface ships, especially bulk-cargo craft, have been considerably enlarged in the past decade to the extent that some cannot pass through the Panama and Suez Canals. Such craft represent environmental threats,

By-passing congested highways to move valuable papers and documents.

Serving as an air platform for selected forestry.

Prototype model for moving heavy loads with short runways.

Emergency ambulance service.

Helicopters A variety of current uses. Clockwise starting upper left: Bypassing congested highways to move valuable papers and documents; serving as an air platform for selected forestry; emergency ambulance service; prototype model for moving heavy loads with short runways. *Reprinted from: Rotor And Wing*

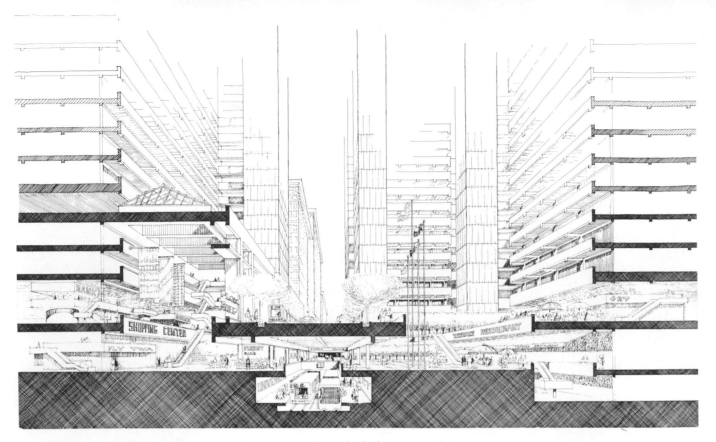

Seattle Subway Proposals Contemporary subway proposals are looked upon not just as transportation solutions but as generators of urban form. The Seattle subway system in the CBD area creates opportunities for distinctive environmental design. The location of the stations and pedestrian circulation systems to and from them are essential parts of the overall city development. *Drawings: Courtesy of Naramore, Bain, Brady & Johnson, Architects*

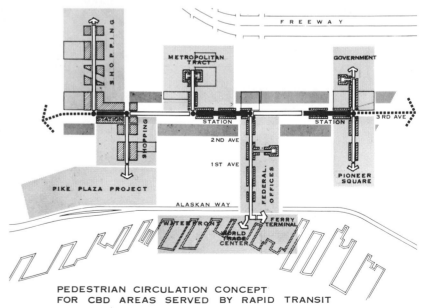

PEDESTRIAN CIRCULATION CONCEPT
FOR CBD AREAS SERVED BY RAPID TRANSIT

as accidents may spill thousands of tons of crude oil or other materials over the water surface, polluting the shore line and destroying tidelands birds, fish, and plant life. Underwater craft may not be as susceptible to such accidents, especially those due to turbulent weather.

Hydrofoil and air-cushion vehicles may enlarge the choices available for urban transportation. The first is a catamaran-like vehicle. The hull rises above the surface of the water as speed is increased, making hydrofoils fast and stabile but prone to damage from flotsam.

Because of engineering limitations, the hydrofoils now in use are not as large or as flexible as traditional waterborne carriers. Accordingly, the cost of operation limits their use of transporting passengers at fast speeds over relatively short distances and usually in areas where the water surface is quiet and clear of debris.

Air-cushion vehicles (AVC) are designed to reduce friction at the interface between water and air and are moved forward or backward by propellers. The air-cushion principle can work on land or sea. The only model now in production is the SK-5 (Bell Aerosystems Company). It can carry fifteen to eighteen passengers at speeds up to seventy miles per hour.

A one-year pilot operation of an AVC vehicle was carried on by the Port of Oakland in 1966. The idea was to link together the San Francisco and Oakland International Airports and downtown San Francisco with high-speed transport. The natural configuration and general urbanization pattern encouraged a direct route across the water rather than travel by ground vehicles along the perimeter of the Bay Area.

During the test period almost four thousand trips were made carrying about twelve thousand passengers. The trips averaged twenty-two minutes at a speed of thirty-five miles per hour. About one out of four scheduled trips was canceled, half of these due to adverse weather conditions.

The results of the pilot program showed the AVC was operationally feasible but more expensive to run than most other types of transport. It was expected that, as larger air-cushion vehicles were constructed, the cost per seat-mile would drop; but at the present stage of development, AVC's appear economically feasible only on routes over relatively calm water connecting places that generate large volumes of passengers who are willing to pay a premium fare for avoiding lengthier and more time-consuming routes.

Railroads, which dominated transportation in the late nineteenth and early twentieth centuries, lost their commanding position as the major carrier in the national economy by 1956. Motor vehicles and pipelines now carry as much bulk as the rail lines themselves. Nonetheless, the competitive position of the railroads may improve with containerization and could rise further with improved methods of handling cargo at the terminals and integration of various forms of transport under a single system of control and management. Most importantly, the major railroad lines have been laid out; areas of expected future urbanization are accessible to existing rights-of-way. Thus new capital requirements are not large in comparison to those of other kinds of transportation. Railroads have the advantage of extensive landholdings at termini and along their routes and have unused capacity in their rights-of-way. With the possible introduction of new power sources (electrical and atomic), the detrimental effects of railroads on the environment may be minimized. Thus their presence in urban areas may be more acceptable now than in the past, and they have good financial potential for undertaking corporate action in development of new towns and redevelopment of the central city.

Passenger service on the railroad dropped so quickly in the 1950's that airlines and commercial motor carriers now have larger numbers of passengers than the railroad in both long-distance and commuter service. Government intervention in recent years has impeded further decline of local railroad service, although cross-country travel continues to worsen, and prestige trains such as the Twentieth Century Limited have been discontinued.

Further public assistance to passenger traffic seems inevitable and desirable, especially through funding of research, purchase of new equipment, and in some cases operation grants. A twenty-eight-month experiment in improved suburban rail service was successfully demonstrated along the Harlem Division of the New York Central Railroad (1964-1966). During the test period service was increased, parking facilities were expanded, coordinated bus service was provided, passenger equipment was improved, off-peak-hour fares were reduced, and patrons were given a second day to use the return portion of cheap one day round-trip tickets. The results were substantial increases in journey-to-work and midday traffic. The key measures in low-density suburban areas proved to be frequent service and substantial and accessible parking facilities at the local stations.

For metropolitan areas, improving train travel may be a cheaper alternative than constructing new expressways and incurring the secondary expenses of policing, pollution, accidents, and the development of parking at the central-city destination. The increased use of railways would require relatively little new investment in rights-of-way and terminals. The use of rail-buses would give the advantages of the flexibility of motor carriers and the volume of rail vehicles. (See description below.) Application of known technology to the railroads could expeditiously return a practical form of transport to common use and subsequently improve the environment.

The proposed high-speed train system for the East Coast corridor is a promising step in this direction. From Washington to Boston speeds of around one hundred miles per hour are expected.

A successful Tokyo-Osaka electric express train has gained world-wide interest for the Japanese national railroad. The new line covers a 515 kilometer route, passing through forty percent of Japan's population and seventy percent of her industrial districts. Long-welded rails laid on concrete sleepers make the journey smoother than on traditional short-rail sections and wooden tie tracks. A centralized traffic control office tied to computer equipment in the trains and along the right-of-way results in safety and efficiency.

In France an experimental train broke the world speed record for track vehicles (December, 1967). The wheelless aerotrain, riding a cushion of air one tenth of an inch thick, is moved along a single concrete track by a turbopropeller. The vehicle is decelerated by reversing the pitch of the propeller and stopped by brakes that grip the concrete rail. A full-size model will carry eighty passengers at 250 miles per hour. In March, 1968, plans were announced to use the trains to connect the major airports outside Paris to the city center. The extension of the lines would turn neighboring cities into suburbs.

At high speeds rail travel from one city center to another will be competitive with the automobile and airplane, both in cost and travel time. Given the overcrowding of air corridors and the expense and difficulty of building new expressways, rail travel of

this quality is not just a good option for bettering the general transportation system but an absolute requirement for an improved environment.

Some forms of rail travel may be placed underground. Experimental studies of gravity and vacuum systems indicate that speeds higher than any known form of ground travel could be accomplished by propelling trains through sealed steel tubes running underground. Preliminary feasibility studies suggest that the cost of boring tunnels would be cheaper than land acquisition for surface routes or the traditional cut and cover methods of subway construction.

Rejected as a popular form of travel, mass transportation may, like railroad passenger lines, be resuscitated by improvements in design, operations, and management. The total number of revenue passengers carried by the transit industry dropped from 13.8 billion passengers in 1950 to 6.8 billion passengers in 1964. During this period the trolley car became a museum piece, its replacement, the trolley bus failed to attract riders, elevated lines were torn down, and subways suffered a drop in patronage. Why did all this happen? The automobile became the preferred means of movement, sizable public investments were made in highway construction to meet the resulting demands, and alternate forms of transport were neglected.

The motor bus is now looked upon by transport experts as the favored form of mass transportation, particularly for those who do not drive and the sizable number of aged, physically handicapped, and impoverished who cannot afford to use an automobile. Although total patronage is well below the peak years of 1950, the motor bus now carries two out of three riders using mass transit. In addition, fixed-rail transit is in difficulty because the older lines were oriented to downtown. Today less than ten percent of

urban travel begins or ends in the Central Business District because of urban decentralization and the use of the automobile for nonbusiness use. As no major additions have been added since 1940 to the number of miles of electrified mass transit in the United States, old rights-of-way and new destinations don't match. Motor buses have filled the travel needs of those who do not drive and are going somewhere else than downtown.

If mass-transit patronage is to be enlarged, however, major changes will have to take place in design and operation of the bus systems. Possible technical advances now under study include combination rail and highway vehicles. The Philadelphia Suburban Transportation Company has equipped a standard GMC bus with a retractile flanged-wheel attachment. The steel wheels can be raised and lowered in one minute without the operator leaving his seat. When leaving the tracks, the flanged wheels retract into the bus's baggage compartment. Called Hy-rail, this prototype rail-bus costs twenty-five percent less to operate than an electric rail-car and has an initial equipment investment substantially less than a traditional electrified carrier. A twenty-mile test in a snow storm in November, 1967, indicated that the rail-bus lost traction on steep grades in icy weather, but a metal cleat wheel has overcome that difficulty.

There are other ways to reduce our reliance on the automobile for urban travel. Engineers have recently suggested the application of automated-control devices leading to prototype passenger carriers (something like automobiles), which can be linked together to form a train and then separated for individual travel. Innovations may also be expected in general flow traffic control, which could include a mass-transport system giving taxicab service. Passengers would call a

Tokaido Express Photo: Japan National Tourist Organization

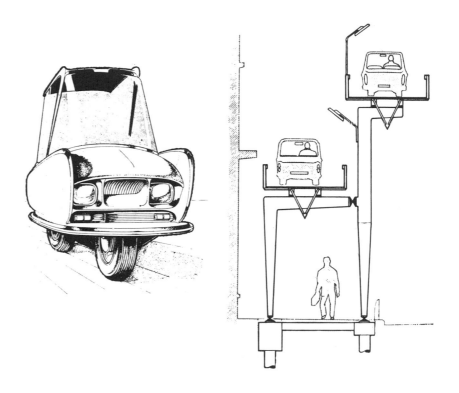

ACV In Operation *Photo: Port of Oakland Authority*

Mini-Cars One approach to city transportation is the reduction of the size of the automobile. Research by the Minister of Transport (Great Britain) suggests that maneuverable individual automobile-like vehicles could be placed in production. Concurrently, an economical segregated street and parking system could be established, and the use of existing urban space for circulation could be made more effective. *Reprinted from: Cars For Cities. Minister of Transport, London, 1968*

central station; the request would be logged by a computer, and several passengers traveling the same route would be picked up by a ten-to-twelve-passenger carrier. This DART system (Demand Actuated Road Transit) would be more convenient than a bus and cheaper than a cab.

Subways, after two decades of decline, are now coming back into their own. Improvements and extensions are planned in Boston, New York, and Philadelphia. Relatively old cities like these simply do not have the environmental capability for accommodating additional highway construction and the concomitant increase in parking. Since the redevelopment and continued use of the central-city core is dependent on large numbers of people using the city during the day, mass transit is the only practical solution.

Encouraged in part by Toronto's recent success, San Francisco, Washington, D.C., Atlanta, Baltimore, Seattle, and Pittsburgh have plans for new subways in various stages of design. These are substantial projects, not token measures. The Washington and San Francisco construction costs alone will exceed all the existing investments in electrified transit equipment in the United States. The encouraging aspect of these designs is that they are not just improved versions of older subway systems but represent innovative operating techniques and are integrated with total urban-development patterns. The engineering solutions have been considered from the point of view of environmental design, using the collective knowledge of physical planners, architects, sociologists, economists, systems analysts, and others with specialized knowledge organized around the problem of modern transportation rather than around professional skills.

Subsurface development has obvious implications for environmental design; yet the least appreciated of modern transport technologies is the growing use of tunnels and pipelines for moving water, power, liquid, and semiliquid raw materials and wastes. Oil pipelines now carry a larger volume of freight than all the inland-waterway carriers combined. The technical feasibility of moving passengers through vacuum-gravity tubes has been mentioned. Similarly, light cargo capsules could be sent along similar rights of way at high speeds.

The construction of an interstate, subsurface transportation system would have more profound influence on the environment and modes of living than further development of rail, public motor, or individual, personal forms of locomotion, such as the family helicopter. For instance, the location of dispatching and distribution centers for freight could significantly affect industrial and manufacturing locations, as well as warehousing and highway development.

Furthermore, transport revolutions have always dramatically altered the size and location of production and distribution. In the age of waterborne freight, the movement of goods depended on vessels whose capacity was measured in thousands of tons. The advantages went to the large producer and the cities with waterfront locations. The railroad allowed the delivery of individual cars with an average capacity of thirty tons, and a linear decentralization of production and distribution points followed. The truck allowed another reduction in the size of freight movement, with a capacity of up to ten tons in each carrier. Freed from fixed rights-of-way, such as the railroad, trucks allowed further decentralization; and a weblike transport network emerged. A further evolution of surface carriers would seem to be spatially limited to a refinement of the known network and the historic carriers. Subsurface transport offers the possibility of an alternative, though perhaps in the beginning it would be no more than a supplementary new movement system.

One possible result would be an improved version of downtown shopping—an immense, enclosed bazaar at the intersection of several subsurface systems, displaying in a single location as many kinds of goods as several large department stores and strings of specialty shops now offer in a typical retail area. After selection and purchase, goods would be shipped by subsurface transport from decentralized warehouses (or perhaps direct from the production line) to designated distribution terminals throughout the metropolitan area. There traditional methods could be used for house-to-house delivery. In instances where the density was high enough, direct pneumatic-tube service from central warehouses to superblocks could take place in a subsystem of the larger network.

One imagines large "pipeline" rights-of-way crisscrossing metropolitan regions with recreation and other open-space activities on top. At special junctures, clusters of industrial and distribution facilities would occur at economic intervals. Obsolete and decaying urban areas could be fitted into the network and possibly reused as employment centers. The typical need for tearing up the surrounding community for highway access, a requisite for central-city industrial parks, warehousing, and shopping centers, would be lessened.

Scientists at the Oak Ridge National Laboratories have discussed the concept of using extensions of the Interstate Highway System as an opportunity for simultaneously constructing a vast cross-continent water pipeline. With comparable investments a subsurface transport system might be introduced at the same time. The use of a limited-dividend corporation such as COMSAT could allow joint public and private investment, funding, and management. Increase and change in technology, industrialization, and urbanization seems to provoke new forms of transport. Since all three are in flux, innovation and invention can be expected. The aerospace industry is less than twenty years old, the automobile industry less than fifty, the railroad about a hundred, the idea of another new transport technology introduced within a relatively short period of time—perhaps subsurface transport is not beyond the borders of credulity. Extended to its logical conclusion, a subsurface system would induce fully automated industrial processes, enlarge the capacity of existing road networks by reducing existing traffic, and free some of the earth's surface and air of polluting machinery. Air, space, and ocean are the popular frontiers for modern transportation technology; but rather than exploit them further, man's habitat and life therein may be more quickly improved, with less economic strain and fewer encounters with incompatible environmental conditions, by simply removing a relatively small amount of industrialization and transportation to a location under the earth.

TRANSPORT PLANNING: A FOURTH REVOLUTION?

Urbanization as a social phenomenon is about fifty centuries old. In the last two of those centuries the most remarkable changes occurred, including three revolutions in transport technology: the railroad, the motor vehicle, and aircraft. The design effects that followed the introduction of these technologies have been more often accidental than intentional, but all the experience to date underlines the fundamental role transportation has had in shaping the environment,

and there are no reasons to think this influence will end soon. A brief summary of trends in urbanization and transportation planning is worthwhile here to illuminate the place environmental designs have in shaping the future patterns of design structure resulting from transportation technology.

From one angle, each succeeding technology has led to the spreading and connecting of urban places. At the metropolitan scale the result has been fairly uniform: heavy densities along the network lines, highest densities at their intersections and termini. This basic design structure will continue into the foreseeable future.

From another angle, many conflicts in environmental design begin when high-volume movement systems intrude into lower density areas and when high-density areas are served by low-volume movement systems. This is a simplistic view, but it states the case historically and currently. Accordingly, the matching of environmental objectives with modes of transportation technology (including the personal vehicle) should be the central objective in transportation planning. This is not the case today.

From a third perspective, recent transportation has been flawed by squeeky-wheel responses: sympathetic public investments for those demanding optimum accessibility and improvements along the networks where congestion has been found or is expected. Central-business-district interests (increasingly less accessible as urban development moves outward) and the transit industries (with a similar concern for survival) have had enough political strength to begin to improve core-area accessibility through mass transit. Industrial management, labor, and others journeying to work in a cross-city pattern have successfully called for other kinds of improvements, largely improved highways located countergrain to the traditional form of the city. In these instances of both mass transit and highway development, the engineering solutions have often come out of projections of past trends in the existing system rather than a creative search for alternative choices through comprehensive planning. In addition, because transportation planning crosses government jurisdictions, the overview was either absent or left to those least qualified for the work because of their special biases.

Due to this background perhaps a fourth revolution in transportation technology has already taken place. Rather than involving a "hard-wave" technology, this revolution concerns an understanding of the nature of transportation systems and their relation to their environment. A transportation system is not an end in itself in the sense of serving merely to get people and freight to their destinations. A transportation system is an instrument for achieving environmental goals by optimizing accessibility and appropriate land use with lowest possible capital and operating expenditures.

If this comprehensive approach to transportation planning is to succeed, four kinds of capabilities are required. The first is management: an area-wide planning and decision-making process and the establishment of a workable method of integrating, controlling, administering, and operating all aspects of the movement systems. The second kind is the ability to conduct an overall identification and evaluation of existing environmental conditions and trends. These would include the qualitative as well as quantitative aspects of land-use, the demands made on all transportation and circulation facilities, a weighing of consumer preferences, and an appreciation of political, social, economic, and aesthetic constraints and opportunities. The third capability is measurement of the engineering and tech-

nical capabilities of known and improving technologies and the consequences their application may have on the environment. The fourth capability is creative speculation, the search for alternatives to the predictable answers that common sense and current planning techniques now produce.

In summary, a better environment will come forth from transportation planning when there is a fundamental comprehension of relationships between transportation systems and environmental goals as well as an extended set of professional and technical skills available for problem-defining and problem-solving, consistent pressure built up for innovation and invention, and political action applied to make the improvements in the environment realized.

Within this theoretical framework present transportation planning is filled with contradictions and paradoxes. One dilemma is where or how to start. Transportation decisions cannot be made independent of land-use planning, but the prediction techniques for anticipating the influence of transportation on land use are inadequate. Improved transportation networks tend to stimulate new land-use patterns, and their emergence is partially a result of yet immeasurable and probably politically uncontrollable human behavior. On the other hand, projection of a land-use pattern cannot be left until after a transportation plan has been set, because the variables in the activities that occur on the land may change the demands made on the network.

Despite these difficulties, there are a number of reasons to anticipate improved transportation planning through policy-making and public administration. The Federal and several State Governments, the prime sources for money, demand a metropolitan-planning context before authorizing grants for any kind of government-supported programs, including highways and transportation. A reasonable percentage of construction funds are being allocated for research. The Departments of Housing and Urban Development and the newly created Department of Transportation have promised to coordinate their various roles and responsibilities with the idea of focusing their efforts on comprehensive solutions.

On the technical side, a new generation of mathematical models is being formulated to account for cause-and-effect relationships between individual location decisions (for a house as well as a factory) and the resultant travel and transport patterns. The techniques may include aspects of human behavior previously unconsidered. A demonstrable theory of how choices are made could offer the necessary clues for predicting land-use decisions. Utilization of cost-benefit factors in sorting out feasible alternatives may encompass environmental variables that cannot be calculated by reckoning only the ratio of initial investment to immediate results. Finally, through successive approximations of land-use models, then transportation models, and the evaluation of the effects of one on the other, a managable number of options may be sifted out, each closing in on a predictable set of actions and consequences—within the limits, of course, of known technology.

What will then remain to be done? We hope a technique will be established so that the quality of human habitation implicit in each alternative can be communicated directly to the general public. Here, environmental designers can play a substantial role in conveying both the design structure created by modern transportation technology and the resulting effects at the microscale. Perhaps for the first time anticipatory planning can reduce the waste, dyseconomies, and injurious effects of adventitious trans-

port design by giving an effective representation of alternatives and options to a public faced with uncertainty, ambiguity, and yet choice.

In the meantime current planning in New York State, Connecticut, Los Angeles, Washington, and Baltimore already provides encouraging examples of how technical problems can be elucidated and how design structure can convey intelligible information to the general public on the choices available for the designed environment at the largest scale of human habitation—a design structure in which lines of circulation, open space, transportation systems, hierarchial centers, their size and location, are the critical elements.

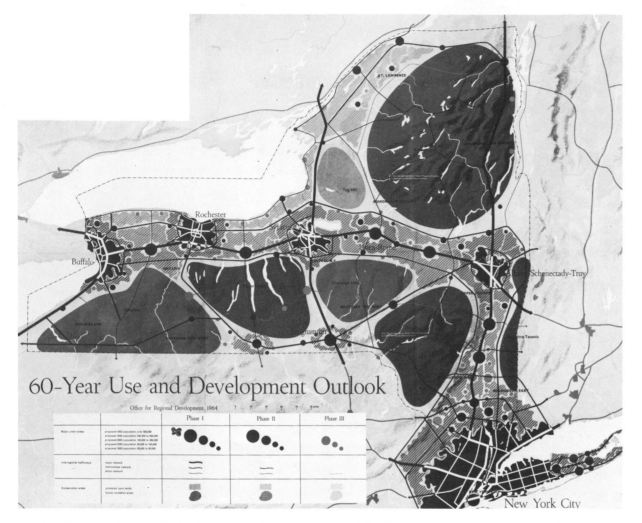

60-Year Use and Development Outlook

Office for Regional Development, 1964

		Phase I	Phase II	Phase III
Major urban areas	proposed 2020 population over 500,000			
	proposed 2020 population 250,000 to 500,000			
	proposed 2020 population 100,000 to 250,000			
	proposed 2020 population 50,000 to 100,000			
	proposed 2020 population 25,000 to 50,000			
Interregional trafficways	major network			
	intermediate network			
	minor network			
Conservation areas	protected open lands			
	forest-recreation areas			

Plan For The State Of New York The suggested sixty-year use and development outlook that comprises the 1964 Plan for the State of New York shows the possibilities of establishing design structure on a large scale. *Source: Office for Regional Development, 1964*

THE DESIGN STRUCTURE OF A STATE: NEW YORK

Planning is a process by which man defines his goals and devises means to attain them. Individuals and families spend much of their lives in planning; businessmen would soon go bankrupt if they failed to plan. Americans would not tolerate a government which did not plan for the nation's security and its economic well-being.

Yet Americans also are very jealous of their private freedoms and of their local responsibility. The dictation of ways of living by a central, autocratic regime is foreign to our history and repugnant to our principles. This report does not seek to impose an arbitrary master plan on the State of New York. It does point out, however, that all the people of the State have a common interest and responsibility in defining goals for future development and making plans to attain those goals.

Change/Challenge/Response, A Development Policy for New York State Office for Regional Development, Albany, New York, 1964

New York State, threshold to the continent, expects its population of sixteen million to double itself by the end of the century. The state is a sprawling mixture of million-people metropolises and villages, thousand-mile state forests and small city parks. In the next forty years it will experience the equivalent of all the physical development accomplished in the past 180 years. For this reason, the State Plan is an administrative device that encourages improved settlement patterns by bringing together in the coming years policies and actions which might otherwise go uncoordinated.

The Plan is essentially an updated version of the one prepared for Governor Alfred E. Smith in 1924 by the New York State Commission on Housing and Regional Planning, and is not a blue-print but a schematic statement of intention. The planners recognize that there are strong economic forces at work concentrating and connecting centers of commerce, manufacture, distribution, and service in the State. Anticipated technological advances in communications and transport plus new kinds of leisure and culture add new dimensions to old patterns. A web of related activities thus takes on a discernible physical shape; and the design structure is established by the location of major urban areas, valleys that link the areas, conservation areas, interregional trafficways, and other transportation lines. Built on trends and conditions that can be projected forward, the Plan gives subtle but strong shape to what otherwise could be a formless environment.

The State Plan is divided into three phases: Phase I will concentrate on metropolitan core-city renewal and constructive guidance of new suburban communities. Many new public facilities will be situated in the linking valley areas. State-wide rural development and conservation of natural resources will be fostered. Phase II will emphasize large-scale residential and social renewal of metropolitan areas, beginning construction of new communities, and organization of high-speed transit in the valley areas. Intense recreation uses and urbanization are anticipated in the relatively low-density parts of the State. Phase III will aim at stabilizing metropolitan growth, balancing urban areas and open space in the valleys, and full-scale construction of new towns in other areas designated for this purpose.

The implementation of the Plan depends on striking a balance between private interests and the 6,846 local governments involved in capital construction in the State. The public share of investment in the physical environment in New York State is expected to be twice as large in the next two

Three basic patterns of growth in New York State, all working simultaneously

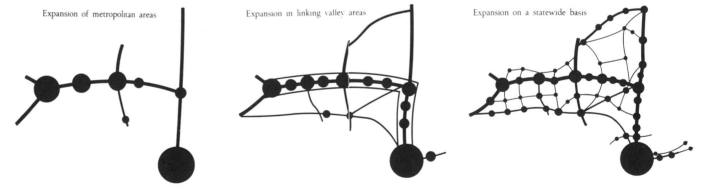

Expansion of metropolitan areas   Expansion in linking valley areas   Expansion on a statewide basis

Growth Interaction *Source: Office for Regional Development, 1964*

159

decades as it was in the last two, growing from $76 billion to around $150 billion. Naturally, the location of this public investment in terms of highways, new universities, hospitals, parks, and so forth will affect the direction of growth and the quality of habitation and environment.

In determining where to locate these investments, therefore, the strategic step was the establishment of ten development regions with boundaries based on local government jurisdictions, natural conditions, labor force areas, and commonality of interests. Through study, education, conferences, and plan preparation, and within the framework of the State Plan, each region will determine its own special life style: ". . . all the people of the State have a common interest and responsibility in defining goals for future development and making plans to attain those goals." The procedural steps in carrying out these measures are shown in the diagram on page 161.

In the New York State Plan, we see an expected intensity of environmental change unprecedented in history and a thoughtful beginning to an era of community-design aspirations rooted in democratic action.

Enormous and frustrating problems become not simple but more manageable as a comprehensive planning framework guides and coordinates public investment so as to sustain the diversity of the past and present and accommodate the demands of the future. Fewer than one out of ten people in New York State will live outside an urban area. The idea of contrasting yet vital environments, a day apart in some instances, is a fundamental environmental objective. In this regard Connecticut's situation is especially acute. The entire state can be traversed in several hours, and there is danger it will become one massive, blurred, undifferentiated urban conglomeration.

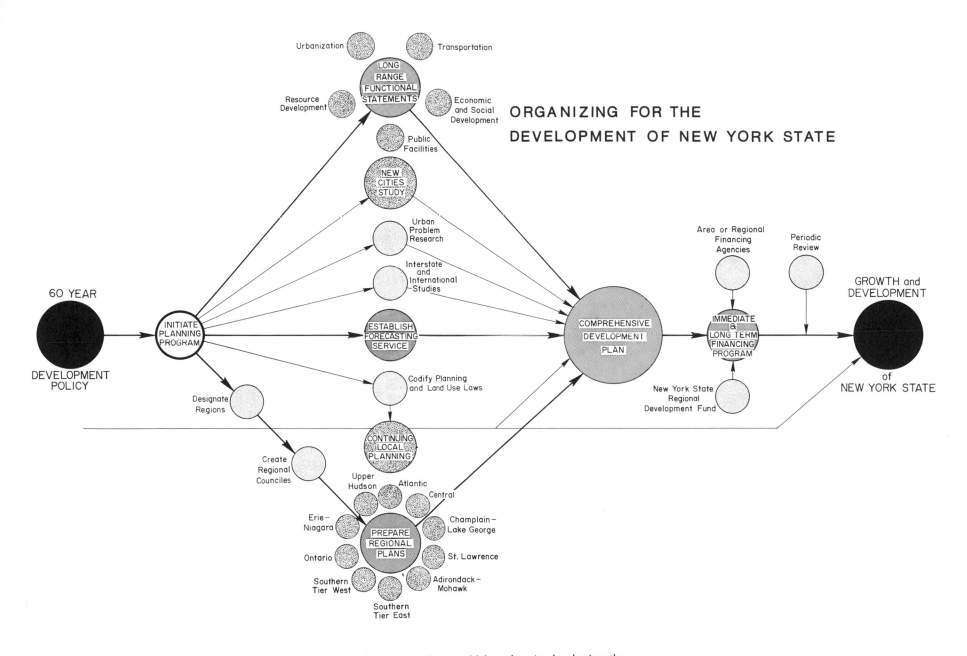

Planning Process This diagram outlines in flow-chart form the basic steps that would be taken in developing the state plan, including the key measures for financing the proposals. *Reprinted from: Change, Challenge, Response. Office for Regional Development, Albany, New York, 1964*

1965 POPULATION DISTRIBUTION
(Cities and Towns)

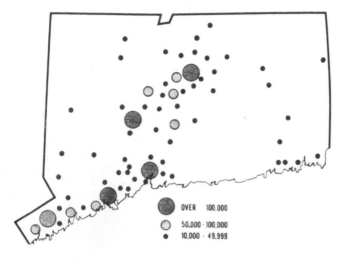

OVER 100,000

50,000 - 100,000

10,000 - 49,999

LAND USE BASE

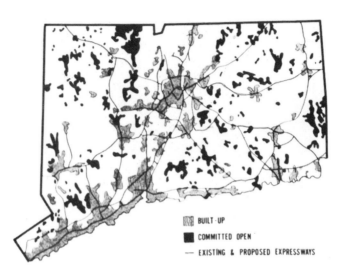

BUILT-UP

COMMITTED OPEN

— EXISTING & PROPOSED EXPRESSWAYS

OPEN SPACE POTENTIALS

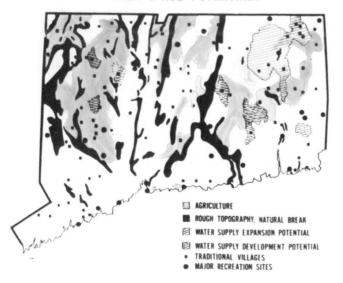

AGRICULTURE

ROUGH TOPOGRAPHY, NATURAL BREAK

WATER SUPPLY EXPANSION POTENTIAL

WATER SUPPLY DEVELOPMENT POTENTIAL

• TRADITIONAL VILLAGES

● MAJOR RECREATION SITES

Open space and natural features acting as constraints or encouragement for urban development.

LARGE TRACTS PHYSICALLY SUITABLE FOR INDUSTRY

Large tracts of land suitable for industrial development. The Connecticut study indicated that the location of job opportunities was a major determinant in the growth of the state. *Four diagrams. Reprinted from: Connecticut Interregional Planning Program, 1966*

THE URBAN STATE: CONNECTICUT

If it continues its present pattern of urbanization, most of Connecticut in thirty years could resemble present-day Los Angeles County: one thousand people to the square mile with dependency on the automobile, a polluted natural environment, and man-made designs that are formless, vacuous, and unrelated to each other.

What may save Connecticut is the logic behind the Connecticut Interregional Planning Program. As with the New York State plan, it responds to the universal dilemma of planning in a democratic society, where future forms cannot be preordained by an authoritarian government. A representative sampling of Connecticut households, about five thousand surveyed from 1961 to 1964, strongly indicated that the general "suburban-rural" appearance of town and state were the most attractive environmental features to its citizens. The realization of this single goal is the overriding design issue, and Connecticut faces the task of designing a plan for 1.2 million acres of land not yet urban.

A favoring factor: The elements to be considered in giving form to the statewide design structure can be easily grasped. First of all, the state's physiognomy is distinctive and varied. Physical features have influenced past growth, but current technologies could overcome most topographic constraints, such as rivers, lakes, ridge lines, and wetlands. Nevertheless, the location of the natural landscape is still of considerable importance to the eventual community design, although the relationship has been reversed: Where the landscape was once a barrier to urban development, it now has to be protected from urban use.

Secondly, urban development follows the seacoast to New Haven and then continues as a linear pattern along the major expressways through the center of the state to Hartford. There is a scattering of historic industrial towns to the west and east, none of which is physically proximate to the major concentrations. One can drive through cities and towns holding the major part of the population in only a couple of hours, as the cities and towns are strung together by a good highway network. This pattern is likely to continue in the future.

The Connecticut Plan shows the benefits of the intentional arrangement of Man's surroundings: The design is rooted in the efficiency, economy, and appearance of large-scale public elements, such as highways, transport facilities, major parks, forests and beaches, and major institutions, such as universities, colleges, and hospitals. The importance of protection of natural landscape and careful location of urban development has already been cited. Of equal consequence, as the plan makes clear, is the location of land for industry, especially those parcels large enough for industrial parks. These singular tracts of land, free from the intrusion of other urban uses, are critically important to long-term economic growth. Large-size, well-located industrial parcels allow optimum efficiency of investment in power, water, and waste disposal and contribute to improvements in the journey from home to work and the movement of goods and materials throughout the state. These environmental features, which will be part of any eventual design for the state, can be seen in the accompanying diagrams. From these physical conditions several alternative design forms of future development can be rationalized. Connecticut has choices, though many conditions are fixed.

No one can predict the future. Seventy years ago few would have anticipated that the automobile would be the common means of transport in the first half of the twentieth century. Similarly, the fact that more money was spent in the aerospace industry than in the automobile industry in 1967 was unexpected. But just as this theme of the unknown shape of the future bears repetition, so also does the theme that uncertainty can be reduced to manageable size and dealt with by the environmental designer. Using credible assumptions, professional judgment, and consideration of past population and urban-development trends, it is possible to make reasonable guesses about the size and location of future urban growth. Such forecasting techniques have been considerably aided by computer technology. Models of possible future development can be statistically derived from existing data through mathematical formulae. The Connecticut Interregional alternative design choices are based on such estimates. It anticipates the most likely form of development, plans government services accordingly, and recognizes the influences of those public services on the total pattern.

The particular prediction technique that Connecticut used is called a "growth distribution model." It works with known relationships between total population, classes of employment, income distribution, holding capacity of land (as determined by zoning, availability of sewers, and other physical characteristics), and accessibility between population groups, income groups, and employment types (measured in travel time). The future relationships between these components are first set by actual conditions found in Connecticut from 1950 to 1960. Since the model produced for 1960 was fairly accurate using 1950 data, it was assumed that 1960 data would yield a fairly accurate 1970 picture, and similarly successive approximations could be obtained to the Year 2000.

This model gives good insights as to future scale of growth and its spatial distribution. Here is a simplified example of one set of interrelationships typical of the model's projections:

Towns having high accessibility to population and employment, having grown relatively fast in the past, having prime vacant land to settle at high densities, and having low land costs will attract more population and thus grow faster than towns having opposite conditions.

The Connecticut model is a pioneering effort, the first time an entire state has been subjected to such forecasts. Models of this kind are neutral, keeping human prejudice and error to the minimum. They do not present a picture in detail and only indicate what the trends might be if no intervening forces were introduced. Such models are the best method for estimating urban growth available today.

After the forecasts are computed, two further steps for planning are taken. The first is the identification of urban places likely to reach their development limits within a thirty- or forty-year period. The second is the mapping, illustration, and interpretation of the Year 2000 forecast. From these studies the current drift can be identified as well as the strategic point for "bending" trends in other directions by studied use of public investments and legal controls on land use. For the Connecticut plan, the land-use trends were mapped on a statewide grid, each square of which covered about 160 acres.

The Connecticut planners used three major land-use categories in their preliminary design: Built-up Land (residential, services, and manufacturing), Committed Open Land (water areas, intensive recreation areas), and Other Open Spaces (flood control zones, State hunting refuges, and so forth). The main limiting factors on future growth were then found to be the amount of vacant land physically available to accommodate new development and the legal proscriptions of zoning. Furthermore, as can be seen in the accompanying illustration, the projected trends take the form of a continuous urbanized pattern occupying a significant portion of the central part of the state and running along the entire coastline. If trends continue, it appears there will be a slight increase in urban land densities, a doubling of suburban densities, and an eightfold increase in exurban land densities. The desired qualities of contrast and amenity, created by closeby open space, would be hard to effect.

The Connecticut Interregional Planning Group presented three alternatives to the trend pattern. Each has a distinctive design structure, but all are compatible with general statewide objectives for variety in housing and job opportunities, optimum space for industrial and commercial development, population groups large enough to economically support local public services, and an environment which is aesthetically satisfying because the parts are complementary to each other and differing landscapes are protected.

Alternative One is a composite plan using existing regional plans and local plans. It is not unlike the trends map except that significantly large public reservations for open space and recreation reduce the width of the linear forms and keep parts of the major urban centers from mixing with one another.

Alternative Two handles future growth in a series of separated urban centers. The pattern would consist of thirty-five to forty individual, self-sufficient centers varying in size and having higher densities than those popularly accepted in Connecticut today. The design would be based on expanding existing places and extending existing highways to tie the communities together. Not all places would be of similar size or equal in economic position. Each center would be separated from the other by open space.

Alternative Three is again a linear scheme, channeling growth into relatively few areas, largely the central spine and five other scattered communities. It would preserve large amounts of open space. The execution of the scheme is dependent on technical solutions for moving high volumes of people and goods through the intensive development. The plan would also require public acceptance of higher density living.

From the point of view of environmental design, the most revealing facet of the Connecticut study is that much of the state's future design is already committed by what exists. Even though large population growth and new land development lie ahead, existing settlement patterns are not likely to be displaced. The question remaining in Connecticut is whether the stated goals of the citizens, rural and suburban environments, are sufficiently strong to bend obvious trends towards undifferentiated urban sprawl to the desired pattern of human habitation through a slightly different design structure than current trends will impose.

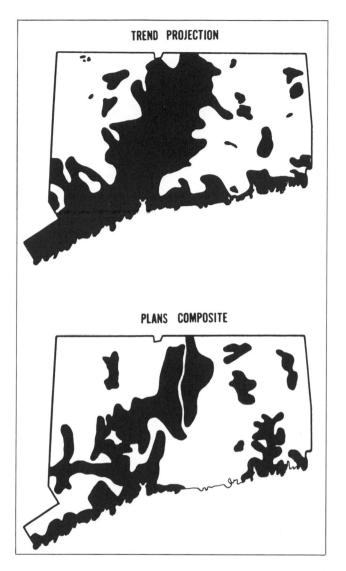
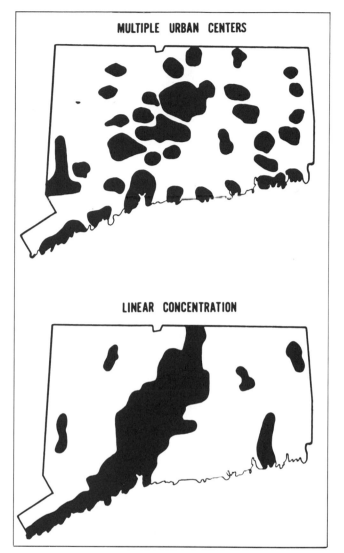

TREND PROJECTION

MULTIPLE URBAN CENTERS

PLANS COMPOSITE

LINEAR CONCENTRATION

Options In Connecticut The design structure of the State of Connecticut by the end of this century will be determined by the selection of options now available for guiding the growth of the state. These diagrams show the differences in design structure if present trends continue, or composite plans of various cities and towns are carried out, or multiple urban centers are created through higher density development and open-space conservation, or if future growth is concentrated along and around the expressway systems in conjunction with open-space conservation. *Reprinted from: Connecticut Interregional Planning Program, 1966*

FINGER AND HAND AND DIAGRAMS: COPENHAGEN

In regional planning every effort has been made to bring the open country as near as possible to the center of the city. By means of the 'finger-plan' principle a large surface of contact between town and country is obtained as well as comparatively narrow built-up areas so that the distance between dwelling and open country is not too great.

Danish Town Planning, Federation of Danish Architects, Copenhagen, Denmark, 1957.

Intelligently determined, a diagram of design structure can communicate the essence of a plan and gain appreciation of its proposed measures without generating confusion from technical detail. The "finger-hand" design for Metropolitan Copenhagen has had good results in keeping the public knowledgeable about the formal development principles guiding urban growth in Denmarks' major metropolis. The concept can be understood quickly and appreciated easily.

Copenhagen spreads over twenty-nine municipalities and extends outwards thirty kilometers or about nineteen miles from the city center. In the past it was often proposed to limit the growth of Copenhagen in favor of other cities, but no agreement could be reached on a national plan within which this limitation could be rationalized. Local communities prized their independence, and overall metropolitan government was not achievable. The 1951 plan was an intelligent planning compromise. It also gave the metropolis a distinctive design structure which it continues to sustain.

In the 1951 plan an "inner zone" was set aside for immediate construction, a "middle zone" for land where plans could be prepared but not yet executed, and an "outer zone" which could be used only for agriculture and other planting purposes. The location of the three zones set the design structure of the region. Subsequently, the boundary between town and country gets set by a joint authority representing the entire metropolitan area. While detailed land uses within the urban area are prescribed by the local municipalities, the joint authority controls the timing of development.

The plan established the three development zones on the basis of mass-transportation systems—tramways, buses, and suburban railroads. All three systems converge in the historic center of the town. As a result, the metropolis has the form of a hand with outstretched fingers. The palm, the "inner zone," contains the heavily used trams and buses, and the fingers the railroads, as well as high-capacity highways. Industrial development and service centers will be kept to the palm area, thus in easy reach of people living in the central city and along the fingers.

The "middle zone" of new residential communities with a full range of cultural institutions, schools, public buildings, and recreational facilities is being developed in the fingers by the individual municipalities. Heaviest densities are near the stations, lower densities at the fringes of the built-up zones.

Forceful measures are taken to keep the "outer zone" areas between the fingers unbuilt. The natural contrast between urban settlement and rural land is being reinforced with the planting of new forests and the encouragement of agriculture. Urban and natural environments are thus kept in balance in accordance with a design structure comprehended and accepted by the citizenry of Copenhagen.

Public Image Regional planners in Denmark have found an effective way of describing the essence of their proposals by suggesting that the future development of the metropolitan area might take the form of an extended hand. Urbanization would be contained within the palm and along the fingers, with the space in between the fingers reserved as agricultural and outdoor recreation zones. *Source: Denmark Federation of Architects*

166

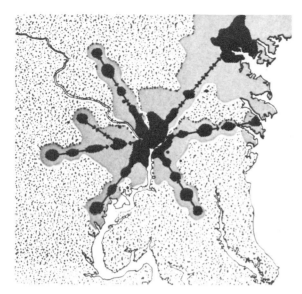

Design Structure, Washington, D.C., Year 2000 *Reprinted from: The Nation's Capital Plan For The Year 2000, Washington, D.C., 1961*

METRODESIGNS FOR THE NATION'S CAPITAL: WASHINGTON, D.C.

By the Year 2000, Washington and her sister city, Baltimore, will form the third-largest metropolitan area in the United States: nine million people. In the Washington sector, recent growth shows signs of "scatteration" and "leapfrogging"—poorly planned, low-density developments, unrelated to each other, with two hundred acres of land used for each one thousand inhabitants, as compared to twenty acres in earlier communities. With development continuing in that fashion, five times the present land will be needed to accommodate a 150 percent increase in population. The probable results: extended and congested commutation from home to work expensive and overburdened public services, a dramatic decline in the visual quality of the region, and the disappearance of any sizable open space.

The policy plan for the National Capital Region is an attempt to reverse constructively these conditions and trends by establishing a strong design structure. This design structure is supported by fundamental social and economic goals and objectives, and these goals set the tone and direction for immediate decisions and actions as well as the necessary perspective for the larger design framework. The first priorities in the Washington plan are these:

A broad range of choice among satisfying living environments: For Washington this means the elimination of substandard housing and the encouragement of variety in housing types and prices throughout the region.

A broad range of employment opportunities: Inevitably, any large metropolitan area generates mixed employment, but in this respect Washington is peculiar, because government is the predominant employer.

The Federal Government accounts for over a third of the total number of jobs. Compared with Detroit, where less than twenty-five percent of the jobs are related to automobile production, and Pittsburgh, with less than twenty percent of the jobs in steel, Washington statistically and psychologically is a one-industry town. The goal in the National Capital Plan is to distribute the major employment centers *throughout* the region and not concentrate them in one area.

Efficiency in transportation and land use: Decentralization of employment centers and increase in variety of housing; that is, the provision of genuine options for choosing where one lives and works, will require *both* land-use decisions *and* circulation decisions concerning how to move mounting volumes of traffic at least cost. To be genuinely effective, one set of decisions cannot be made independently of the other.

A healthy and visually satisfying environment: The giant steps needed to achieve a healthy environment are the elimination of slums and water and air pollution. These measures will be costly. To be effective, they will require a firm view as to where overall development patterns should occur.

For visual distinction the National Capital Region is fortunate in having both sea and mountains within reasonable driving time from the central city. Strong local design can be anticipated because of local support for regional architecture and the natural landscape and beauty of community sites. The problem area is the land that lies between the well-planned small communities and the large natural assets (sea and mountain).

Opportunities for participation in decision making: The planners believe that the above goals can be achieved only if there is widespread involvement in the many de-

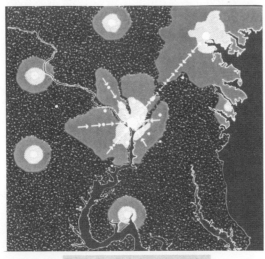

Expansion by independent cities

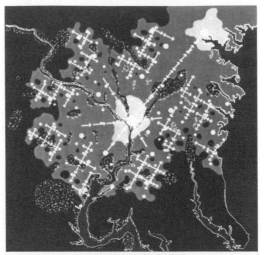

Peripheral communities

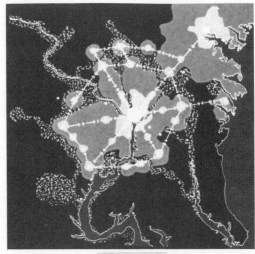

Planned sprawl

Alternative Development Patterns *Reprinted from: The Nation's Capital Plan For The Year 2000, Washington, D.C., 1961*

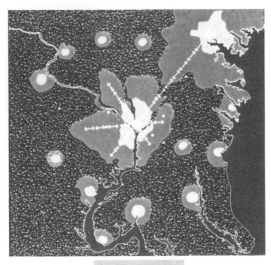

Satellite new towns

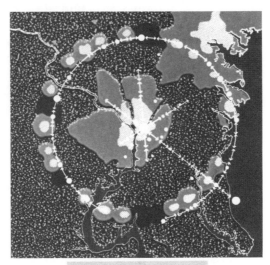

A circumferential ring of towns

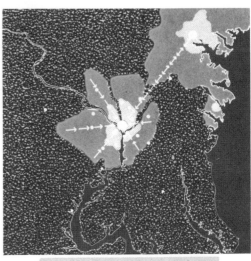

Restriction to present size by a green belt

cisions that have to be made to execute the regional plan. This participation will have to come from *supra-government* agency, not a *super-government*. The strongest influence will be Federal decisions on the quality of the central city (Washington, D.C.), on the location of Federal employment centers, and on Federally aided roads and transport. Unlike other metropolitan regions in the United States, metropolitan Washington's design will largely be determined by Federal intentions, commitments, and performances. It is a showcase opportunity of national importance.

In giving form to the National Capital Region, the planners approached their work in two ways. First, they thought in terms of four elements: the patterns fashioned by residential areas, the general disposition of major employment centers, the regionwide lines of transportation, and the open spaces outside the built-up areas. The current design, more adventitious than planned, consists of a dominant city, with residential and employment densities highest at the center; scattered, small suburban job centers; weak and unconnected regionwide road and transit lines; and countryside ten miles or more from downtown Washington but rapidly receding.

Second, the planners thought in terms of design choices that anticipated continuing growth. The alternative to growth is restricted metropolitan development through strong controls on employment opportunities. Such a policy is in force in Moscow and in Tokyo, but both regions today have twice the population expected in the Washington-Baltimore region by the year 2000. Since no out-migration is expected, natural increase alone in Washington would result in a population of four million people within four decades. Accordingly, the most likely trend is towards guided development rather than restricted development. The design options available are these:

New independent cities: An alternative to the metroregion of today is the creation of large independent cities at the edges of the region. These urban areas would have approximately 500,000 people and be relatively self-sufficient in the sense of affording as wide a range of activity, service, job, and housing choices as any metropolitan city.

Full realization of this design concept would require not only construction efforts of unprecedented magnitude but also accompanying improvements in political and administrative management. Under presently foreseeable conditions, the development in a relatively short period of time of large independent cities with very distinctive designs is not a likely occurrence for fiscal and legal reasons. The idea does give clues, however, as to how our national public expenditures might be pointed in better times to come, when the cost of national defense is substantially reduced.

Planned sprawl: This alternative guides the present trends towards a *designed* sprawl. Average densities would be higher than those of today, allowing a more efficient and economic development pattern. New communities would be formed by accretion with the building of new employment and commercial centers. Each community would have a range of housing types, jobs, and services. None would be large enough to compete with the central city. The low density of development might present an attractive appearance, but there would be no real contrast between urban areas. The journey-to-work distance would be extended, and the automobile would be the predominate form of travel. Technology would triumph over amenity.

Dispersed cities: Regulated sprawl and designated, concentrated suburban centers suggest the possibility of making these communities somewhat denser than would be expected and separated from each other by the open space thus saved. This concept differs from the independent-cities alternative described earlier with respect to size and number of the cities and travel time to and from the central city: The dispersed cities would be smaller in size, more numerous, and closer to the central city.

Location of the dispersed cities would be determined by topography, water supply, present trend of development, and other practical considerations. The overall regional design, however, is a central city circled by an urban area and then open space spotted with strongly-defined small cities having 100,000 or more people.

Ring of cities: Another possible design form, not dissimilar to the 1948 Plan for Greater London, is a ring of cities of less than 100,000 people about thirty miles away from central Washington and at considerable distance from the present urban area. This arrangement has several advantages over the dispersed-city design.

The new towns would be in close contact with each other, thus allowing a sharing of activities, particularly cultural and service activities. The remoteness from the central city would reduce the chance of eventual pressures to fill in the intervening space. Controls over the central urban area would, therefore, not have to be as strict as in the case of the dispersed-city plan. To succeed, as with the London Plan, the design would need national government subvention for physical development to an extent unknown in the country's recent experience.

Peripheral communities: This alternative might also be called the Stockholm approach. New and intensively developed communities are placed at the very edge of

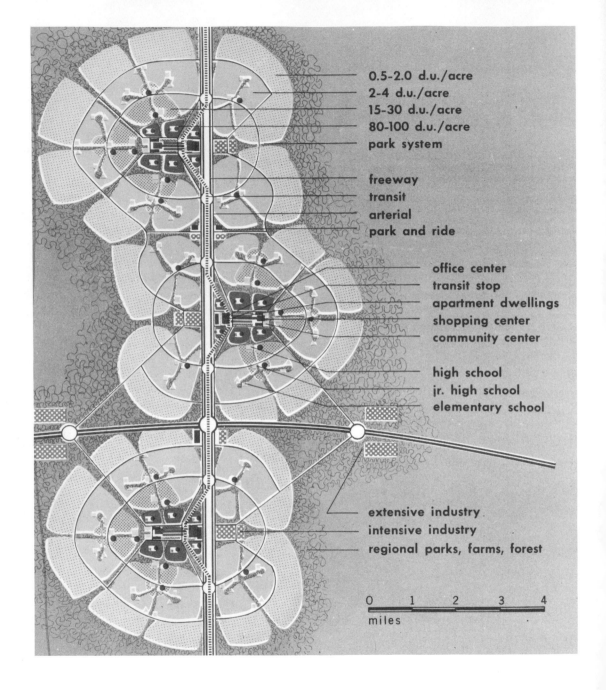

0.5-2.0 d.u./acre
2-4 d.u./acre
15-30 d.u./acre
80-100 d.u./acre
park system

freeway
transit
arterial
park and ride

office center
transit stop
apartment dwellings
shopping center
community center

high school
jr. high school
elementary school

extensive industry
intensive industry
regional parks, farms, forest

0   1   2   3   4
miles

Concept Diagram, Satellite Communities Corridor development would encourage the construction of satellite communities along major transportation lines, both freeways and mass transit. The adjacent diagram is a further refinement of the design structure suggested in the illustration on p.167 just as the Illustration of Reston, Virginia (p.49) is an executed example of the concept diagram. All these drawings and photos illustrate how a strong design idea can be established in general principle, yet allow diversity in the execution of the details. *Reprinted from: The Nation's Capital Plan For The Year 2000, Washington, D.C., 1961*

the urban area and connected to the central city by corridor-like developments. The design is compact and places more people and communities within easy reach of each other than the dispersed-cities or ring-of-cities alternatives. The peripheral plan, however, requires a sizable network of freeways and reduces the amount of countryside immediately available to the central city. The Copenhagen Plan (page 166) solved this latter problem by increasing the densities in the corridor and widening the dimensions of the open space.

The radial corridor plan: The selected design is called the radial corridor plan because it strengthens and extends the existing development trends along known and probable future lines of mass transport and highway corridors from the central city to large communities to be strung together in star-like fashion.

This scheme has many of the advantages of the previous five designs and fewer of the disadvantages. It accommodates considerable regional growth in new communities, offers the necessary variety in service and employment opportunities, maintains the traditional strength and position of the central city, optimizes mass-transit solutions for moving heavy traffic, saves much of the regional countryside, and suggests the possibility of distinctive designs for the communities that make up the regional form.

To create the corridors, there are three immediately realizable steps: location of mass-transit lines and expressways, establishment of employment centers, and preservation of open spaces. Within that framework any number of alternative community designs can be planned, each having components that reflect individual site differences, entrepreneurial skills, and community preferences.

As shown in the model scheme to the right, the high-density employment and residential areas would cluster around the mass-transit stops. Government, retail, and cultural buildings, of low density, would extend away from the center into the community fabric. Residential areas surrounding these facilities would decrease in density towards the edge of the community. Local shops and community facilities would be connected to the center and to each other by parks and greenways. These would also serve as the site for recreation and leisure close to family life. Studies by the National Capital Region Planning Council indicate that space for one million more people is available in new communities along the corridors within the presently committed urban area. Three million additional people could be accommodated in metropolitan communities on land not yet developed, again along the corridors.

The radial corridor concept lends itself to phasing of development. Construction of both the total corridor and the communities along it can be staged over time.

The obstacles to bringing the Washington plan into fruition are not those of substantiating the many advantages of the plan but rather of devising a set of incentives that will encourage people to voluntarily use their private and personal decisions for a common goal. While Federal action will help, the quality of the future in Washington as elsewhere is dependent on forceful local measures; and in Washington the present picture is bright. The central city has been given a new form of government, and privately-sponsored new towns such as Columbia, Maryland, and Reston, Virginia, are on the horizons or in the early stages of development.

The use of historic and new techniques in producing design structure at the state, regional, and metropolitan level is evident in the schemes for New York, Connecticut, Copenhagen, and Washington, D.C. All these examples show the beneficial effects of establishing visual and functional order. All hint that variety and embellishment can follow, once the skeleton of the design is laid down. These former design objectives involve something more than design structure. They can be summed up simply: the creation of a sense of place.

171

A Sense Of Place? *Photo: Louis B. Schlivek, Regional Plan Association*

A SENSE OF PLACE

East Side, Manhattan *Reprinted from: Lower Manhattan Plan, 1966, Whittlesey, Conklin, and Rossant, Architects; McHarg, Wallace, Todd and Associates, Planners*

A SENSE OF PLACE

A sense of place is first the sum of all those environmental characteristics that distinguish a part from the whole. For primitive man the encampment differed from the savannah. In preliterate cities the sacred precinct was unlike the *urbs* around it, just as today downtown differs from the rest of the city, and a city differs from the suburbs. But of more importance than functional differentiation, a sense of place implies physical qualities which allow one to separate the fresh from the stale, the sweet from the sour, protecting, enhancing, cultivating amenity and ambience, stimulating variety, banishing monotony, creating tranquility from raucous insistence.

A sense of place is man-made, for even the contrast between ruralscape and cityscape is not an accident of nature. While a few square miles of primeval forest, swamp, and ledge may exist within the immediate reaches of urban communities, this countryside is largely a contrivance of Man. True, the wilderness, sea, and desert have unique aesthetic interest, but only because Man sees in them the quality of art.

A sense of place often comes about, in Thomas Sharp's words, from "the superimposition of the works of succeeding generations upon earlier achievements and upon the unequal incidence of these accretions on the separate parts." While continuity plays a role in creating a sense of place, as in a New England village, a sense of place can be developed quickly and completely. The town center of Reston, Virginia, page 49, is a good example.

In this section we review the present state of the art of environmental design as it is reflected in the establishment of a sense of place. Techniques which can be used in both small and large-scale designs are described, using case examples of government centers and neighborhood and citywide development plans. The influence of climate and regional architecture is explained as a special factor in environmental design. The utility of historic preservation as a stimulus for design action is reviewed. Finally, the role of a sense of place as a prime value in a world of facts, is discussed with reference to advocacy planning, traditional views of city design, and the implications a sense of place has in a world increasingly subjected to impersonalized planning.

BEYOND ARCHITECTURE

In the *Design of Cities*, Edmund N. Bacon says that "each generation must rework the definitions of the old symbols which it inherits from the generation before; it must reformulate the old concepts in terms of its own age." Bacon defines architecture as "the articulation of space to produce in the participator a definite space experience in relation to previous and anticipated space experiences." The methods of doing so are many. As Bacon so well illustrates in his book, each culture and each age produces a special vision, the total of which is our design heritage.

Environmental design is concerned with the art larger than architecture but contains parallels to the definition of architecture which Bacon reveals in his book. Environmental design is not a set of formulae about how to design the space occupied by man but a set of attitudes about how to arrange the physical elements in the environment to achieve a satisfying and rewarding human habitation. There are many ways of reaching this objective.

Each approach involves apprehending the environment and representing it through tangible designs. The realization of the designs depends on the degree of precision intended, the time span required for executing the design, and the size of the enterprise. Some design approaches are of recent origin, especially those involving the creation of the urban world at a scale greater than that which can be comprehended in a single glance. The reasons these new approaches came into being and the ways they are used are indicative of the evolving art of environmental design and the importance of creating a sense of place.

The 1950's saw fundamental changes in the complexity of design opportunities. The automobile became the dominant means of transportation, and the metropolis "exploded." New highways opened new land for urban development. The central city was subjected to dramatic redevelopment. The environment was viewed not as a collection of parts but as a total fabric susceptible to design at a regional as well as a local scale. Despite unprecedented construction, the visual quality of the environment was not noticeably improving. Out of a critical examination of what post-war Europe was accomplishing—especially the townscape and new towns in England and housing and community development in Scandinavia and Switzerland—professional and public interest in urban aesthetics was revived.

One line of inquiry followed the Beaux Arts notion of applying theories of building design to large urban areas. The vocabulary of scale, balance, proportion, color, harmony, and unity was considered appropriate for urban design. But as larger areas became subject to design and as the dynamic factors of movement and staging were introduced, the static Beaux-Arts approach became noticeably inadequate for analyzing design problems and representing design solutions. Further, the classical approach, suitable for the work of a single designer, seemed out of place in a process that involved many designers and other professionals, many clients, and many forms of design simultaneously interacting on each other. Subsequently, a fresh look was taken at what the other arts—particularly the film—might contribute to a new vocabulary. The sciences and philosophies of perception and behavior were raked over for contributions and insights. Infused with the intuitions of critics like Gyorgy Kepes, Susanne K. Langer, and Hans Blumenfeld, young designers such as Sydney H. Williams, Philip Thiel, and Kevin Lynch embarked on exciting research, reorganized professional and other educational curricula to disseminate the information and techniques discovered, and began to alter for the better old ways of viewing and designing the environment. The relevance of these new modes for the development of a sense of place will seem obvious in view of its function of distinguishing the part from the whole.

Williams was among the first contemporary designers to recognize the basis for a new aesthetic suggesting three specific reasons for improving the visual character of urban areas: first, "to give the city dweller the opportunity to comprehend and orient himself to the city as a part of his daily life"; secondly, to provide "a visual emphasis" for those functions which are socially, culturally, and economically important; and thirdly, to stimulate civic consciousness and pride. Williams outlined a procedure for classifying urban sites on the basis of land forms and the significant man-made features. Drawing on Gordon Cullen's townscape vocabulary, Williams also suggested that the environment could be perceived in several ways: as a panorama, a skyline, a vista, an urban open space, or an experience of design in motion.

In his contributions to environmental

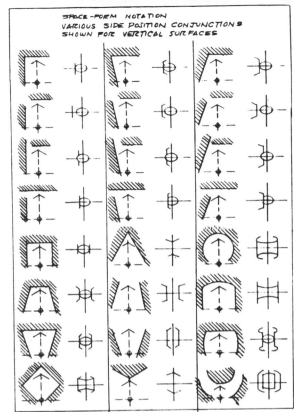

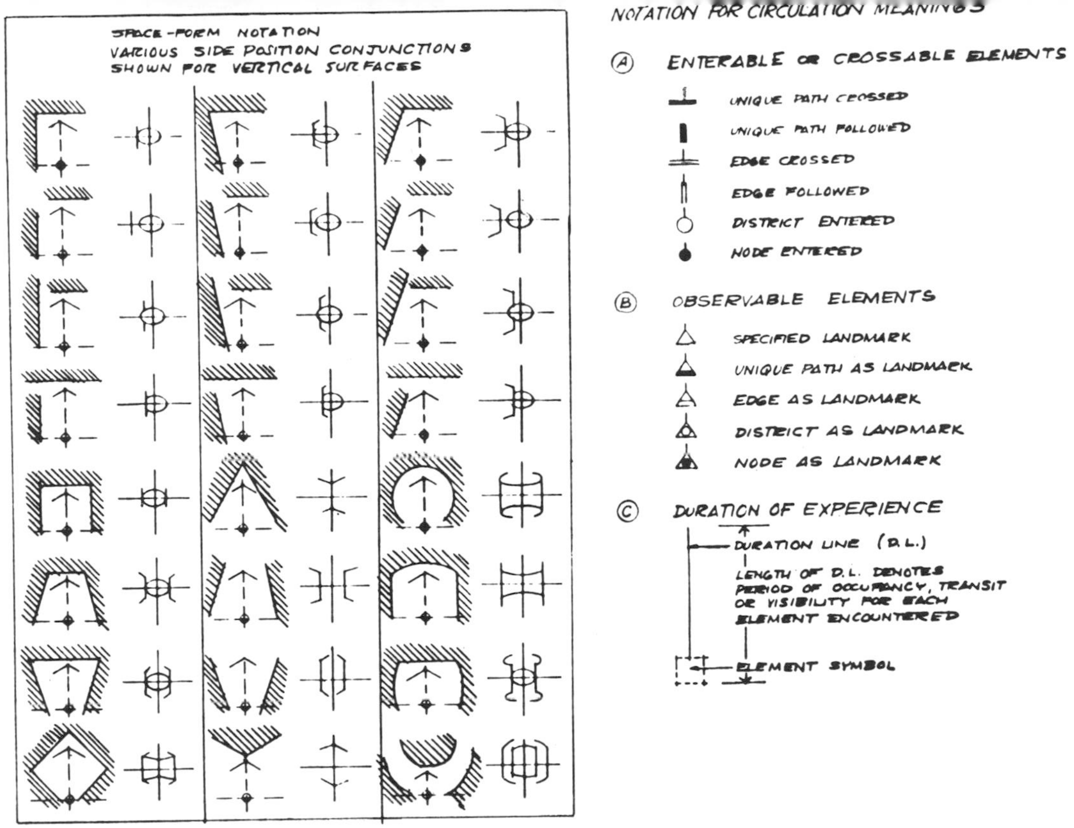

SPACE-FORM NOTATION
VARIOUS SIDE POSITION CONJUNCTIONS
SHOWN FOR VERTICAL SURFACES

Thiel's Notation System *Drawing: Philip Thiel*

NOTATION FOR CIRCULATION MEANINGS

Ⓐ **ENTERABLE OR CROSSABLE ELEMENTS**

- UNIQUE PATH CROSSED
- UNIQUE PATH FOLLOWED
- EDGE CROSSED
- EDGE FOLLOWED
- DISTRICT ENTERED
- NODE ENTERED

Ⓑ **OBSERVABLE ELEMENTS**

- SPECIFIED LANDMARK
- UNIQUE PATH AS LANDMARK
- EDGE AS LANDMARK
- DISTRICT AS LANDMARK
- NODE AS LANDMARK

Ⓒ **DURATION OF EXPERIENCE**

- DURATION LINE (D.L.)

  LENGTH OF D.L. DENOTES
  PERIOD OF OCCUPANCY, TRANSIT
  OR VISIBILITY FOR EACH
  ELEMENT ENCOUNTERED

- ELEMENT SYMBOL

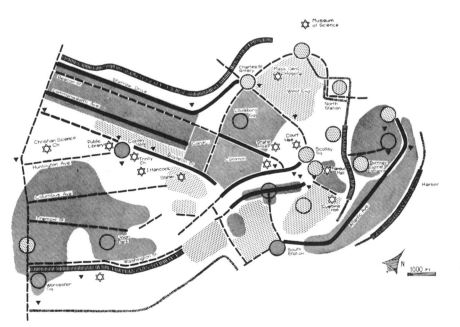

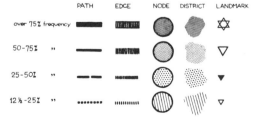

	PATH	EDGE	NODE	DISTRICT	LANDMARK
over 75% frequency	▬▬▬	▥▥▥	◉	◆	✡
50-75% "	▬▬	▨▨	◯	░	▽
25-50% "	▬ ▬	▦▦	◌	⣿	▼
12½-25% "	•••••	⫶⫶⫶	◎	⫽	▽

N 1000 FT

Lynch's Shorthand For Describing The Visual Form Of The City Lynch condensed the important form-giving elements to five features: path, edge, node, district, and landmark. Using these elements, the visual form of Boston (as seen by a trained observer in the field) could be reduced to a simple diagram. *Reproduced from "Image of the City," by Kevin Lynch by permission of the MIT Press, Cambridge, Mass. Copyright*

177

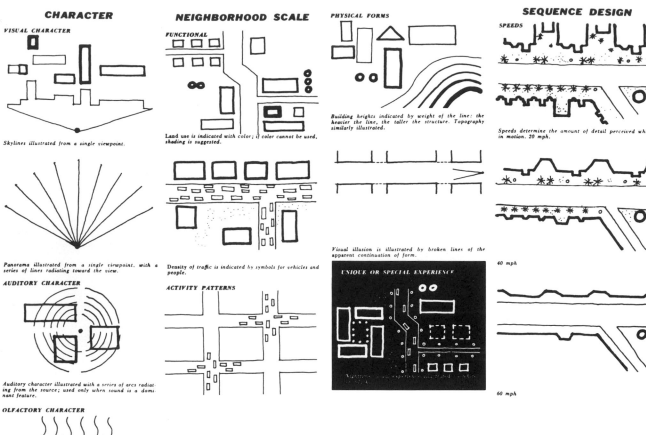

CHARACTER

VISUAL CHARACTER

Skylines illustrated from a single viewpoint.

Panorama illustrated from a single viewpoint, with a series of lines radiating toward the view.

AUDITORY CHARACTER

Auditory character illustrated with a series of arcs radiating from the source; used only when sound is a dominant feature.

OLFACTORY CHARACTER

Olfactory character illustrated by means of undulating rays; pungence is illustrated by degree of wave proliferation; odor by line frequency.

NEIGHBORHOOD SCALE

FUNCTIONAL

Land use is indicated with color; if color cannot be used, shading is suggested.

Density of traffic is indicated by symbols for vehicles and people.

ACTIVITY PATTERNS

PHYSICAL FORMS

Building heights indicated by weight of the line: the heavier the line, the taller the structure. Topography similarly illustrated.

Visual illusion is illustrated by broken lines of the apparent continuation of form.

UNIQUE OR SPECIAL EXPERIENCE

SEQUENCE DESIGN

SPEEDS

Speeds determine the amount of detail perceived while in motion. 20 mph.

40 mph

60 mph

Urbanography Thiel and Lynch's work in notation systems continues to intrigue designers for two reasons: as a way of becoming more sensitive to the subtle aspects of the designed environment and as a method of communicating design proposals without the constraints of specific recommendations. *Source: Dept. of Architecture, University of Cincinnati*

178

design, Thiel stressed the experience of design in motion, believing that the "spaces, surfaces, objects, events, and their meanings, which are associated in such varied combinations to constitute both our natural and man-made landscapes, cannot be seen simultaneously but must be experienced in some temporal sequence." Finding a lack of theory and tools for representing design in motion, Thiel developed a system of graphic notation for the continuous recording of architectural and urban space-sequence experiences. Intended to be used by architect, planner, critic, historian, and student, the technique required no skill in sketching, no mechanical devices, only paper, pencil, and common sense. Thiel's shorthand is similar to choreographic notations. The experience of moving through space is charted by registering the principal features along the route and the spatial dimensions of the territory passed through. Indeed one can draw analogies between the art of the dance and the art of arranging civic space for sensuous pleasure, which is an idea central to Lawrence Halprins' concept of *motations*, a further refinement of Thiel's work.

Kevin Lynch's research has gained widest acceptance as a simple way of recording the design image of the city. After several years of investigation on the form of the city, Lynch reduced the vocabulary of large-scale design to five elements: paths, edges, nodes, districts, and landmarks. The image of Boston is thus made up of distinct districts such as the Back Bay and the North End, edges such as the Charles River, paths like Commonwealth Avenue and the Washington Street Elevated train lines, nodes such as South Station and Louisburg Square, and landmarks—reference points of distinctive imagery—such as the State House and Trinity Church.

By being able to record the existing image, designers can sensitively place new elements either to reinforce the old or to contrast it. Both measures equally create a sense of place. Design-analysis techniques such as Lynch's do not inhibit the design process or the creation of a sense of place but serve them. Lynch and others are not sure anyway that the form of the city or the metropolis can "exhibit some gigantic, stratified order." A master design may be constraining if it does so, but the organization of the image can be accomplished so as to suggest sensuous form "through clarity of structure and vividness of identity." Urbanography, the notation of visual city forms, thus continues to intrigue designers and planners. Theoreticians such as Christopher Alexander are working on computer systems to dissect, measure, program, and project environmental designs in mathematical languages. In research projects at the University of Cincinnati, sensory stimuli other than visual elements are being recorded and the graphic vocabulary of describing a sense of place extended to produce a ready method for interpreting additional dynamic and static features in the environment.

Lynch is to environmental design what Giotto was to painting, sharing the intuition that Bernard Berenson ascribed to the thirteenth-century master of pictorial art: ". . . the given quantity of atmosphere is sure to contain other objects than those the artist wants for his purpose. He is free to leave them out, of course; but so far as he does, so far is he from producing an effect of reality." Lynch explored dimensions of the environmental design that others, using Renaissance explanations of how people view their physical world, left unsettled. With other members of his research team, Lynch raised the issue as to whether or not special groups viewed the environment in special ways, and if they did, what did they share in common and how did they differ in their environmental needs. Although his notation system gave a synoptic rendering of the image of city, true for the largest number of people, by indirection his urbanography contended that differentiation was equally important. Lynch and his research group explored the use of varying methods for presenting the environment, including random and scheduled analysis of the urban scene by trained and untrained observers, the use of the camera for still and motion records, the analysis and categorization of the content of both interviews and discussions of the constituent elements in the environment as presented by untrained eyes, and the investigation of ways of presenting in two and three dimensions, singly and simultaneously, selected aspects of the research. Lynch and his staff of experimenters were discursive, occasionally discordant; and the studies sometimes were discontinuous. But the body of work fairly reflected young designers' disenchantment (circa 1955) with old views of how the environment should be designed and heralded a wave of insight and thought only recently beginning to be appreciated as a new environmental aesthetics.

The underpinnings of this new aesthetics are threefold: program (an elaborate examination and description of goals, objectives, and needs), performance (the description of standards and criteria for design), and perception (the description of perceptual dimensions of the design, largely spatial and visual but not exclusively so). Programmatic approaches to design seem to work best at the micro-scale, where the number of variables to be accounted for are manageable, the time span for executing the design is not great, and the process of community participation and decision-making allows ready contact between the user and the designer. College campuses, model-city neighborhoods, and private development projects are all susceptible to the programming approach.

The theory is that a programmed design is tailor-made—cut to fit. Thus one advantage is that programming allows the introduction of human foibles, traditions, and the contrariness that animate a community. The environment is not imposed by the designer on the basis of archaic notions of scale and perspective but surfaces from a regard for how people live in their environment, affect it, and are conditioned by it. Technological coercion, the possible degradation of human choice, because neither user nor client understands each other nor shares a common language, is eased by programmatic design. A special place is becoming to special people, and it can be strengthened by cross-referencing from function to symbol and image through programming, allowing the combining of behavorial engineering with the variables of art.

On the technical side, programming reduces risk-taking and dependency on a single-minded, preconceived solution and encourages task-force efforts. The general level of intelligence and information available to a group is likely to exceed that of a single client and a single designer. Expert knowledge may be readily introduced in reaction to seemingly imponderable questions, for programming is also a rational method for obtaining, organizing, and applying information. Finally, programmatic techniques encourage applied experimentation: trial and error designs; mock-ups, directed speculation, prototypes tested and refined before execution. Some of these experiments may be carried on at full scale and in three dimensions, or as mathematical simulations and models, or by graphic and verbal communication and review.

Performance criteria constitute the objective reckoning of the quantitive and qual-itative levels the design should reach. Originally, these criteria included lighting, thermal comfort, spatial dimensions, level of pollutants in the air, noise, and other physical characteristics that could be machine-controlled. Then cost factors become recognized as an influence on design choices—original construction costs versus maintenance costs, installation costs versus operations costs. The concept of performance has been expanding through eventful research. Physical anthropologists have enlightened designers as to how products can be best fitted to the human skeleton. Sociologists and psychologists have measured the relationships of environmental situations to ascertain, for instance, what are the best seating arrangements for group conferences, how maximum social communication can be carried on through the siting of buildings and the arrangement of corridors and hallways, what is the impact of movable furniture on learning, how and why colors soothe or abrade the sensorium, and whether or not windowless rooms lead to mental-health problems. These are samples of voluminous research awaiting not answers but application. Currently, systems analysts are also creating performance standards—evaluating the effectiveness of an overall activity by critically examining the component parts with the view of improving key and critical elements so as to raise the level of performance in the entire system. The systems analyst views a neighborhood park, for example, not just as a community open space, filled with equipment, of which the use is directed and maintained by a special branch of local government. Rather, he sees the park as part of a recreation system, formal and informal, with discernible functions that are meant to satisfy different age groups, at different times of the day, in different seasons. Designs must therefore vary in accordance with the num-ber, age, social characteristics, and life styles of the people who may be using the playground. The design is not an abstraction but a set of special conditions and responses.

All such systems approaches today measure and predict action and reaction in terms of past behavior and as such seem primitive in light of recent research. An environmental issue related to performance standards, based on known human behavior, is the alternation of such behavior through psychological conditioning or more drastic measures. Rather than change environmental conditions, some believe it may be easier to change Man's reaction to them.

B. F. Skinner and his followers have adequately demonstrated how the pairing of stimuli with rewards and punishments can modify responses to exceed or fall below achievement standards, regardless of some environmental conditions. The interaction between human behavior and the physical environment is also being recast by new research in the biological sciences, where chemical and neurosurgical modifications and stimulus control are being introduced to intervene in patterns of response previously thought unalterable. This raises the specter of conflict between civil rights and human engineering and the choice of who picks the standards and values against which performance is measured.

The emergence of conditioning through biological, chemical, physiological, and psychological intervention will raise as many political questions and moral issues as the introduction of nuclear fission and fusion two decades ago.

Ultimately, the acceptance of the performance-standard approach in environmental design to its fullest potential is dependent on cultural acceptance. If an innocuous measure such as fluoridation is rejected, where will the community stand on serious environmental interventions? Despite these ethical issues and uncertainties as to application, the idea of a performance standard as a guideline for design is a promising one, because it involves rationality and comprehensiveness and yet allows differentiation.

Though program and performance are the hottest areas of research in environmental design, interest in the perceptual qualities of the environment has not abated. Kenneth H. Craik believes that the continuation of the work of Lynch and others may produce the first legitimate marriage of science (with its objectivity) and social improvement (with its inevitable taking of sides). Craik would enlarge perceptual design investigations to include an understanding of how stereotyped social attitudes affect one's sense of place. He would encourage experimentation in total sensory immersion (sound, color, smell) to gain knowledge of environmental preferences and use such testing devices as adjective checklists, activity and mood checklists, Q-sort descriptions, and adaptations of Thematic Perception Tests to find and give order to the cues and clues of environmental stimulation. With these insights the environment can be made rich, increasingly varied, and comfortable.

The new aesthetics is still in the laboratory, not fully tested; its contributions to date are largely the awareness and sensitivity it provokes among those struggling with environmental issues. But an improved environment is not entirely dependent on techniques that lie on the horizon. A sense of place can be created and animated by carefully positioning key elements in the design structure, by responding to the natural conditions suggested by climate and regional influences, and by preserving and enhancing the best features of the existing environment. These are old tricks of the trade which, brought up to date, continue to work well.

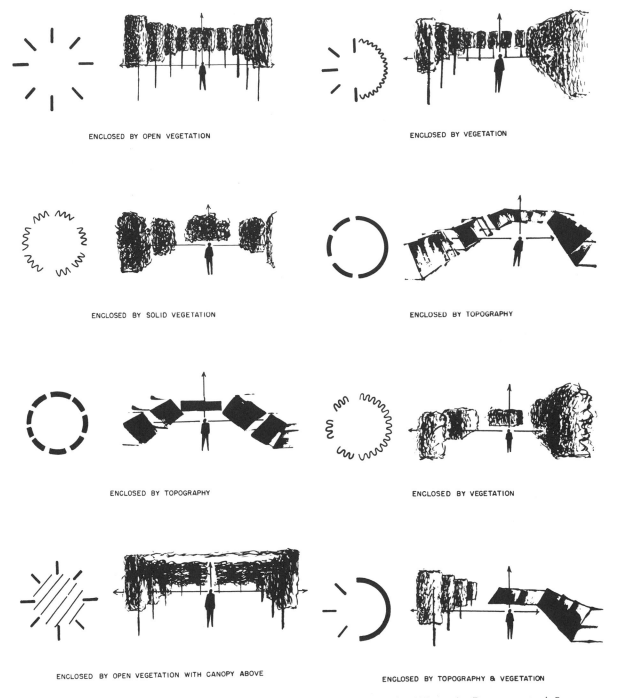

ENCLOSED BY OPEN VEGETATION

ENCLOSED BY VEGETATION

ENCLOSED BY SOLID VEGETATION

ENCLOSED BY TOPOGRAPHY

ENCLOSED BY TOPOGRAPHY

ENCLOSED BY VEGETATION

ENCLOSED BY OPEN VEGETATION WITH CANOPY ABOVE

ENCLOSED BY TOPOGRAPHY & VEGETATION

Recording The Perception Of Open Space *Source: Recreation In Wisconsin, Wisconsin Department of Resource Development, Madison, Wisconsin, 1962*

COMMUNITY SCALE	I	ii	iii	I	II	III	IV	V	VI	VII	VIII	IX	X	XI	XII
EKISTIC UNITS	1	2	3	4	5	6	7	8	9	10	11	12	13	14	15
	MAN	ROOM	DWELLING	DWELLING GROUP	SMALL NEIGHBOURHOOD	NEIGHBOURHOOD	SMALL TOWN	TOWN	LARGE CITY	METROPOLIS	CONURBATION	MEGALOPOLIS	URBAN REGION	URBANIZED CONTINENT	ECUMENOPOLIS
NATURE															
MAN															
SOCIETY															
SHELLS															
NETWORKS															
SYNTHESIS															

POPULATION
T (Thousands)
M (Millions)

1 2 4 40 250 15ᵀ 7ᵀ 50ᵀ 300ᵀ 2ᴹ 14ᴹ 100ᴹ 700ᴹ 5000ᴹ 30.000ᴹ

Ekistic Logarithmic Scale

EKISTIC ELEMENTS

Nature
1. Water resources and pollution
2. Air circulation, temperature and pollution
3. Mineral resources
4. Land forms and landscape
5. Agriculture and fishing
6. Recreation and preservation of nature

Man
1. Biological needs (space, light, temperature, etc.)
2. Sensation and perception (the five senses)
3. Emotional needs (human relations, beauty, etc.)
4. Moral values

Society
1. Population composition
2. Social relations
3. Cultural patterns
4. Economic development policies
5. Industrial development
6. Education, health and welfare
7. Law and administration

Shells
1. Housing
2. Community services (schools, hospitals, etc.)
3. Commerce and recreation
4. Industrial equipment
5. Transportation facilities

Networks
1. Public utility systems (water, power, sewerage, etc.)
2. Transportation systems (water, road, rail, air)
3. Communication systems (telephone, radio, television, etc.)
4. Physical layout (land use system)

Ekistics Grid Ekistics is a systems approach to the science of human settlement. The grid is a useful device for organizing information about the elements that comprise all scales of human habitation. *Reprinted from: Ekistics, May 1968*

Design Research The computer now makes it possible to deal with large amounts of information in design research so as to set performance standards for environmental designs. *Drawing: Courtesy of Caudill, Rowlett, Scott, Architects*

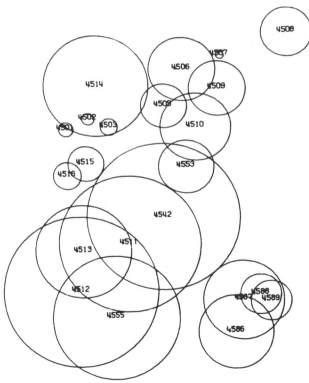

Computer drawings facilitate the study of relationships on campus. Drawing above shows how intensively particular buildings are used. It describes the person-hours spent in each building during 7 consecutive days, 24 hours per day, as recorded in the diaries of 100 students. Number in the center of the circle identifies a building; the diameter of the circle is proportional to the amount of time spent in the building.

The computer could be ordered to refine this information for particular student groupings, i.e., to show how intensively buildings are used by "female liberal arts majors living off campus," or by "male sociology majors in residence," and the like.

6.30 p.m. People do not always want to watch T.V. when it is on, and need a place to sit away from it. The children need quiet when they are being settled down to sleep.

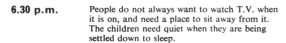 

7.00 p.m. When Father makes or repairs something, he needs to be out of Mother's way in the kitchen and where he will not disturb sleeping children.

8.00 p.m. Sometimes visitors are being entertained while a child is watching his favourite T.V. programme.

Performance Criteria A simple example of how performance criteria can be organized for the designer's use. *Source: Ministry of Housing, Great Britain, 1966*

The site

The pond in the Boston Public Garden.

Key:
o   Underwater lights
●   Speakers

Specifications

Fifty-five xenon strobe lights, 50 watt/seconds nominally, Sylvania Π4310 flash tube.

Sixty 20 watt Amperex audio amplifiers PCA-8-36.

Fifty-two poly-planar ERA P-20 speakers.

One Electro-Voice 643 Super-Directional Microphone.

Punch paper tape reader, one inch asynchronous, 100 characters per second maximum read rate, Photologic Model Number 100, loaned by Friden.

Hybrid-digital-analog-signal synthesizer and sequence generator, developed by Pulsa.

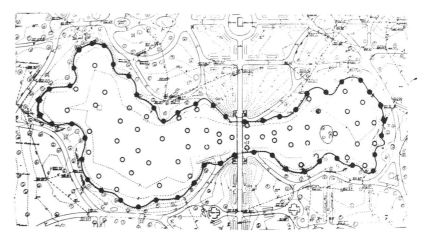

Environmental Research Drawing showing location of underwater lights and speakers for an experiment in recording and transmitting abstract patterns of light and sound in an urban environment. Public reaction during the demonstration would be observed to help determine environmental standards for the design of signs and lights in the city. *Drawing: Pulsa, 1968*

The Undesigned Environment *Reprinted from: Long-Range Plan, 1966. Capitol City Planning Commission, Dober, Walquist and Harris, Inc.*

STRATEGIC DESIGN NETWORK

The environment is a demeaning experience when the small-scale elements that comprise the urban scene are neglected, and the synergism of that neglect becomes a shabby, uninspired, unattractive urbanscape. The State Capitol Plan (Springfield, Illinois) is an example of how a long-range view, a strategic design network, can guide public and private investments to overcome urban blight and revive a sense of place.

Early in the study for this Plan, it was evident that there could be no grand plan for the expansion of State Government facilities unless there was a conscious effort to improve their general environment. No buildings, however impressive, could overcome poorly designed and badly maintained urban spaces around them. The Capitol area had many buildings which at first glance seemed inappropriate to the place. Service stations, hamburger stands, and commercial buildings with hanging signs or gross advertisements painted on their sides surrounded the State Capitol. In nearly all instances the structures were not objectionable in their function, as in the case of the commercial buildings which were convenient to State employees and others who worked nearby. It was their appearance that hurt. The display of garish signs, banners, and bright lighting robbed the Capitol of its dignity. The planners noted that:

A gas station need not dominate everything in view, and a shop need not use the entire side of a building to announce its presence. General restraint in outdoor advertising would result in an upgrading of the quality of the area. Views to major landmarks such as the State House could be enhanced and the Capitol Buildings once more made the important visual elements.

Public agencies were at fault as much as private groups. Directional and information signs lacked distinction or character and added to the visual clutter. Street-furniture elements were uncoordinated, and there was little pleasing about their appearance or intelligent about their location. The Springfield streets cried for better maintenance. Drains needed repair. Grass was growing through the sidewalk. There were cracked and crumbling curbs, bricks missing, and paving surfaces patched with a contrasting material—all of this giving the streetscape the look of endemic civic scrofula. Equally displeasing were the alleyways throughout the Capitol area and adjacent downtown—filled with overhead wires and transformers, drab with the unpainted backs of buildings, trashy, unkempt, suggesting complete decline of the central city.

The design erosion evident in Springfield came about because the city's downtown was in transition from one dominant economic role (retailing) to another (commerce and professional services). Most of the recent private construction was limited to modest renovations and rehabilitation in the process of changing retail establishments to office and service buildings. Population growth in the city was modest. Any significant additions depended on the enlargement of State Goverment, the city's major employer. But each expansion move the State made congested traffic, compounded the parking problem, hastened the move of marginal retail stores to other parts of the city, and accelerated blight.

As the cycle of decline continued, underused buildings were demolished and turned into surface parking, giving the center of Springfield a "bombed-out" appearance. By 1966 parked cars occupied the major portion of the land surface in downtown. Automobiles were jammed in every open space. In the Capitol area some homeowners turned their lawns into parking lots, while surface lots set up by the State showed indifference to appearance. In summer the large expanses of cars and asphalt with few trees to give shade or contrast made the streets and sidewalks between the Capitol and downtown inordinately hot, empty of people, and desolate. What had once been an attractive town was recognizably a slovenly city.

In reaction to these events, the Capitol City Planning Commission was formed by joint State and City legislation to guide the inevitable expansion of State facilities and simultaneously improve the Capitol area and adjacent districts. But good intentions were not of themselves sufficient. The development of a sense of place had to be tied into the economic and political realities of Springfield.

The Commission's studies indicated that the centralization of most of the State's expansion in Springfield in a Capitol Complex was desirable for efficient government. In addition a high-density complex on the edge of downtown would create opportunities for self-renewal in the core of the city. Plans under way for the central business district could be coordinated with the State's plan.

The Commission noted that the downtown probably would not grow in acreage. Existing low-density, obsolete blocks were likely to remain for years ahead. To give incentive for private investment, it was decided that the Capitol Complex would grow in checkerboard fashion towards downtown. New State buildings would be located so that their appearance, density of population, or function would encourage private entrepreneurs to refurbish the nearby urban scene. Because it was uncertain how or when the obsolete blocks between the Capitol Complex and the center of downtown would be redeveloped, several of the worst connecting streets were selected for special planting, paving, and lighting. By refurbishing these streets, pleasant and easy pedestrian connections could be established between the Capitol Complex and shops and offices downtown.

The plan called for the center of downtown to be resuscitated by restoration of the Old State Capitol Building and the park around—both of which had been scenes of Abraham Lincoln's life in Springfield. Because of Springfield's association with Lincoln, several City and State buildings are tourist attractions. In the plan a special street, Lincoln Way, was designed to connect these landmarks. The Old State Capitol was taken apart stone by stone and replaced and landscaped in the fashion of Lincoln's time but with an underground garage and archives below. Two adjacent streets were closed to form a pedestrian mall tied into other street improvements. The Capitol Complex and downtown were further strengthened by a proposed convention and hotel center to the north, immediately adjacent to a new major thoroughfare that replaced an underused railroad right-of-way.

Despite the low-grade design quality of the center of Springfield, the Commission appreciated that the overall design of the city is quite distinctive. The urban area is surrounded by countryside and not extended in wasteful sprawl along the major highways. Not only is there a boundary between city and country, but the skyline of the central district delineates the symbolic center, with the dome of the present Capitol as its chief feature, visible thirty miles away. To preserve this quality, the Commission passed a zoning ordinance so that new construction would not be any higher than the eaves line of the Capitol, thus holding on to an unusual visual feature and maintaining a historic sense of place.

The key to Springfield's revitalization

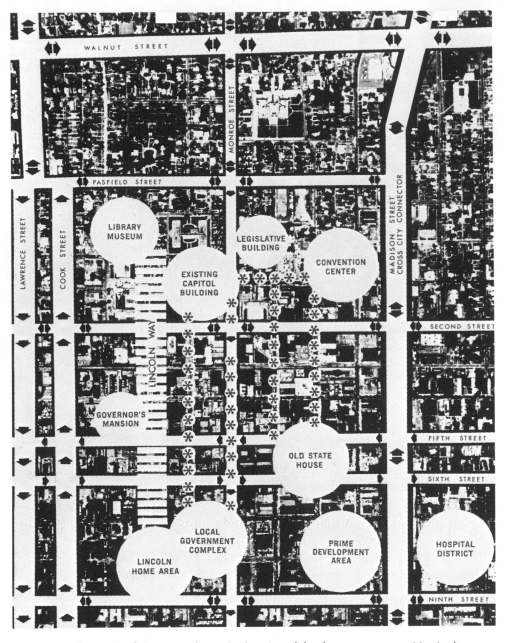

Strategic Design Network Diagram shows the location of development opportunities in downtown Springfield. The organization of these opportunities into a strategic design network is meant to establish a sense of place and to correct environmental deficiencies as quickly as possible. *Reprinted from: Long-Range Plan, 1966. Capitol City Planning Commission, Dober, Walquist and Harris, Inc.*

was the strategic design network: New development was concentrated in selected locations appropriate to functional requirements but also fitted into the form of the city in such a way that the smaller elements and designs that give a sense of place could be refurbished and augmented with new construction. Design structure and design content reinforced one another. This set the stage for private and public groups to carry out individual measures to improve the environment, including tree-planting along the downtown streets, aesthetic controls on new development in the Capitol Complex, and a vigorous effort on the part of the downtown business groups to make the center of the city an attractive place.

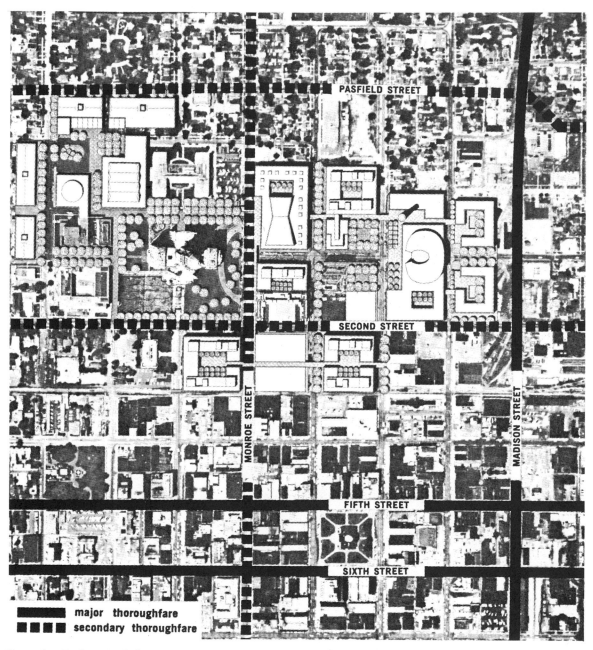

Illustrative Design, Capitol Area A further refinement of the strategic design network in the State Capitol area, showing the scale character and design interrelationships of the various proposals. *Reprinted from: Long-Range Plan, 1966. Capitol City Planning Commission, Dober, Walquist and Harris, Inc.*

Capital Web Birdseye view of the major proposals that constitute the capital web in the 1975 plan for Boston. *Source: Boston Redevelopment Authority*

CAPITAL DESIGN/CAPITAL WEB

Capital Design/Capital Web is a short-hand term to convey a design process. From the outset of the overall design concept through the various stages of realization in detail, large expenditures for major public construction establish and build an overall design structure. This general design focus can encourage economic and social progress. Within it, Capital Web refers specifically to the inventive placement of municipal facilities, related open space, and important streets in a continuous physical form that links significant community landmarks and centers of activity. This form acts as a unifying "seam of services" for the common use of adjacent neighborhoods and districts.

The application of the concept will depend on local conditions; but the general idea is to realize municipal economies, social vitality, and environmental impact on adjacent private areas. Economy can occur through sharing of sites and buildings by municipal and other public agencies, efficiency in maintenance and policing, and the built-in advantages of flexibility in reassignment of building interiors and outdoor spaces to other uses if population needs and service boundaries in the neighborhoods change. Social vitality comes about through a broad range of public facilities available on the site for many social and age groups and for all hours and seasons. This concentration of public facilities may have the additional advantages of reducing nuisances on private property, such as noise and traffic generated by municipal facilities, and bringing about increase and some stabilization in residential real-estate values. David A. Crane observes, "Public agencies can prime the pump of private investment by making use of public development and design tools, such as the design of public buildings, streets,

and open spaces, and the harnessing of the natural environmental assets, all of which have considerable influence on the quality and amount of private construction." *The 1965/75 General Plan for the City of Boston* illustrates both the concept and its design vitality.

Boston prides itself on having one of the largest and most effectively organized Federal urban-renewal programs in the country. This ability to renew and improve the urban fabric through city enterprise is not new to this city. Boston's environmental achievements in the past include land-filling projects which produced a third of the city land, the country's first metropolitan-wide system of open spaces, water and sewage disposal, distinguished civic and institutional architecture and parks, pioneering air-rights development, and a river basin impounded from a salt marsh—a basin whose design continues to serve as a model for similar undertakings all over the world. The 1965 Plan, however, adds measurably to this custom of civic vision and bold design. It sets the guidelines within which an estimated $455 million of Federal funds (for renewal grants, highway, and Federal facility construction) augmented by $416 million in City and State expenditures will be used to develop a physical environment that can encourage $2.4 billion of private investment in new construction and rehabilitation.

The price the individual must pay for the improved environment is calculable—divide the total estimated expenditures ($3.6 billion) by the projected 1975 population (680,000 people), and the answer is $10 a week per person for ten years. The value is immeasurable, for in the decade preceding the 1965 Plan, the city was losing ten thousand people a year in resident population. Despite excellent historic examples of designed environments, blight and decay

were everywhere and expanding.

As might be expected in a city shaped by past events, the blight itself could be traced to historic occurrences. Boston began as a series of individual, dispersed communities situated along valleys, streams, and shorelines where the natural channels could be used for taking people and goods back and forth from the major seaport. With industrialization and urbanization, these channels became filled with rails and roads as well as housing and places of work. This mixture of incompatible land uses, poorly related to the traffic systems that spilled through them, and such local phenomena as changing tastes in housing that left good neighborhoods prematurely abandoned, all combined into environmental deficiencies more widespread than those seen in many other larger and more densely developed American cities.

The situation was difficult to reverse. Dramatic population losses heightened the feeling of lack of confidence in Boston's future. The local tax base was in slow transition from manufacturing to commerce and the professions. Public services declined—magnificent public parks became public eyesores.

Fortunately, the adverse physical conditions were concentrated in pockets of blight. The topographic constraints that shaped the decentralized city pattern also preserved many sound neighborhoods and districts. Strong political leadership from a new mayor, a renewal administrator skilled at gaining Federal funds, a business community that could not afford further decay, especially downtown, the historic physical assets that remained whole and sound, and the blight that qualified Boston for Federal aid were all points of departure for the 1965 Plan.

All these happenstances were brought

into a singular design expression by rigorous design ideas. The environmental-design structure in the 1965 Plan is composed of strongly defined centers of activity connected by highways, streets, rail transportation, and intensively developed public and private buildings and open space all in hierarchical relationship to each other. A three-dimensional image of this structure, and the sense of place that follows, can be seen in the perspective opposite.

Presenting a comprehensive view of design opportunity, the 1965 Plan sets out major large-area design concepts which will enable small-scale design actions, both short and long term. With the large picture in view, designers can proceed with the details of capital construction to create, through the control of geographic distribution, site selection, architecture, and landscaping of municipal facilities, a web of design elements fostering social and economic progress and civic and cultural identity in a sense of place. This process constitutes the long-range and day-to-day scheduling of public improvements summed up as Capital Design/Capital Web.

Where it is environmentally desirable to do so, the Boston planners would like to draw private development and quasi-public uses into the Capital Web: shops, churches, historic places, special housing (for the elderly and multi-family apartments), local off-street parking, and other landscape and urban-design features favorably related to large, public facilities.

In meeting the design objectives in the Boston Plan, careful consideration is given to the environmental qualities of urban renewal projects. Design objectives are determined from the beginning of the renewal planning, not introduced later as a secondary factor. The public agency uses three-dimensional drawings and models for analyzing and presenting possible renewal solutions to neighborhood and district citizen groups. In this way the communities at large instill a local flavor into the schemes, and new construction is intelligently fitted into the design fabric of the city. But the Boston Redevelopment Authority also assures improved design through formal design controls on other new development as well as on designs for large-scale rehabilitation and area renewal.

The execution of design is not left to chance. In theory, Boston has made design a municipal function as important as licensing, firefighting, and waste disposal. In practice, this function is actually carried on by encouraging developers to select talented architects, establishing design criteria for development areas, assisting the developer and his architect in preparing proposals and securing necessary official approval, advising on the design quality of specific schemes, and surveillance to assure that designs agreed upon are carried out in actual construction.

The official agency's administrative staff concerned with design is backed up with independent and objective advice from the Boston Redevelopment Authority Design Advisory Committee. This Committee is composed of distinguished local architects serving on a voluntary basis. Occasionally, experts from outside the region are brought in to give additional design advice.

Because allocation of rehabilitation funds in renewal areas is the immediate responsibility of the Authority, it can support teams of qualified architects and engineers directly in the field as well as carry out prototype rehabilitations so as to stimulate home-owner interest and participation in residential renewal.

Finally, as a public body, the Authority can take responsibility for design standards used by other City agencies concerned with low-cost housing, schools, recreation, open spaces, and streets. Using Authority staff and outside consultants, prototype designs are created. These show how good designs can be practical and economical. Further, these designs challenge private and other public agencies to improve them or use them.

Good design in Boston thus combines bold and grand designs and the small-scale, everyday measures which together form the sense of place which is Boston. The New Boston is not a rigid, formal, committed architectural statement but rather an adaptable, comprehensive system (Capital Web/Capital Design) of using public and private actions to preserve and enhance the best of what exists.

The system is not without conflicts and strains, however. The City's expenditure of political and administrative energy and innovation to encourage private construction for a stronger tax base has not yet found equivalent success in improving job opportunities for low-income groups, better housing, a school system as good as that in the suburbs, and a park and recreation system anywhere near the levels of accomplishment in the historic past. Even public design measures are a mixed lot, despite a strong public commitment to design. The New City Hall is one of the outstanding pieces of American architecture in this century, but the new civic auditorium is an aesthetic and environmental disgrace. Despite the strongest urgings of the official spokesman of the design professions in Boston, the skyline of Back Bay has been twice injured by competition between commercial interests seeking to gain symbolic dominance over one another by constructing the highest and most visible landmark.

Capital Design Elements Examples of capital design elements whose careful placement within a larger design structure creates a distinctive sense of place. *Source: Boston Redevelopment Authority*

191

A District With a Sense Of Place Perspective view of community-scale design elements which were organized to create a sense of place in the West Philadelphia District Plan. *Reprinted from: The West Philadelphia District Plan. Philadelphia City Planning Commission, 1964*

The Process A diagram of the steps taken to organize and define the significant features and proposals to be included in the District Plan and the translation of these items into a design concept and a program for carrying out the plan. *Reprinted from: The West Philadelphia District Plan. Philadelphia City Planning Commission, 1964*

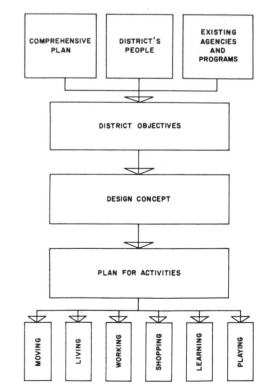

DISTRICT PLANNING PROCESS

COMPREHENSIVE PLAN

DISTRICT'S PEOPLE

EXISTING AGENCIES AND PROGRAMS

DISTRICT OBJECTIVES

DESIGN CONCEPT

PLAN FOR ACTIVITIES

MOVING

LIVING

WORKING

SHOPPING

LEARNING

PLAYING

PUTTING IT TOGETHER

A building with courtyard, a handsome street, an urban park, all these may be designed to create a memorable scene and a sense of place. A more difficult design task is to give a neighborhood, district, or entire city a similar aesthetic order and visual delight. The problem is complicated, because the design form makes sense only if it serves social purpose and responds to utilitarian demands and functions. Health, safety, and economy are prime concerns. But within these constraints there are ways to achieve a communal expression—a civic design. The plan for West Philadelphia shows how the physical elements that comprise the image of a whole district can be put together to create a sense of place. Philadelphia holds an important place in environmental design, not only as one of the historic planned cities in North America but also as a contemporary example of effective community planning. The Comprehensive Plan for Philadelphia, adopted in 1960, established the broad framework for city development. The West Philadelphia District Plan serves as a vehicle for fine-grain planning and design.

The Philadelphia planning process is charted to the right. Using the general scheme of the Comprehensive Plan, the attitudes of the people in the district towards it, and existing commitments of public agencies, the planners draft a statement of public policy concerning the area. This statement is then translated into design concepts and these in turn are described in terms of land uses, transportation, and other key elements necessary in effectuating the Plan.

West Philadelphia by itself is as large as many American cities and illustrates the impact and challenge of urbanization. Up to 1900 West Philadelphia was a series of villages and small communities, many well planned and noted for their environmental design. The district attracted several educational institutions, including the University of Pennsylvania, and was the site of one of the largest and most famous urban-park systems in America, Fairmount Park. But in the next thirty years the population tripled, largely due to annexation and the opening of new streets and mass-transportation lines; and West Philadelphia's spacious qualities were lost. Today the district is still predominantly residential, but in many parts obsolescence, age, and the effects of an automobile-oriented way of life superimposed on a pedestrian-scale district are evident. Many of the old blocks of housing have been penetrated by industrial and commercial enterprises which change the scale of the area as well as cause environmental conflicts. Parts of the district are thus susceptible to blight and urban decay. The worst housing now "shelters" the lowest-income groups, largely Negro migrants from the South, who have little choice in deciding where they live.

The general goals of the District Plan are to reverse these trends and ameliorate these conditions, and to develop an environment of quality, safe and secure, with freedom of location and mobility, so that the physical form of the community enhances opportunities for education, employment, variety, and individuality in the lives of the people living there.

The elements of the physical form of the district—open space, major streets, and clusters of specialized buildings—are shown on page 195, some of which uses Kevin Lynch's design vocabulary. Thus the triangular symbols represent major generators of human activity—the larger the symbol, the greater the importance of the activity. As can be seen, the principal generators lie on the eastern side, close to the center of the city. These include such region-wide facilities as the Central Post Office, the major city newspaper and railroad terminal, and several large manufacturing firms. Westward, other generators include the institutions of higher education, hospitals, and shopping centers.

Landmarks—star-shaped symbols—include visual focal points such as the sky, architecture so large or unique in shape that it stands out from the general background, and small parks and landscaped areas that contrast with the dense urban development surrounding them.

Connectors are the major rights-of-way, along which people and vehicles move from one part of the district to another. The width of the line expresses the volume of traffic. The frequency with which the line is broken indicates speed: Solid lines suggest fast and continuous motion.

Barriers and edges are places where district design is broken off by topographic features, major highways and railroads, or abutting and different land uses. The most significant visually are the Schuykill River to the east and Fairmount Park to the north.

"Greenways"—landscaped pedestrian walks—will tie together the various parts of the design. They will connect living areas to key places such as schools, shops, playgrounds. Many parts of the system are already in place, such as the University precinct and selected renewal areas.

It becomes apparent through the design analysis that the West Philadelphia District Plan established a strong district design, but planning for the area did not stop there. The overall planning aimed not only at visually relating the people to their district but also at visually relating the district to the city and region as a whole. An opportunity lay in the design of Philadelphia itself.

In the original plan William Penn gave Philadelphia (1683), Broad and Market Streets were placed at right angles to each other so as to form the axis of a gridiron street system. The intersection today is crowned by the City Hall, with its memorable statue of the City's founder. The length and design scale of the streets approaching this center are such that the heart of the city is readily recognizable—the City Hall and adjacent buildings are a dramatic and familiar reference point.

Market Street connects Philadelphia's center with West Philadelphia. Accordingly, this important street will be reconstructed as a boulevard and its elevated transit line will be replaced by a subway. As a green parkway the street will serve as a design spine to which many of the communities in the District Plan can be related functionally and aesthetically.

Points of interchange between "district-level" and "community-level" transportation systems along Market Street will be used as locations for high-density housing and commercial services. These activity points become more closely spaced and increase in size as one approaches the central city, giving order and rhythm to the street design. What otherwise could be an undifferentiated string of mixed uses, visually unappealing, becomes a strong design, orchestrated for those in motion as well as those at rest.

The District Plan exploits yet another chance for relating the part to the whole. The Schuykill River is an impressive natural boundary to the West Philadelphia District, and along its edges lie good opportunities for creating identifiable gateways—those clusters of intensive-use buildings or areas, special architecture, and landscape that mark the point of arrival at a district. The major threshold to West Philadelphia will be on Market Street.

The sense of place in West Philadelphia will thus consist of a strongly defined street pattern, the strategic location of major buildings and intensive-use areas as gateways, the development of a set of greenways that internally connect the communities of the district, and the redesign of Market Street as a symbolic and functional link to the city center.

GENERATORS
BARRIERS AND EDGES
LANDMARKS
CONNECTORS
PARKS

DISTRICT STRUCTURE

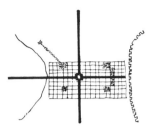

CENTER CITY—FOCAL POINT OF THE REGION AT
INTERSECTION OF MARKET AND BROAD STREET

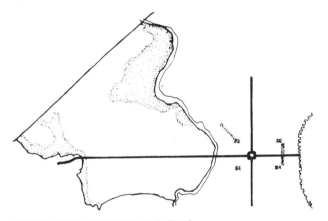

MARKET STREET DOMINANT AXIS TYING
WEST PHILADELPHIA TO CENTER CITY

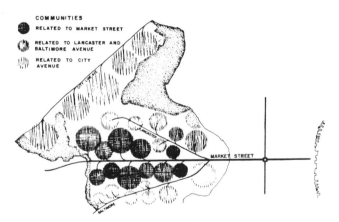

COMMUNITIES
RELATED TO MARKET STREET
RELATED TO LANCASTER AND
BALTIMORE AVENUE
RELATED TO CITY
AVENUE

MARKET STREET THE UNIFYING ELEMENT
FOR WEST PHILADELPHIA COMMUNITIES

Structure And Content An example of how structural elements and design content are combined to establish a sense of place.
Reprinted from: The West Philadelphia District Plan. Philadelphia City Planning Commission, 1964

Fisherman's Wharf Location and activity translated into a regional design. *Source: Embarcadero Development Plan, 1961. The San Francisco Port Authority, Boles and Born, Architects*

TAYLOR WHARF AND CRAB FLEET BASIN. *As piling deteriorates in the older part of the restaurant group at the Wharf, a system of structures with restaurants on upper floors is recommended. A café or two might be located at ground level, along with entrances and service areas for restaurants, but circulation at ground level would be maintained on all sides to permit direct access to wharves and views of the mooring lagoon.*

Regional Expression Consistency in the use of building materials and a respect for the influences of climate create memorable design effects. (The Village of Crail, Fifeshire, Scotland.) *Photo: Copyrighted by the British Travel Association*

Public and private architectonic elements constructed over four centuries in this Swiss village south of Basel again demonstrate how a sense of place can be established by regional design expression. *Photo: Richard P. Dober*

CLIMATE AND REGIONAL DESIGN

Climate strongly influences the creation of a sense of place. Economic and political decisions and technological development may create, preserve, and enlarge the designed environment but only to the extent that the natural climate permits. The state of agriculture, public health, the ranges of human comfort, all are set by the degree of heat, cold, and humidity present through the year. Thus poverty and prosperity arise in part from climatic conditions, as do historic and contemporary migrations involving the search for better climate and the establishment, disposition, and growth of urban places. The first American moving away from glacial intrusions thirty thousand years ago and the population growth in the southeastern and southwestern United States today bear witness that Man feels neither efficient nor safe in areas of extreme climate.

Furthermore, climatic conditions—unlike topography, soil, natural resources, and the use of the land—are not readily susceptible today to man-made changes of long duration. Despite continuing progress on short-term weather modification and a willingness to invest large sums to overcome the effects of climate, Man's patterns of habitation remain bounded by climate. Fortunately, human ecology is not as strongly tied to climate as animal and plant life. Energy produced by coal, gas, and now atomic power can cheaply heat and cool interior environments. Transportation provides mobility in almost all kinds of weather; durable building materials to some extent have emancipated Man from climatic impediments.

Though year-long freedom from the extremes of climate, such as in the Arctic and low-lying equatorial zones, seem beyond immediate practicality, the beginnings of climate-based architecture for environmental extremes are visible. In Artesia, New Mexico, an underground elementary school has successfully overcome the limitations of hot, arid climate. The subsurface building can more economically maintain a stimulated environment than a similar structure above the ground.

Buckminster Fuller, grandfather of all the megastructures, has demonstrated how simple engineering principles can be used to span large spaces and close out weather. In theory, Fuller's tetrahedral construction can even be put together to form satellite earths, launched into space for colonization. In practice his designs have been used for cheap and portable buildings, but in the American exhibition at the Montreal World's Fair (Expo '67), he and the Cambridge Seven created a memorable structure. Another example of design for climate is the well-publicized Astrodome (Houston, Texas, 1965). It provided an all-weather indoor arena for games traditionally played out of doors. The dome has altered the conduct of professional athletics and aroused considerable journalistic conjecture on the state of spectator sports across the country affected by the uncertainties and hazards of inclement weather. Others viewed the Astrodome as a mechanical feat long overdue and are more upset about the intrusion of synthetic grass than the introduction of a synthetic sky. Both the Astrodome and the Expo '67 structures in any event showed the advantages of climate control, allowing other projects to gain public and private attention. In his designs for the Hyatt-Regency Hotel (Atlanta, Georgia), John Portman created the most exciting interior environment since Grand Central Terminal—enclosing street, lobby, and twenty-two stories of balconies with an all-weather cover. The artifacts of pleasure and play yielded support for experiments long called for by visionary architects and planners, ideas earlier labeled "impractical."

The University of Minnesota launched an Experimental City, a dome-enclosed environment for 250,000 people (1968). An extremely cold winter site was selected to give incentive for technological innovations and to insure sufficient removal from conventional urban areas so as to afford true experimental conditions. The use of the dome will mean a city free from pollutants and noise. Industry, commerce, residence, and culture can then exist side by side, thus reducing extravagant commuting and air-burning transportation. Athelstan Spilhaus sees the Experimental City as "being designed backwards, starting with innovations in the newest engineering systems conceived for a certain number of people and no more." Urban space and technical systems will be leased by a private corporation to the city-dweller, not owned outright. Architectural designs will be free from the rigid geometry of plat lines. Diversity of forms will follow. The environment will not be entirely uniform, as enclosed portions will be subject to variations in temperature, humidity, fumes, and light. The differences will be used to stimulate both mental and physical well-being. In a very different climate for a very different location, yet another climate-based architecture is being born. Oceanographic engineers in Southern California are preparing plans for underwater communities: sea farms, sea mines, and resorts, all climate-proof. Many of the engineering solutions proposed in the Minnesota and California experiments are by-products of spacecraft construction and operations, leading some observers to believe that the next frontier is not the space above the globe but the land not yet inhabitable by man because of poor climate and the territory just beyond the water's edge.

197

By modulating the face of the continent, climate determines the major landscapes and opens up many possibilities for climate-based architecture in response to regional variations. Climate dictates the length of the growing season, the types of vegetation, their expanse and composition. True, without oceans and mountains, bland bands of similar vegetation would cross North America, for the only ecological energies available would be solar heat and light. The interaction of ocean winds and mountains gives rise to a mixture of tundra, evergreens and deciduous forests, grassland, desert, and tropical vegetation. Local soil conditions further influence plant ecology, as do birds, insects, and animals. But climate spins the web of life.

Climate differences resulted in the Ice Ages, which sculpted the upper United States and Canada. Glaciers pockmarked the earth, eventually allowing small ponds and great lakes to hold rainwater and establish a distinctive and picturesque regional setting. Glaciers ground away the old mountains, carved out valleys, and carried rock and soil southward to build up the landforms of New England, Long Island, and the northern tier of the midwestern states. Glaciers rearranged river beds. When the ice melted the seas rose, defining along new lines the edge of land and ocean—an edge which still subtly changes from day to day. Less precipitous has been the gradual erosion of the land by wind and water, all climate induced, producing the profile of the Rockies, the broad expanse of the Mississippi and its tributaries, the Grand Canyon, the Badlands of North Dakota, external and internal carving away of the surface and subsurface in Arizona, New Mexico, Virginia, and Kentucky. The result is a diverse national landscape, of remarkable variety and beauty. Each region instills a unique sense of place to be built upon or simply enjoyed in its natural state.

Geographers such as Vidal de la Blache have even classified civilizations by example of regional architecture, suggesting that in non-technical cultures Man built largely in response to natural conditions. In the view of de la Blache, successful anonymous architecture heralded the advance of kin and cult, thus serving as a visible prelude to a literate society. In the more recent past, in any case, climate has been a strong influence on indigenous regional architecture in Europe and the United States, especially prior to the widespread use of standardized building materials cheaply transported thousands of miles and uniform construction methods maintained by labor practices and building codes. Examples of regional architecture survive in the United States, but the widest range of building types influenced by climate can be seen abroad.

Traditional forms of housing in Europe show a clear relationship between terrain, the use of local building materials, and landscape. Rural architecture particularly illustrates how climate and geography have produced an art in which fashion and style had no important part. Economy and availability of materials were an overriding consideration and thereafter the trial-and-error approach to handling isolation, prevailing winds, rain, snowfall, and temperature ranges were the dominant design factors. Both house and settlement were climate-oriented. Man's innate sensitivity to these conditions produced a distinctive and visually appealing sense of place. For example, to prevent high winds from uplifting the roofs, both Scottish fishing villages and German farm towns placed the gabled ends into the direction of the prevailing winds, thus organizing the community design. An architecture of patios, courtyards, flat roofs, thick walls, and verandas cool the interiors of houses in high-temperature zones. In addition to these measures, narrow streets provide shade and comfort to those living in the Mediterranean climate. Where rain and snowfall is excessive, as in Alpine regions, the roofs are pitched over 60°. The walls here are kept low in relation to the roof area, the roof overhanging to protect the walls and the walls themselves set on thick stone bases so as to be dry and free from ground moisture. All these models of regional architecture, appearing at first glance quaint and arbitrary in calendar art and on travel posters, were rooted in function and a common-sense regard for climate.

Paradoxically, the delights of regional expression, while missing from mass-produced contemporary housing, are a common feature in today's custom-designed buildings in Europe and America. In his picture book *Modern Houses of the World*, Sherban Cantacuzino shows three-dozen examples of recent domestic architecture—modern in the choice of materials, the mechanical systems, the life style of its owner, and the outlook of its architect. In Scandinavia, France, Spain, Italy, and the United States, what stands out immediately in these designs is a site arrangement that fits into the regional landscape and an articulation of views, vistas, and internal and external spaces that reflect climate. One of the designers represented, Paul Rudolph, calls regional sensitivity one of the main determinants of architectural form, the others being (in his view) function, the setting created by other nearby structures, the psychological demands of the building, and the spirit of the times. Rudolph's aesthetics are appealing because they offer a verbal hook upon which anyone can hang a subtle or mundane plea for diversification over monotony.

Interior, Hyatt-Regency Hotel. Atlanta, Georgia, 1966, John Portman, Architect. *Photo: Hyatt-Regency Hotel*

Houston Astrodome. *Photo: Houston Astrodome Inc.*

U.S. Pavilion, Montreal, EXPO '67, Buckminster Fuller and Geometrics, Inc., Architects. *Photo: General Electric Company, Nela Park, Cleveland, Ohio*

Climate-Controlled Advances in technology are encouraging designers to free themselves of traditional building forms. These recent structures give clues as to how man-made environments can overcome the constraints of climate.

199

The builder's house—the anonymous architecture of the United States—occasionally shows differences in design from coast to coast. But this is a relatively recent event and not widespread enough to merit acclaim. The common experience is senseless repetition of a limited design vocabulary. Historically, one could say that this is an American tradition.

The initial immigrants and migrants in the United States brought memories of many kinds of "old country" architecture with them as they pushed the continental frontiers westward. However, their traditional architecture proved to be unadaptable or difficult to construct in the newly conquered wildernesses, so a degree of uniformity became evident through wide use of an English model, itself modified by the climate of New England.

In their panoramic work, *The Architecture of America*, John Burchard and Albert Bush-Brown described how the English Medieval house remained a prototype in the original Pilgrim and Puritan settlements along the Atlantic seaboard "perhaps no longer than the first winter. The exposed Elizabethan struts, stuffed with bricks or lath and plaster, simply let in too much air, and the American tradition of heavy siding was urgent." The successful adaptation was soon used everywhere. Despite good reason to develop designs indigenous to the region, the New England house "marched across America," suffering depreciation as it "became less and less suited to its surroundings." In several sections of the country French, Spanish, Dutch, and Scandinavian influences can be traced in domestic housing and public buildings, but none of these designs endured or extended themselves as the New England images did. Note Burchard and Bush-Brown, "Farmhouse, barn, elevator, school, church, store, town house, town

street, all came from this one bolt of indigenous cloth."

The history of American architecture shows a tug of war between those searching for a national style (by international acclamation) and a minority voice calling for regional architecture. The records of both camps are uneven, the motivations unclear, the achievements to date uncertain. Frank Lloyd Wright's Prairie Houses, Santa Fe's adobe, and California's Bay Area style, stand out as expressions of climate in architecture but not as forms having countrywide acceptance. If diversity is embraced, a singular national architecture may be beyond the reach of a continental nation. Let this be recognized, and the follies of spoiling the American landscape with ill-suited repetitions of imported mediocrity can be reversed and regional styles, climate-based, can be encouraged.

To design in response to climate and a regional setting is not to rob architecture of inventiveness and innovation or to downgrade function. On a fairly intimate scale, environmental design studies for the expansion of research and educational facilities at the Woods Hole Oceanographic Institution show how a sense of place arises from the interaction of program, site, and climate.

Underlying the design concept for the Institution was the idea of creating a precinct of maximum opportunity for contact and communication among all members while fitting a large building program to a small irregular site, suggesting a fusing of building and landscape. Accordingly, the program called for a number of structures to be connected, though other buildings would be free-standing. The design concept maintained the contours of the land by creating a profile of roof and ground-floor lines that stepped up and down the slopes. Buildings were edged with hard paving which allowed

space for occasional service and emergency vehicles, but beyond the paving, the design called for natural landscape—the existing vegetation pattern supplemented by additional evergreen material for winter screening—to grow to its fullest extent. The highest elevations on the site were used for the heaviest density of buildings, lifting them above the surrounding streets and allowing views and vistas to the nearby ocean, bay, and ponds. The simplicity of the architectural expression recognized the functional requirements of research buildings as well as the quality of the site but was not neutral. Site design captured the continuity of purpose the Institution represented, while the details of interior design expressed the personality of individuals and of groups within the organization.

A sense of place oriented to regional design can be equally well achieved at a large scale of design. Planning studies for the Rocky Mountains and Great Plains area show how awareness of design influences, especially climate, can generate policies and programs that lead to an improved environment.

The resolution of environmental-design issues often begins best with simple statements about the planning area which may then throw the design problem into relief. The map opposite suggests an opportunity on a grand scale for the distinctiveness in urban form and habitation that arises from the natural situation. The plan covers one third of the land area in the United States and one twelfth of the population—the Rocky Mountains-Great Plains region. It is a major biological zone with boundaries determined by the Rocky Mountains and a semi-arid climate where the average rainfall is less than twenty inches a year.

Within the zone three settlement patterns can be seen. The Northern Plains is predominantly rural with a few large cities. Basically a farming and livestock economy prevails. The Front Range of the Rocky Mountains runs from Cheyenne, Wyoming, to Trinidad, Colorado. It is dominated by several metropolitan areas—Denver, Colorado Springs, and Pueblo, Colorado—which may eventually merge into a megalopolis. The Southwest Plains Region which extends from the southern end of the Rockies to the Mexican border, is the most culturally diverse, reflecting influences of Indian, Spanish, and Northern European settlers.

All three areas today show country-to-city population movements. Urban standards are deteriorating because of the influx, while genuine values and purposes of rural life go neglected. While the overall regional density (calculated by dividing numbers of people into acres of land) is low, a handful of cities holds a large percentage of the population in the individual states, and these concentrations of people pose the threat of environmental disaster.

As a special study committee convened by Colorado College noted,

Man has confronted the forces of nature [in this Region] with practically no comprehension of their meaning for his life, and his adjustment to the realities of his environment has been seriously incomplete. For the central fact is that the area is a partially developed, semi-arid hinterland, while for the most part the local human settlements have borrowed their ways from quite different environments—from the humid East Coast, the interior manufacturing areas, the corn belt, and the agricultural South.

Carrying their environments westward to the region, the urban settlers brought institutions, politics, laws, and design attitudes clearly unsuited for the area. Despite periodic droughts, trees and lawns were planted that had difficulty in surviving or wasted the most precious regional resource:

Region And Climate Design studies for proposed research facilities in which both climate and regional expression were used to suggest the criteria for the design of future buildings. *Source: Long-Range Plan, 1966. Woods Hole Oceanographic Institution. Dober, Walquist and Harris, Inc.*

water. The architecture of wood was used instead of an architecture of indigenous materials, such as rock and earth. In general and in detail Man's adaption to the land in the Rocky Mountain-Great Plains Region was ill conceived and poorly executed, one of fashion, not intelligence.

In attempting to sort out a direction for guiding changes and improvement in the region, the study committee concluded that a general respect for the special character of the large-scale environment was the first step towards effective small-scale solutions.

This respect must begin, the study committee said, with the protection and enhancement of individuality of the three major urban complexes in the region. While they all share common difficulties—water shortages, tax inequities, rural-urban political conflicts, frustrated community planning, uncoordinated economic development —they possess characteristics that make them visibly different from each other. These include cultural and ethnic attitudes, microclimatic conditions, population composition, and urban and regional design opportunities. It is the preservation and encouragement of these differences that are valued.

As a long-range development policy within the lines of respect for the character of the area, the committee recommended that the region should *not* encourage large-scale manufacturing or a diversified, production-oriented economy. Rather the area should capitalize on its unique location and natural advantages as a specialized service area for the major populations located to the east and to the west. The Rocky Mountain-Great Plains Region should be the locus of large new cities built for education, the fine arts, advanced scientific research, finance, and recreation. Concentrated populations with this kind of economic base use less water than other kinds of communities

and are easier to fit into the natural conditions in the region; in turn such populations can ease the pressures on natural resources in the area which then can be creatively mined, forested, farmed, and used for aesthetic enjoyment and private and public profit.

By encouraging functions that the character of the region can support economically, the plan can maintain the dividing line between rural and urban settlement. By not straining the natural resources, urbanization can enhance the region's attractiveness to visitors. With climate as an orientation, the architecture and landscaping need not be a transplanted aesthetic but can grow out of the regional setting.

Since sizable portions of the world— including the Pampas of South America, the interior deserts of Australia, and a band of territories from the Mediterranean through Turkey and Soviet Russia to the Indus Valley—have natural conditions similar to the Rocky Mountains-Great Plains Region, successful large-scale environmental design can serve as an inspiration or model for those areas abroad in efforts to avoid the weary reiteration of undifferentiated building and landscape that robs Detroit, Dallas, and Denver of any sense of place. But what about areas with other types of natural condition?

Surely, for advanced industrial societies anywhere, regional planning has become a critically important device for containing megalopolitan growth. Designs for regional development have been rooted in dimensions set largely by the time and distance scales of economic transactions. Yet it is clear that the extension of existing inventions for transmitting power, water, materials, and ideas can free regions from dependency on geographic proximity. From primitive savannah encampments to modern cities,

economic development has determined settlement patterns and modes of transport, but new concepts of community can follow a different line of development, for the logistics of commerce are no longer as compelling or commanding as they have been in the past. Further, the major portion of the work force need not be employed in the production and distribution of goods. Freedom from locational constraints and old work patterns may lead to increasingly varied forms of human habitation. Technological advancement can free people from a dependency on the metropolis. A mindless accretion of population beyond the limits of economic necessity can be replaced with planned urban decentralization. Of course, the technical studies for accomplishing this require examination of the interdependencies of economic and social classes in the post-industrial world, as well as consideration of how to run smaller-than-present systems of food and fiber production, how to manage natural resources better than heretofore, and lastly but equally important, how to sustain cultural and intellectual stimulus.

Intellectual discourse about regionalism is often tinged by a fear that regionalism will lead to ruralism with the bleak prospect of people being out of touch and thus out of action. But as the Rocky Mountain-Great Plains study hints, that prospect is an illusion. Advancing systems of communication can give ready access to all kinds of stimuli including information. High-speed transport can cross the continent in four hours and might halve that time through supersonic travel. Ninety percent of the population does not have to live on ten percent of the land in the United States—the current national prospect. Regionalism offers an alternative to gigantic megalopolitan areas. Given the expenditure of $100 billion a year for housing and community development, the means for a new national pattern for community development are available. Some of these communities can be *synoptic*, a term we use to describe a balance of population and employment. Others can be *singular*, of special purpose; in these, regional character can especially animate the design. Such a community is Flagstaff New Town (Maine), a good example of a community where leisure-time pursuits and regional design are blended into a special place.

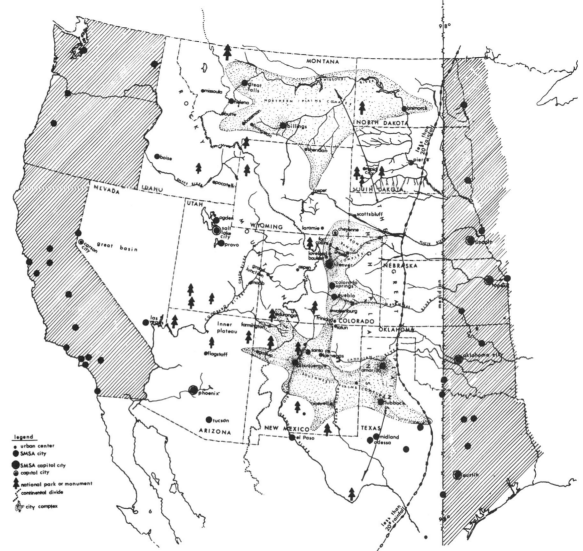

A Sense Of Place For a Third Of The Continent The adjacent map shows the extent of climatic influences on the Rocky Mountains and Great Plains area. Within this region special ecological and environmental conditions deserve a distinctive design response from those concerned with large-scale and small-scale community development. *Reprinted from: Civic Design Study, Colorado College, Colorado Springs, Colorado, 1962*

Flagstaff Town Center

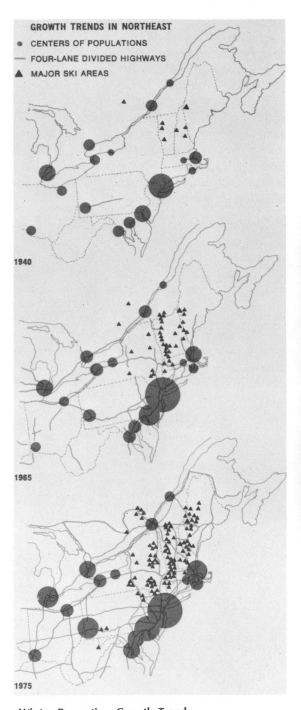

1940

1965

1975

Winter Recreation Growth Trends

COMMUNITY FUNCTION AS A DESIGN INFLUENCE

Americans are presently spending upwards of $40 billion annually on recreation and related leisure-time activities and will spend more with each passing year. Consequently, recreation has become a major industry providing increased employment, high incomes, increased retail sales, and increased tax revenues to the states and local communities catering to the recreation market.

While the growing demand for recreation facilities has been pronounced throughout the country, it has been particularly dramatic in the densely populated Northeast where twenty-five percent of the nation's population must share only four percent of the recreational land. The resulting pressure on recreational resources has particularly affected New England. From all over the Northeast people come to New England for weekends and longer vacation periods in both winter and summer. The region's unexploited recreational resources, its untouched mountains and lakes, are becoming scarce. Most of the mountain and lake areas in Vermont, New Hampshire, and Massachusetts are already crowded, and only the interior of Maine remains to be fully developed.

The proposals for building Flagstaff New Town (Maine) recognizes these trends. The town is intended to serve the largest, richest, fastest growing, and most highly educated recreation market in the world, consisting of the eastern United States and Canada. Including the dense urban strip lying along the Atlantic seaboard from Virginia to Maine, Flagstaff's primary market area contains over 100 million people.

Behind the design of Flagstaff is the thought that economic and social trends in America need not lead to an ill-defined megalopolis and the exploitation and overuse of regional assets. Constructive guidance of these forces can increase available choices for community development, and the character of a region can animate the spirit of that development.

In these objectives recreation is a particularly favorable industry, since its development does not deplete any natural resources or produce undesirable side effects such as air and water pollution. The economic impact of recreation is particularly significant in regions like Central Maine where the mountainous terrain and heavy snow conditions (the heaviest snowfalls and longest snow retention on the East Coast), so favorable for recreational development, generally would not be conducive to other types of industrial operations. The creation of new communities for leisure activities, such as the proposed Flagstaff New Town, along with concurrent development of highway and air transportation to the area, would inevitably bring the benefits of economic growth.

Economic analysis by Paul Hendrick has indicated that for every tourist dollar spent in a community like Flagstaff, at least $1.55 will be generated by the multiplier effect in the state and local economy. Of this amount, 93 cents will go into new salaries and wages, while at least 20 cents will be generated in the form of state and local tax revenues. On the basis of estimated tourist expenditures, this means that fully developed Flagsaff could eventually produce some eighteen hundred to two thousand new, full-time jobs with annual state and local tax revenues of some $2 million.

The planners of Flagstaff anticipate it will be more than just a recreation area; it will be a permanent town, embracing a full, year-round community life, a town oriented toward recreation as its principal industry. That industry will foster the growth of many supporting businesses, and institutional developments may include a private school, junior college, conference center, and other community-enhancing facilities. The atmosphere and social character of the town of Flagstaff will be in keeping with the best traditions of New England towns, yet Flagstaff will have a flavor uniquely its own.

The focal point of the town center will be the main plaza, with ice skating in winter, and exhibition space in summer. Surrounding the plaza will be restaurants, hotels, shops, and condominium or apartment houses. Beyond them will be the main residential area of row houses, guest lodges, and other types of housing and the institutional area containing a cultural center, schools, churches, medical and other community facilities. All these areas will be within walking distance of the main plaza. Public parking space will be ample and easily accessible, but located outside the town center, and all buildings within the project area will be required to provide off-street parking. The center will be exclusively for pedestrians.

The appeal and interest of Flagstaff's town center will be enhanced by the development policy of encouraging service operations on a personal scale and small-scale business enterprises, individually owned and operated. Specialty shops, art galleries, and a variety of colorful and unusual restaurants will contribute to the village atmosphere. While some large hotels will be developed, much emphasis will be on smaller inns and lodges, each with its own special character. This type, variety, and scale of commercial facility is sometimes found in areas surrounding other resorts, but Flagstaff's unique feature will be its compact arrangement into a highly urban scene.

In the design of Flagstaff, regional functions, regional architecture, and regional landscape combine to create a sense of place, pointing to a future for urban development less gloomy than a prognosis of present trends in urbanization would first suggest.

Flagstaff Marina The organization of recreation activities into functional areas will result in sub-centers that possess a sense of place.

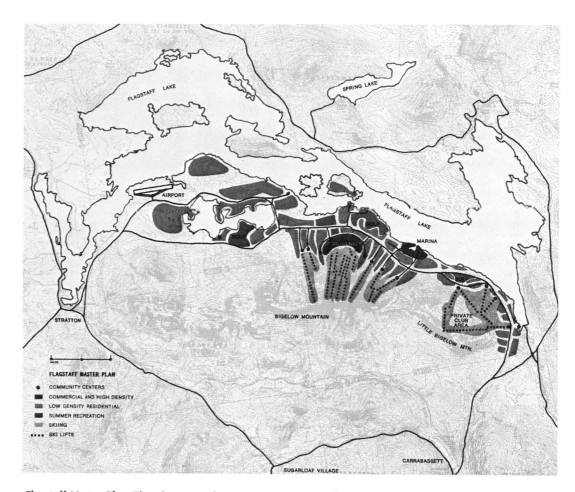

Flagstaff Master Plan The elements of a contemporary special-purpose town. The location of the activities reflects site decisions in which both the natural conditions and the requirements for carrying on the activities are joined in a singular design concept. *Illustrations on pages 204-207. Reprinted from: Winter Spring Summer Fall Flagstaff, Dober, Walquist and Harris, Inc., Planners, Richard C. Stauffer, Architectural Consultant*

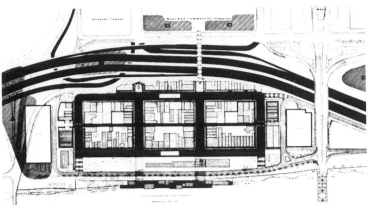

LEGEND: Circulation and Parking Diagram

- FREEWAY
- LOCAL STREETS
- HISTORIC STREETS & WAYS (LIMITED TO HISTORIC VEHICLES OF PERIOD AND SERVICE VEHICLES DURING CERTAIN HOURS)
- PROJECT ACCESS & SERVICE ROAD
- SERVICE VEHICLE ONLY
- PEDESTRIAN
- PARKING STRUCTURES
- PARKING UNDER RECONSTRUCTED BUILDINGS
- POTENTIAL PARKING AREAS OUTSIDE OF PROJECT AREA

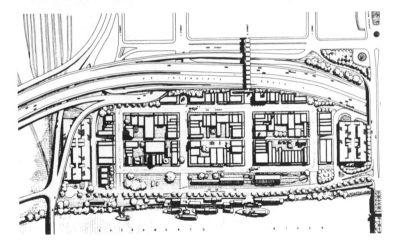

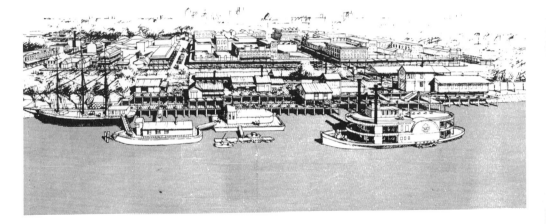

Old Sacramento, 1965 Proposals for the conservation and rehabilitation of the historic beginnings of Sacramento, California. The planning aims at recreating the physical appearance of the structures and district with maximum authenticity. The area would be developed as a location for professional offices, specialty shops, motel-hotels, restaurants, and entertainment facilities — using existing streets and structures. The district would be tied back to downtown Sacramento by a pedestrian bridge and moving sidewalk. The area today is a Skid-Row. The carrying out of the plan can be considered successful only if the relocation and improvement of the lives of the unfortunate people in the area is accomplished. *Source: Sacramento Redevelopment Authority. Plans prepared by Candeub, Fleissig and Associates.*

HISTORIC PRESERVATION

Wholesale addition of new construction offers one opportunity to create a sense of place, historic preservation another. Not every old building, street, neighborhood, or district deserves restoration and maintenance, but sizable portions of the environment are not only worthy of preservation but essential to advance the cause of good design. Historic preservation can be used to restore individual buildings to economic life, to animate renewal and. redevelopment schemes, to save in their entirety, unique districts, landmarks, and urban settings. America has only just begun to value its past, but this new-found interest is a welcome support for environmental design.

Many reasons can be advanced to support historic preservation as a means for achieving a sense of place. Historic structures can often be brought up to modern residential standards or modified to serve some purpose other than their original design. Often the design and craftsmanship of an old structure are better than that of many more recent buildings of the same kind. The older buildings display a craftsmanship and discipline of construction that deserve emulation. In America the rise of disposable income (what economists used to call discretionary spending) seems to bring with it a rise in the incidence of disposable communities (meretricious, tasteless, and readily obsolete). But firmness of structure, commodity, and delight are ageless values. By displaying these characteristics, even the humblest of historic buildings gives three-dimensional testimony to the unsoundness of our architecture, especially in the scale and details of the everyday environment. Historic structures are needed to show the individuality and identity of a time and a place, not so much to be imitated and mir-rored in kind as to instill a desire for qualities of integrity that transcend style and period.

In addition, historic buildings and districts are fun to see and to live in, not only because they are three-dimensional representations of history but also because they are likely to fill the urban scene with colors, textures, and landscape different from the present fashions. Novelty, say the biologists is Man's fifth need for survival after air, water, food, and love.

Furthermore, forces of economy and social purpose are at work in historic preservation. Entire districts can be recaptured from the past at less cost than new construction and with dramatic consequences to the life of the cities in which they are situated. Charleston, South Carolina (1924), and two dozen cities since then have set up formal legal boundaries around historic districts to preserve and protect them from encroachment and destruction. These measures have not only arrested the spread of blight and obsolescence but have reversed such trends, helping to avert a one-class, one-race, one-age-group city. Historic districts such as Beacon Hill, Georgetown, and Old Annapolis have changed from incipient or actual slums to highly valuable real estate because of preservation. They are certainly not multi-class neighborhoods today, but nationwide when the principles of urban renewal through urban conservation have been applied to non-fashionable neighborhoods in the middle- or upper-class sense, after years of neglect old structures and old neighborhoods, have re-engaged themselves in the everyday traffic of life.

The redevelopment of Old Sacramento by conserving and using historic buildings is a splendid example of enlightened self-interest in California. The project area covers about five blocks between a freeway and the Sacramento River. Fifty-three historic buildings remain from the days of the Gold Rush. Now serving as a Skid Row—an area of flophouses, bars, diners, and pawnshops—its strategic location adjacent to the central business district (across the freeway) provides potential of reuse for professional offices, specialty shops, restaurants, and entertainment. Plans prepared for the Development Agency show how restoration of the area can recreate a semblance of the past and allow the old structures to be returned to a purpose other than the ones they now serve. An unusual feature of the scheme is the proposed utilization of moving sidewalks and escalators to connect Old Sacramento to the central business area. Along with the presence of wagon trains, riverboat, stagecoach, pony-express station, and railroad, the sidewalks should give visitors an unusual view of transportation in America and its impact on city life and development by having these modes of transportation in a living museum.

Historic preservation is also a way of avoiding clearing and crushing a sense of place in the haste to redevelop neighborhoods and communities. The physical features of the past can be revitalized not only for themselves but also for enlivening new environments which may be inserted into the urban fabric.

New Bedford, Massachusetts, affords a good study of how a city can use fragments of the past to renew itself. New Bedford grew in the nineteenth century as a wealthy seaport with an unusual role in American maritime history. For a hundred years the community's fleet ranged the oceans to gather whale oil and bone for lighting and manufacture. The enterprises based on these sea industries created great wealth, large mansions, and unusual cultural institutions. With the discovery of petroleum and the naval disasters of the Civil War, the city's prosperous period ended. In its struggle to develop a competititive role in the twentieth century, officials in New Bedford have used the artifacts of the past to animate new city growth. Old civic buildings such as the Seaman's Bethel, memorably described in Herman Melville's *Moby Dick*, the Mariner's Home, and the New Bedford Whaling Museum are open to the public, drawing visitors from around the world and adding to the income of the city. Johnny Cake Hill with its many excellent old buildings is being restored as a downtown neighborhood, and badly needed contemporary housing, complementary in design, added there. New and old are brought together into a sense of place.

Historic preservation is important in still more ways. The pulse of the past has given designers and planners clues and cues for undertaking large-scale new construction as well as preservation. The Society Hill area of Philadelphia is a notable example of combining urban conservation, historic preservation, and new development. The threads of historic buildings and open spaces have been used to establish the design structure of the area. High-rise buildings, row housing, and community facilities have been integrated into a distinctive design in which the scale and building materials from the past have served as points of departure for contemporary design—all this producing a sense of place.

In Columbus, Ohio, a preservation measure for a unique district—German Village—has engendered rehabilitation over 150 acres. Originally settled by a German immigrant group, the Village's scale and architectural distinction, though modest and moderate, have captured the interest of enough people to generate sizable private investment in restoring and using homes and business. The proximity of the city's central business dis-

trict has made it possible for the Village to draw specialty shops and professional services serving the entire metropolitan area. Location in German Village is now recognized as an economic advantage, but whether a carnival air and heavy traffic caused by the renewal of its commercial sections will be to the district's long-range advantage as a residential area remains unanswered.

Historic preservation has been shown to work. A review of the efforts on College Hill, Providence, Rhode Island, revealed the effectiveness of community awareness and action in using historic structures and open spaces for community renewal and development. The original study and proposals for historic preservation (1958) generated considerable confidence in the district's future. Mortgages for rehabilitation became easier to obtain, and private companies were organized to profitably carry on reconstruction. Historic-district legislation was enacted, a national-park site established, and accommodations arranged for tourists. The upgrading of College Hill had a strong effect on an adjacent, four-hundred-acre urban-renewal project. City housing inspections showed that forty percent of the units showed "measurable improvement" over a ten-year period; large-scale clearance became unnecessary.

The economic advantages of historic preservation are not inconsiderable. There are approximately 100,000 acres of historic sites in urban areas directly under Federal and State control and half that much again in private hands. Historic buildings and sites are an important part of the tourist industry; and if there is anything Americans share in common, it is the urge to travel and "see the sights." In Virginia, for example, tourism is the state's largest industry, bringing more than a half-billion dollars into the Commonwealth each year, with much of the commerce attracted by the sites and artifacts of

colonial America and the Civil War.

But economic advantages are not the only benefits of tourist and sight-seeing attractions. Museum villages, such as Williamsburg, Virginia; Greenfield Village, Dearborn, Michigan; Mystic Seaport, Mystic, Connecticut; Old Sturbridge, Massachusetts; and New Salem, Illinois, have proven effective in dramatizing the events and occurrences of the past and by giving people a realization of the richness of our heritage. Such places are excellent depositories for period artifacts and furnishings. They place common objects in a setting of the past recalled, an environmental setting not possible in the frame-and-case format of older museums. There is an important difference, as André Malraux hints in his *Les Voix du Silence*, for the marble palace museums "estrange the works they bring together from their original functions and . . . transform even portraits into pictures." In-place examples of Shaker furniture and Georgian tableware expose the state of contemporary industrial design, pointing to things that have to be done to raise the quality of the everyday environment.

In a radically different spirit, bits and pieces of the past can also be effectively sown for environmental betterment. Five lighthouses in the Hudson River, ". . . splendid witnesses to the golden age of American navigation . . ." wrote one architectural critic, had been abandoned, boarded up, and left to the ravages of the elements. Each was a unique momento of nineteenth-century architecture and commerce, but all were planned for demolition. At the request of the Hudson River Valley Commission, work was postponed. As studies for preservation revealed (1967), all could be used as special design features in imaginative schemes for outdoor recreation.

America's attitudes about the value of the past have varied but are worth tracing

For all those people who thought they would miss the old Imperial, here's what is going up in **its** place: The new Imperial. A majestic, 17-story high rise scheduled to open in March, 1970, on the site of our old wing. With a thousand spacious new rooms added to the 600 in the present building, the Imperial will be the largest hotel in the Orient. And then, as now, you can be sure the Imperial's famous personal attention to service, and its matchless facilities in the heart of the world's biggest city, will live up to the name. Imperial. ...and the legend continues.

IMPERIAL HOTEL
T. Inumaru, President and General Manager   TOKYO

An Eighteenth-Century Townhouse Rehabilitated For Residential Use In Philadelphia's Society Hill Area
Photo: Philadelphia Redevelopment Authority

USING HISTORIC BUILDINGS

Desecration World-wide historic masterpieces are in danger. The tearing down of Frank Lloyd Wright's Imperial Hotel and its replacement by a bland, anonymous architecture is a case example of destroying a sense of place. *Source: Imperial Hotel, Tokyo*

Seamen's Bethel Described in *Moby Dick*, the building is now used as a focal point in the renewal of old sections of New Bedford. *Photo: The Whaling Museum, New Bedford, Massachusetts*

New Salem, Illinois Authentic structures were brought together to recreate a pioneer village. The living museum tells better than words how life was carried on more than a century ago. *Photo: Richard P. Dober*

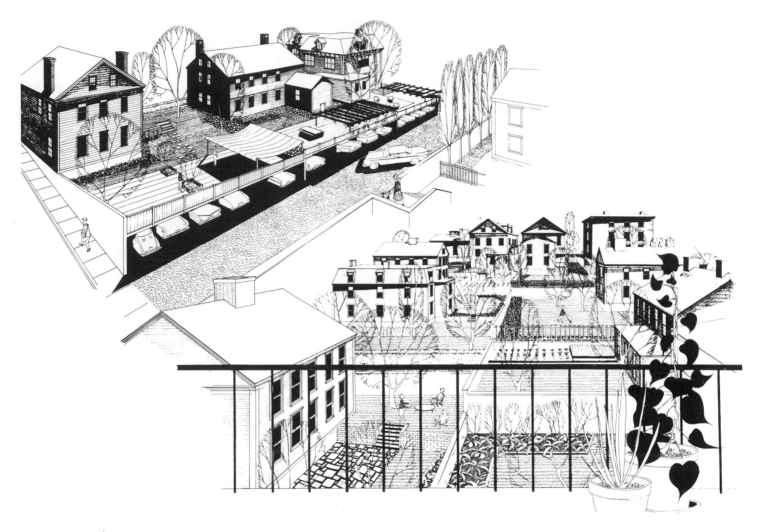

College Hill, Providence, Rhode Island Design proposals by Lachlan F. Blair and his associates for the renewal of an important historic district.
Source: Providence Redevelopment Authority

briefly for the light they shed on changes in national policy and cultural instincts. Certainly, the last century's regard for history was mixed, for at the same time that the graft-ridden Pennsylvania State Legislature was offering the old National Capitol and Garden (now called Independence Hall) for sale, the Federal Government was enacting laws to acquire generous open space around the City of Washington's monuments so as to preserve them in perpetuity. It wasn't until 1850 that any American government, local, state, or national, acquired a historic building specifically for public exhibition. Thus New York State's purchase of the Hasbrouck House (George Washington's headquarters in Newburgh) was an unusual public action, although a private group had previously saved Andrew Jackson's home from demolition, and the Fairbanks had earlier formed a special trust to keep up their family's first home in Dedham, Massachusetts, probably the oldest house in North America. Civil-War internment sites became national memorials in 1863 and 1865, and in 1880 eight additional Civil War sites and three Revolutionary War battlegrounds became National Military Parks under the aegis of the War Department. Concurrently, the Army's role in preservation took an unusual turn, largely unremembered and certainly unsung, for it served as police and groundskeeper for the Department of the Interior. For thirty years, Federal troops occupied, protected, and maintained Yellowstone and Yosemite Parks, effectively preventing encroachment and destruction by illegal lumber companies, hunters, and other intruders.

The Year 1891 marked a turn in the approach to preservation measures: The General Court of Massachusetts authorized the chartering of the Trustees of Public Reservation with the right of acquiring lands "deemed worthy of preservation for the enjoyment of the people." Scenic places as well as buildings and sites were covered by the legislation. The charter became the model for the British National Trust and later the National Trust for Historic Preservation in the United States.

The next forty years were nevertheless times of indecision with periodic public cries for government action, agitation by muckrakers and private preservation associations, and seeming slowness on the part of Congress to do anything more than meet the immediate crisis. Almost reluctantly, if newspaper accounts are indicative, the Grand Canyon was saved and a National Park Service established. In its first annual report (1916) Stephen Mather wrote, "We have only just begun the solution of many questions that have to be dealt with . . . and for the handling of which we have no coherent policy."

Not until Franklin D. Roosevelt's administration was their a unified national program. Among the highlights of his terms in office was the Historic Buildings Survey (1935). It used unemployed architects and draftsmen to measure and classify five hundred special structures, almost all of which have since been preserved and labeled as National Landmarks. Thus the chartering of the National Trust in 1947 crowned a century of many personal efforts and the work of hundreds of private and state historical associations to confirm the value of the past to present and future generations.

Clearly, the urge for historic preservation today is not a matter of spontaneous patriotic combustion but the end result of individual initiative of many generations, education, the solicitude of associations, and, finally, government leadership. The evolution of concern may be due to the emergence of pride in a common heritage or a reaction to the blandness and monotony of contemporary life. In either instance, present-day attitudes are encouraging in what they mean for the creation of a sense of place and the possibility of a growing, widespread appreciation for the natural environment and the subtle manifestations of environmental design.

Twenty years ago it would have been political folly to use zoning laws to limit the heights of buildings for aesthetic considerations—not just because overly tall buildings threw unhealthy shadows but also because they blocked the views and vistas into and away from an attractive and historic urbanscape—but San Francisco did just that in 1964.

Few American cities have a more coherent sense of place than San Francisco. One of the last twenty-four-hour downtowns in North America, a strong topography, moderate climate, spectacular panoramic views in all directions into a landlocked bay make up a memorable place. Since 1927, special height-control districts have been used in San Francisco in response to special circumstances; but they were positioned haphazardly, sometimes overlapped other types of control areas, and left several critical areas unprotected. The holes in the coverage could permit wiping out of historic views which other laws protected. Timely studies by the San Francisco Department of City Planning concluded that no exceptions should be allowed in the city's forty-foot height limitations on the northern flatlands and, in addition, that precise limitations were necessary over a great part of the city slopes upon which high-rise buildings had not been constructed. (A sampling of the study technique and the ensuing district is shown on the next page.) The advantages of height limitations were known to be more than aesthetic. Without controls, land values would drop; traffic, parking, and urban congestion would mount; and the ratio between open space and building density would diminish. But it was the possible loss of one of San Francisco's greatest single assets—the historic panorama—which galvanized public support to find legal precedence for using aesthetics to restrict the use of private property.

In using historic preservation in environmental design, the ethics of authenticity cannot be neglected; for the boundary line between the real and the fake is more than a date, the degree of imitation, or the resemblance to the original. In the biography of Walt Disney, Richard Schickel asks about the relationship between "authentic experience" and "the impoverishment of our national mental life." Does a clean version of the past—such as the historic recreations of Disneyland—offer in place of reality escapism, quick and cheap thrills through nostalgia, a mental appreciation no deeper than the immediate stimulation? As a self-proclaimed proprietor of the American spirit, Disney found a way to exploit history; and others will profit by his example. This is the danger of historicism.

José Ortega y Gasset calls for a history that is like "an elaboration of the cinema . . . reality, which for one moment seemed an infinity of crystalized facts, frozen in position, liquifies, springs forth, and flows." There is a newly demanding role for historic preservation: the need to leap the boundaries of infatuation with a historic building and site and search out some larger perspective and idea of contemporary human habitation. Florida's beaches are historic, a singular natural resource upon which Florida's economic livelihood is heavily dependent. They are slowly eroding, growing unstable, losing their visual and functional qualities. Can love of historic man-made objects be extended to nature? Nantucket Island has vigorously enforced Historical District zoning. Its past need not be recreated; it exists. But outside the District boundaries, an unusual ecological pattern is continually injured by uncontrolled urbanization. More than nature is threatened. The charm of the Island comes from contrast between grey shingle and green moor, town and country, all in one piece. The reasons why the endearing quality emerges from the contrast can be made explicit. It does not involve a mysterious art but an art of search and sensitivity. The qualities of the special sense of place which is Nantucket can be extrapolated. The next pages show how. But extrapolation and preservation only resolve the surface problems. The fundamental issue for Nantucket is whether it can find a way for creating in this generation something worth preserving in its own right by the next. Can locality, the sense of place, yield new designs, or is the eventual fate of Nantucket and the mainland a gradual decline to large Disneylands, modern-day Roman circuses passed off as "the designed environment"? World-wide the dilemma must be faced, as under the guise of service and improvement, bland and scaleless buildings replace distinctive masterworks.

Hudson River Lighthouse These special historic structures, abandoned and obsolete, are now being used as landmarks for new recreation areas along the river. *Source: State of New York*

San Francisco The impact of height on a sense of place in San Francisco can be appreciated from this design study of the Golden Gateway Redevelopment Project. The well-designed and spaciously sited towers present an entirely different visual impact when seen from the southeastern slope of Telegraph Hill. To preserve the traditional views to and from the Bay, an imaginative zoning ordinance has been established to restrict the heights of buildings in selected areas. The location of the restrictions was based on environmental design studies. *Photo: San Francisco City Planning Department.*

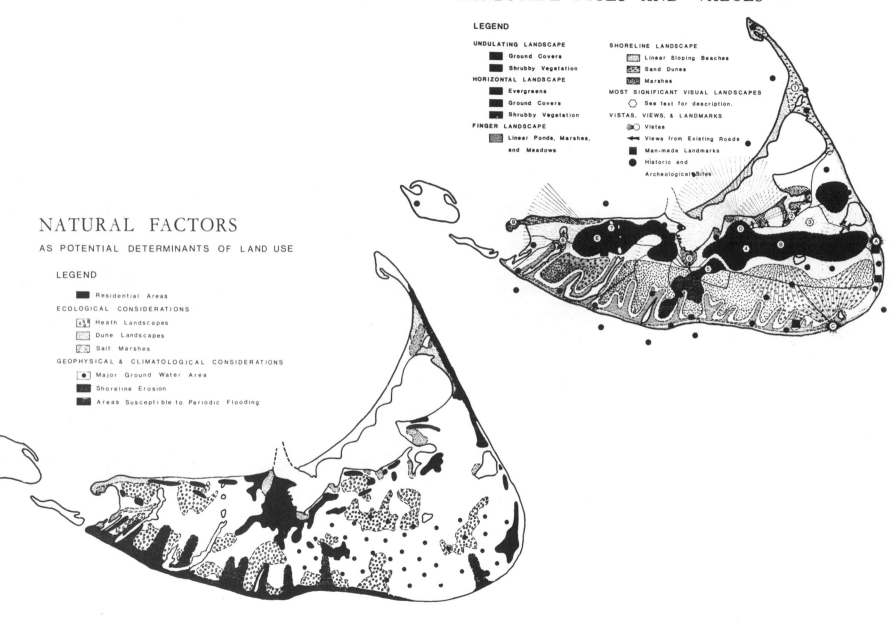

LANDSCAPE TYPES AND VALUES

LEGEND

UNDULATING LANDSCAPE
- Ground Covers
- Shrubby Vegetation

HORIZONTAL LANDSCAPE
- Evergreens
- Ground Covers
- Shrubby Vegetation

FINGER LANDSCAPE
- Linear Ponds, Marshes, and Meadows

SHORELINE LANDSCAPE
- Linear Sloping Beaches
- Sand Dunes
- Marshes

MOST SIGNIFICANT VISUAL LANDSCAPES
- See text for description.

VISTAS, VIEWS, & LANDMARKS
- Vistas
- Views from Existing Roads
- Man-made Landmarks
- Historic and Archeological Sites

NATURAL FACTORS

AS POTENTIAL DETERMINANTS OF LAND USE

LEGEND

- Residential Areas

ECOLOGICAL CONSIDERATIONS
- Heath Landscapes
- Dune Landscapes
- Salt Marshes

GEOPHYSICAL & CLIMATOLOGICAL CONSIDERATIONS
- Major Ground Water Area
- Shoreline Erosion
- Areas Susceptible to Periodic Flooding

The Elements That Comprise a Sense Of Place A sense of place arises out of site and situation. The above drawings indicate the variety of design features that combine to form a distinctive environment. Man-made and natural conditions play unique and complementary roles. Through time, development has responded to ecological, geophysical, and climatological considerations. Nantucket may lose its sense of place if urbanization runs counter-grain to what nature has traditionally encouraged. *Drawings: Department of Landscape Architecture, University of Massachusetts, Amherst*

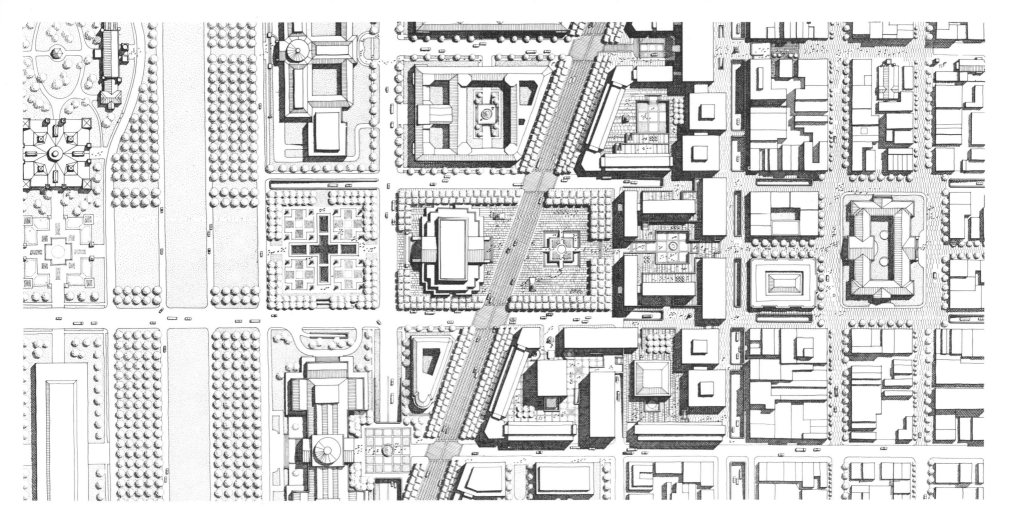

Extension Without Artifice The heroic qualities of the historic city can be extended without artifice, and without compromising contemporary technical requirements, as indicated in the proposals for new designs along Pennsylvania Avenue, Washington, D.C. *Source: Report of the President's Council on Pennsylvania Avenue. 1964*

217

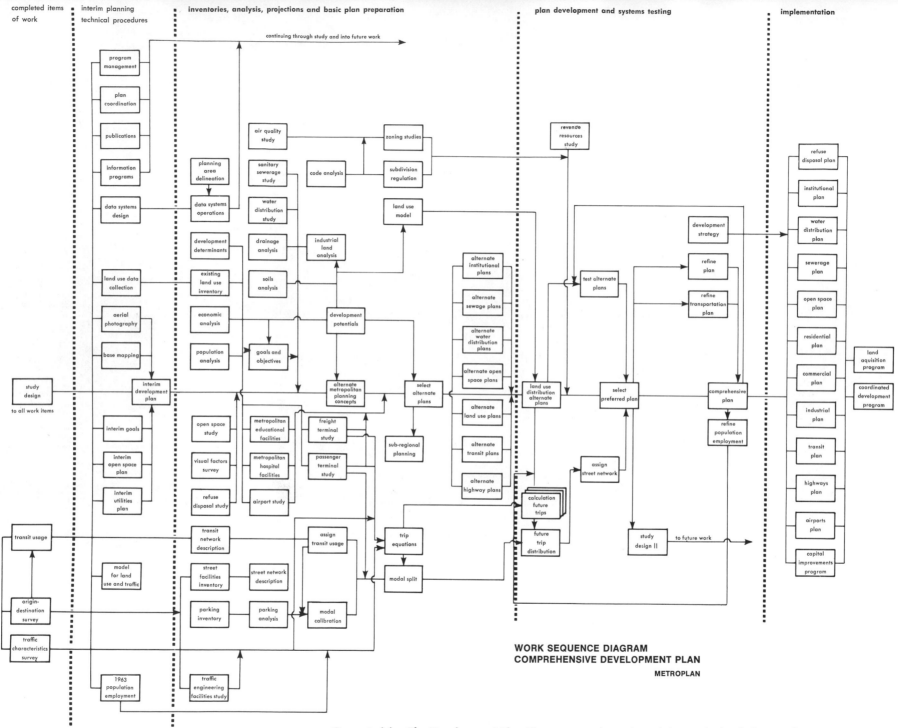

WORK SEQUENCE DIAGRAM
COMPREHENSIVE DEVELOPMENT PLAN
METROPLAN

Computerizing The Development Plan The computer has enlarged the methods of planning but not simultaneously the general public's understanding of the process and the product. In the latter respect, environmental design can play a important role in describing the general meaning of various proposals. *Redrawn from: Metropolitan Planning Commission, Kansas City, Missouri, 1967*

THE SEARCH FOR VALUES IN A PRINTOUT OF FACTS

A sense of place can be created through many devices ranging from an order imposed by the process of urban development to a common-sense respect for natural conditions or the preservation of what exists. This idea of a sense of place may seem a romantic notion at first glance, out of place in the twentieth century. But the concept has utility, purpose, and value well beyond visual satisfaction and pleasure. A sense of place can serve as a counterforce to what otherwise could be a contrived and expressionless world, through the avoidance of impersonalization, as computer-age technology brings about a return to the ideas of comprehensiveness and uniqueness in large-scale design.

Significant improvements in environment are dependent on the management of technology and political processes. Neither the quantity nor quality of the designed environment will emerge spontaneously, episodically, and unregulated. Intuitive response, the traditional tools of the design professions, are no longer sufficient. The scale of the problems and the scale of the solutions are beyond simple elucidation. Pope, king, or magnate can no longer impose by decree a designed environment, at least in the sense that it encompasses twentieth-century machines and social organizations effectively at work to produce and maintain it. This is a worldwide condition.

Reforestation about Chinese villages and traffic control in Belgrade succeed or fail not because of political philosophy but because they follow a course of action that is apolitical, a science of management itself. And the same is true in the immense efforts of recent years in the United States to improve the human condition through public welfare, public education, public health, and public design. These measures may start with simple actions by elected representatives, but their effectuation requires institutions and bureaucracies which are complex, expensive, and skillfully managed. Industry, commerce, institutions, and government now share common features: complexity and interrelationship. While it may be convenient to see different points of emphasis by suggesting a private sector and a public sector, the lines between are blurred. This has happened for many reasons, all of which can be summed up as growth and change in every dimension of human life at a double-time pace.

The discovery, accumulation, and application of knowledge increases at a rate faster than the capacity of the human mind to absorb the facts and values of even the most specialized technical fields. Accordingly, technology comes close to the edge of a self-destructive cycle. The gigantic efforts to overcome the inherent limitations in materials, energy sources, and natural phenomenon demand extraordinary expenditures that seem to throw perspective and moderation out of kilter. A nation prepares for a voyage to the moon, leaving half its rivers and streams filled with sludge and stench. An uninhabitable terrain is occupied, while a precious piece of earth is left uninhabitable.

In the long-term view, however, the investments in technology in the past, and most likely in the future, may eventually produce more important benefits than those which give rise to the immediate impulse for innovation. Techniques discovered in spacecraft engineering, for example, may allow farming of the seas, improvements in transportation, advances in medicine, and efficient control of environmental pollution. Further, the very complexity of techno-

Life Styles Common sense suggests that environmental designs should respond to everyday problems, a number of which are humorously summarized in this drawing, prepared by a group of Dutch housewives to remind public officials what is important in new housing. *Source: Ministry of Housing, The Netherlands*

logical change itself has stimulated ideas for organizing and applying the process of invention and production. Technology is now viewed as a system of causes and effects which lends itself to management through a set of procedures that is rational and comprehensible.

The presence of a system further suggests the possibility of human intervention. Through strategic deflections of money and manpower, a seemingly anonymous technology may be guided for social purposes and human satisfaction. Furthermore, the pervasive feeling that the technology is in a runaway condition today may not be so much a flaw in technology as it is indecision and uncertainty about national priorities. Thus a consensus on goals and priorities expressed in political action is the missing element in widespread environmental improvements.

Technology needs to be challenged, but it also needs to be commanded. The setting up of political programs and institutions for environmental improvement has not, however, kept pace with the growth and change of technology and urbanization. National committees have reported time and again the desirability for metropolitan government to give order and efficiency to methods of water distribution, sewage and other waste disposal, transportation, and outdoor recreation. While intercity compacts and special authorities have been established to provide these services, the concept of government structure acceptable to central-city and suburban interests alike founders time and again when tested at the polls.

Voters are not unreasonable, nor do appeals to rationality go unnoticed. In the last decade approval has been given to consolidation of school districts to carry out educational reforms. In fact the number of school districts has been halved, from 47,600 to 23,500, in a quiet but dramatic shift in local politics and an assertion of common sense. The rejection of metropolitan government in the same period may be due to partisanship, lack of political leadership, traditional conflicts between local groups, and a fear of large and anonymous government. The specter of population growth underlines the last issue: How do people stay in touch with the institutions that govern their lives? How do communities remain small in feeling while growing large and complex in actuality? As we will show later, the creation of a sense of place is the environmental designer's contribution to this issue.

The anonymous and impersonal environment that evokes student revolt, community disorder, and political disenchantment has also encouraged counteractions of importance to design. Experimental procedures have encouraged formation of small urban units wherein people have some degree of self-determination in choosing, shaping, and managing their environment. Conceptually, the subunit is not measured by scale but by the quality of its relationship to a larger part. Microscale planning and microscale design are convenient descriptions of this reaction against the meaningless environment.

The Federal Government's approach to renewing cities now reflects this thinking. Under Title I of the Demonstration Cities and Metropolitan Development Act of 1966, special grants have been set aside to plan and design model neighborhoods and districts, not through the imposition of a formal set of guidelines, but rather through "flexible performance standards, responsive to local conditions and capabilities." An unusual feature of the program is that the residents of Model Cities areas are not just consulted on community development; they actually determine the alternatives for improvement and—through their elective committee—vote on the options presented. Because of limited Federal funding, the competition for Model Cities grants has been intense, resulting in fresh and novel views about how the environment can be planned and designed through community participation. The first results of the Model Cities programs show that administrative impediments can be overcome to draw together previously fragmented government programs, reducing the anonymity and bureaucratic insensitivities often found in large-scale operations, private as well as public.

In addition, the movement against monolithic, anonymous government structures, stressing the microscale, has been advanced on the American scene by advocacy planning. Special-interest groups have engaged skilled professionals to plead their causes in the political chambers. Campus planning for example has been carried on by public institutions of higher learning in part to persuade their legislatures that their capital-construction requests were not arbitrary but were rooted in an intelligent estimate of past events and future needs and tailored to a special situation. Neighborhood groups have reversed official decisions on highway locations, on the equipment selected for public playgrounds, on the construction of shopping centers, and on the filling in of estuarine tidelands.

Advocacy planning can produce exciting environmental designs, because it is carried on outside the normal bounds of official planning and is thus easily freed from cant, ritual, and formula.

Often advocacy planning is charged with a social conscience. Under consultant Walter Thabit's direction, citizens and businessmen in the Cooper Square District of New York City were successful in repelling

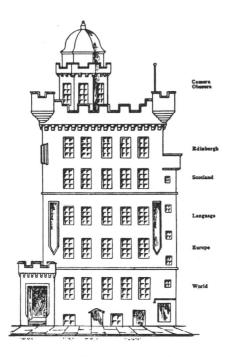

Patrick Geddes' Outlook Tower Geddes proposed that each town have an Outlook Tower in which the history and environmental features of the particular place would be available for comparison to official planning and development proposals. *Reprinted from: Cities In Evolution, Patrick Geddes, London, 1915*

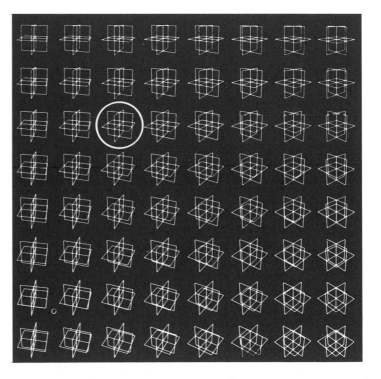

Computer Graphics A series of computer-drawn perspective views of three coordinate planes used to divide space. Given one set of three-dimensional coordinate locations for the end-points of the line, the computer produced on request the sixty-four views shown above. *Source: Harvard Laboratory for Computer Graphics*

the City's official plan for redeveloping the area. The City's scheme called for the construction of 2,900 units of middle-income housing through the displacement of five hundred businesses, twenty-four hundred tenants, 450 furnished rooms, and four thousand beds used by homeless men. Thabit's report showed the unbelievable inaccuracy in the statistical compilations that were made by the City to justify Federal funding for renewal. The plan clearly was of no benefit to those displaced by the scheme. It would also have disrupted a subtle pattern of interdependencies and economic associations among ethnic and social groups that together created the distinctive character of the Lower East Side.

The Cooper Square area was the kingpin in the relationships holding together a community of working artists as well as serving as the cultural center for Italians, Slavs, and Puerto Ricans. The massive clearance proposed by the City would have removed unique shops, entertainments, and organizations dispersing, dividing, and emasculating many microcommunities. *An Alternative Plan for Cooper Square* (1961) is a landmark in advocacy planning, not just because it successfully fought City Hall, but also because it alerted the design professions as to how the techniques of environmental planning can be applied to muster evidence and support for maintaining and improving a sense of place instead of destroying it.

Around the world, many institutions are addressing themselves to the task of microplanning. Charles S. Ascher observes that in the 1966 reorganization of the Archdioceses of Paris, the division of centralized ecclesiastical administration into smaller dioceses, zones, and regions than heretofore brings the process of decision-making on religious matters closer to the people, revealing "the flexibility that has enabled an institution to survive for two thousand years." Ascher sees in this action to create subunits of a scale to increase individual participation, a move that parallels events in the secular world where immediacy and participation are the order of the day.

Microdesign as a conscious objective is not new, though dependent in realization upon the political context. The working ideal of comprehensiveness, which is basic to modern systems approaches and not without social and political consequences, is also as old an ideal in planning and as fundamental a part of the environmental-design process itself as the creation of a sense of place which it fosters. Patrick Geddes with his admonition *Survey Before Plan* in 1906 voiced the basic planning tenet of comprehensiveness in a manner still valid today. He said:

A town plan is a forecast—one of the most vast and comprehensible forecasts imaginable. How then shall anyone consent to, much less draft or criticize such great general schemes of his town's future as a whole without investigating into the growth of its parts and elements in detail.

In Geddes' mind planning was the active agent for social reform and environmental improvement. Thus he encouraged the idea of Civic Surveys as a prelude to plan-making. He called for details on site and situation, land and water transportation, the prospects for industry and commerce, and a knowledge of the population that went beyond mere numbers. Geddes saw planning not as a design "constructed by a municipal official and inspected by a government one." For him a plan was a unique expression of all the elements, all the processes, all the aspects of a community's individuality. Planning must either express progress toward varied city life, he suggested, "or else embody, as our present towns so largely do, tendencies which are the very opposite of all these and so reproduce the past's decay."

To ensure that planning was comprehensive and reflected the special character of the city and town, Geddes proposed that each community maintain a permanent exhibition of survey material. The exhibitions would gain the active interest and cooperation of the citizens and give continuity to all civic changes. These Lookout Towers, as he called them, would house all the plans for community design and development. Criticism and ideas would be solicited. School children would use a camera obscura to view their environment, so that they might grow up as "caretakers of its beauties and renovators of its structure." The *genius loci* would be expressed in simple charts, diagrams, maps, and photographs. Geddes wrote (1908):

The problem for solution which each city has increasingly to face is thus to conserve and to express its local individuality, its uniqueness and character, yet to reconcile this with a full and increasing participation in the material appliances and immaterial advantages of other cities—in short, at once to live its life and this more and more intensely; yet to be also in the great world, and this more and more fully also.

Today unperceptive critics are too quick to suggest that because the environment is ragged and reiterative, someone planned it that way, or even worse, an authoritarian body imposed a mechanical solution and somehow this is part of the planning tradition. This generalization arises from poor planning, but is not necessarily implicit in planning itself; and Geddes' life work and that of other historic figures who shared his concern for a complehensive approach that took not only details but relationships into account, refute such balderdash.

Frederick Law Olmsted recognized that his designs could not be isolated from the social purpose which they served. Writing in 1877, he commented:

Our country has entered upon a stage of progress in which its welfare is to depend on the convenience, safety, order, and economy of life in its cities. It cannot prosper independently of them; cannot gain in virtue, wisdom and comfort, except as they also advance.

In a typical statement, *Town Planning in Practice* (1909), Raymond Unwin wrote that it is the function of the designer to find artistic expression for the requirements and tendencies of the town, "not to impose upon it a preconceived idea of his own." Later he said, "the designer's first duty then, must be to study his town, his site, the people and their requirements."

The discovery of uniqueness is not enough, however; the fact is that the translation of environmental planning into executed designs cannot occur without public approval and enthusiasm. No plan, however meritorious its objective or stimulating its design, can survive without public support.

Thus, the involvement of the community in environmental improvements has long been recognized as a critical step in the planning process, carrying along the current objective of microdesign but very much a part of the concern for comprehensiveness. As Robert A. Walker well asserts, to cast a plan "unheralded and unsupported upon a disinterested city would be to condemn it to early oblivion."

Periodically, versions of Lookout Tower have been promoted to gain citizen interest and backing. Recent examples have been the Greater Philadelphia Planning Exhibition of 1946 and the American Institute of Architects Guide Lines for the Visual Survey, written by Paul Spreiregan (1963). An earlier and perhaps the most successful continuing effort in America were the publications prepared by Walter A. Moody in support of the Chicago Plan.

The enthusiasm for civic improvement sparked by the Chicago Exposition (1893), which gave thousands of Americans their first glimpse of urban design, was kept alive by the Merchants and Commercial Club of Chicago, of which Moody was a member. They succeeded in making the exposition grounds a permanent park, turned their energies to developing the lakefront, and extended that idea to a plan for all of Chicago and its immediate environs. The task was assigned to Daniel Burnham and his associate, Edward H. Bennett. It was in Burnham's own words a "big plan," concerned with the central area of Chicago and the adjacent towns. Burnham advocated that each community create a public improvement commission "to bring about orderly conditions within the town itself, and especially to act in cooperation with similar bodies in neighboring towns so as to secure harmonious, connected, and continuous improvements."

Among the major recommendations made when the Plan was issued in 1910 were these far-seeing features: a system of regional highways extending up to sixty miles outside the city, an orderly arrangement of streets and boulevards for moving traffic to and from the center district, the design of the entire shoreline of Lake Michigan within ten miles of Chicago, the development of an outer-city park system, including a forest preserve of thirty-seven thousand acres and the construction of cultural and civic centers in Chicago and many of its suburbs.

The Chicago Plan is important as an early example of advocacy planning. It was initiated, conceived, and prepared by private individuals. It was the first comprehensive plan prepared for a modern American city. Though continually attacked as impractical

and unrealistic, it survived because its supporters countered every argument by stressing its practical aspects, not its design philosophy. Under Charles H. Wacker's and Walter A. Moody's guidance, a simplified version of the plan became adopted as an official eighth-grade Civics text in the public schools. Wacker and Moody started a lecture bureau and were the first to use movies for propagating environmental design. Through their efforts the Plan guided $300 millions of public investments along the lines set out in the report, giving Chicago a memorable city setting along the lakefront, a sense of place all its own.

Parenthetically, the success of the Chicago Plan inspired civic leaders elsewhere to produce plans for cities beautiful and sometimes cities functional. In a short time the initiative to plan was accepted as a formal government activity. A federal government task force produced model planning and zoning ordnances to encourage local communities to plan comprehensively. By 1928 the planning process was professionalized. The new respectability was crowned when Harvard University introduced graduated studies in planning at about the same time.

Some of the defects in these early approaches to community design, for all their comprehensiveness, are now clear. The social consequences of planning acts were not easily recognized. Planners prevented the worst environmental disasters from occurring but did little to stimulate or understand new forms of urban habitation. Some of the plans were static, inflexible, and insensitive to some of the fine-grain distinctions of community life that Geddes and others believed in; but many of these plans suffered because there were no reasonable ways to accumulate cheaply and evaluate easily the amounts of information needed to make realistic environmental-design decisions. In a chain of events too lengthy to show here, these shortcomings often led to the dehumanization of official planning, a disregard for a sense of place, and the almost criminal neglect such as Thabit's fine critique of the official plan for Cooper Square sadly revealed.

In the past decade several things have happened to modify the techniques of planning, to encourage community participation, and again view comprehensiveness as a useful context for environmental design. Research and application of program, performance, and perception techniques have been described as manifestations of rekindled interest; the ideas of community participation and microplanning are of equal consequence. These concepts have been supported by the potential of scientific management to organize the process of planning and the advent of the computer to analyze and present clearly extraordinary amounts of data. Though use of these tools is not widespread, it has been successful.

The assets and implications of computer science for environmental design now go far beyond data analysis and can be quickly appreciated. Through a computer program called OTOROL, perspectives and other drawings can be produced of any number of views of an object in space, at any scale, and as seen from any point in space. By producing a series of drawings, the computer can simulate the visual effect of moving around or through a building or group of buildings. The same program can produce orthogonal projections as well as two-dimensional plans and elevations. Thanks to fast, accurate, and inexpensive three-dimensional graphic portrayals of objects in space, environmental design alternatives can be pretested and costs and benefits evaluated. Furthermore, computer programs can be written so that statistical information

may be displayed in the form of charts and maps. As changes are made in the statistical inputs, subsequent changes can be printed on the maps. Mathematical models of environmental designs can be constructed and manipulated. Some researchers believe it is now possible to devise a machine with which the simulation of a proposed environment can be recorded and felt as if it were a real-life experience and modifications in design concepts made accordingly.

These are unquestionable contributions to environmental design. There are other aspects of the computer age however that are less certain in value and more deserving of comment and concern.

The preparation of a comprehensive plan for metropolitan Kansas City, Missouri, calls for a five-year study (1967-1972) and an expenditure of $2.5 million. The computer data bank will contain forty-seven items of information on every parcel of land in seven counties containing a total population of over a million people. The planning methodology consists of over a hundred different steps and requires 132 pages of descriptive text to explain the process. Since the eventual result is a composite of many different plans, each piece of the work has to proceed in an orderly and economic fashion. To control and manage the study, a critical-path work system has to be used (PERT) so that various subelements can be completed on time and in sufficient depth to allow the next stage of the work to proceed. The process is routinized and rationalized.

In this kind of planning, aesthetic contemplation, the nightmares and dreams of dawn which are preludes to design, sketchy ideas, impressions, the uncertainty that precedes commitment, all the vicissitudes and energies of art seem alien to the computer process. In the computer age at hand, time,

budget, and schedule would seem to dominate—the triumph of process over product. And there are problems other than aesthetic ones.

Few observers of government and planning deny the utility of the computer and scientific management in resolving environmental problems; but many, like Don K. Price, raise the question of how to avoid having the new techniques diminish leadership roles in translating technical studies into political values, policies, and in turn feasible programs for action. "The real danger," Price says, "is a body of techniques that can be mastered only by a highly educated and highly professionalized corps of officials, who will be content to serve only if they can make impact on the major problems of the age; and we are by no means prepared to create or maintain such a corps."

On matters of environmental design, a sense of place has a special role in the judicious utilization of computer technologies. A pictorial representation of the future can offer a statement of values in a world of facts. Pounds of statistics and yards of data can be reduced to essential issues: *What options are implicit among the choices presented?* A sense of place can be discerned because it exists, but it also can be created. Skillful graphics—environmental images—can describe options and choices in terms suitable for fluent public response and in turn political action.

The importance of a sense of place as a planning tool is strengthened further as one observes how environmental designs were used to plan the city which many said could never be planned.

CHAPTER XIII

PERFECTING OUR STREET SYSTEM

In the early days of Chicago, as we have seen, press of business and lack of wealth made the people of the city intensely practical in their daily lives. Every public action proposed was subjected to scrutiny as to whether it would pay financially and

One of the first needs of the future city is a perfect street system. There must be enough streets to easily accommodate the traffic of the millions who are to live in the city. They must be wide enough to insure comfort in traversing them, and they must run in the right directions to enable the people to go from place to place quickly. We must realize that lifetimes are made up of minutes, and that to save minutes means to lengthen life. Thus we can

CHICAGO. Proposed Twelfth Street Improvement at its Intersections with Michigan Avenue and Ashland Avenue. It is planned to make this a Broad Business Street and not a Boulevard, as Shown in this Picture. The Proposed Railway Terminals are Shown Fronting on Twelfth Street at its Level, Which is Raised to Allow North and South Traffic to Flow Underneath. Access to the Street is Provided at Alternate Streets. The rise Presumably will begin at Wabash Avenue and end at Canal Street. At the Intersection of Twelfth and Canal Streets a Diagonal Thoroughfare is Shown Extending to the Proposed Civic Center. Between this Diagonal and the River is Shown the Beginning of the West Side Railroad Stations.
[Copyrighted by the Commercial Club.]

practically before it was undertaken. It has become, therefore, a habit among the people to give first importance to the matter of direct gain from contemplated changes. It was this spirit which ruled in the minds of the architects in working out methods of perfecting the street system for the Chicago of the future.

justify the spending of millions of dollars today if it means saving time for millions of people in years and centuries to come.

In all cities which have grown up without being governed by a proper plan, it has always been necessary to make large expenditures of money and labor to correct the error of planless building. We have

Wacker's Manual Of The Plan Of Chicago Typical page of an influential publication used for over forty years in the Chicago school system in support of the Plan of Chicago. *Source: Wacker's Manual Of The Plan Of Chicago, Walter Dwight Moody, Chicago, Illinois, 1912*

THE FOUR ALTERNATIVES

	CENTERS CONCEPT	DISPERSION CONCEPT	CORRIDORS CONCEPT	LOW-DENSITY CONCEPT
HOUSING CHOICE	Emphasis on low & high densities	Emphasis on medium density	Emphasis on all density types	Emphasis on Low density
EMPLOYMENT DISTRIBUTION	Large centers of employment	Dispersed employment	Employment concentrated in central core and along transit corridors	Continuation of existing employment patterns
OPEN SPACE	Large scale open areas	Use of existing publicly owned land & creation of new small parks	Large scale open areas	Use of existing publicly owned land & creation of new small parks
MOVEMENT	Both private car and public transit	Private transportation	Public and private transportation, emphasis on public transit	Private and public transportation

Alternative Choices For The Design Of Los Angeles, 1967 *Drawing: Los Angeles City Planning Department*

226

COPE, COP, OR OPT

All the conflicts, contradictions, disparities, and despairs in contemporary environmental design can be summed up in two words: Los Angeles. A county of seven million people, a city of three million, Los Angeles possesses a beautiful site, a unique climate, and a complete dependence on the automobile for its Angeleno style of life.

By reputation the Angeleno is either very young or very old, eccentric or exuberant. Statistically, the Angeleno is Mr. Average: Caucasian, Protestant, thirty-two years old, married, has a high-school diploma, a white-collar job, and an annual income of $7,000 a year. The Angeleno's city is larger than the combined land areas of St. Louis, Cleveland, Minneapolis, Milwaukee, Boston, Pittsburgh, San Francisco, and Manhattan. Every acre of this 450 square miles is shaped by life in the sun—and life behind the wheel. Most people live in single-family houses, with generous back yards and patios which they use obsessively when not driving ten to one hundred miles a day for work and pleasure. One could live in the auto from womb to tomb. Los Angeles had the first drive-in clinic, church, bank, theater, and funeral home. It is the first drive-in city.

Two out of three acres in Los Angeles City were undeveloped in 1940. Today, less than a third of the land remains unoccupied, and most of this is mountainous terrain. Low densities and widespread urbanization were made possible in Los Angeles by extensive freeways, a mile of high-speed expressway for every square mile of land. To some observers the resulting environment is a shattering experience. Harrison Salisbury's reaction (1958) is not untypical:

Here, nestled under its blanket of smog, girdled by bands of freeways, its core eviscerated by concrete strips and asphalt fields, its circulatory arteries pumping away without focus, lies the prototype of Gasopolis, the rubberwheeled living region of the future. When Lincoln Steffens went to the Soviet Union just after the Bolshevik Revolution, he proclaimed, 'I have seen the future—and it works.' Today's visitor to Los Angeles might paraphrase Steffens and say, 'I have seen the future—and it doesn't work.'

Ten years have passed since Salisbury's comments. His viewpoint is still shared by many, but Los Angeles continues to grow. Twenty thousand people migrate each year, and the natural increase (births over deaths) adds about thirty thousand more. Every week a thousand people arrive one way or another in the city and a hundred previously unoccupied acres of land are taken from the fast diminishing stock. Significantly, the urbanization outside the city limits is growing at an even faster rate. The obvious solution of adding more land to the city is severely constrained by existing urban growth surrounding the borders as well as undevelopable land. The economic problems typical of all central cities politically bars voluntary association with the city's suburbs. Adjacent communities are willing to share in the assets that make Los Angeles the nation's second wealthiest urban place, but no one wants the poor, the disadvantaged, or the minority groups.

Within a future close enough to stir up immediate anxieties and uncertainties about environmental conditions, Los Angeles City is expected to add two million to its population, the county around it about three million more. Spread over an area a third the size of Rhode Island, the city faces development problems more extensive than any one of thirty-eight of the fifty states. Not only must Los Angeles build for expected population growth, it must simultaneously replace extensive sections of the city—jerry-built and socially gerrymandered, to a degree peculiar to the city's history.

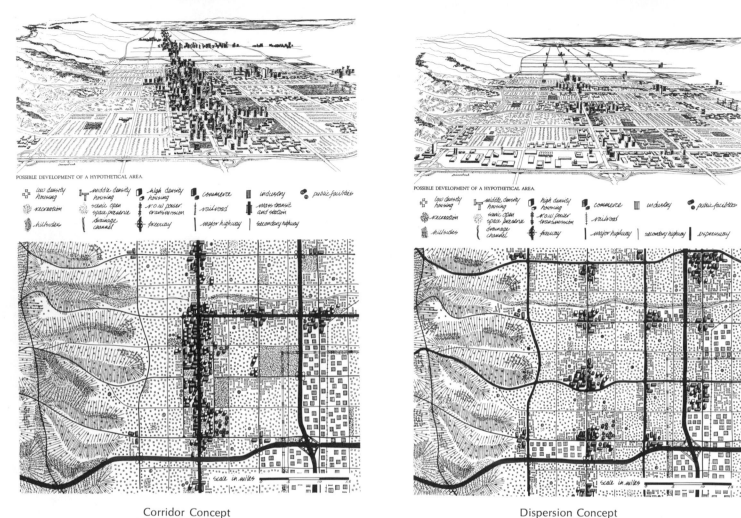

POSSIBLE DEVELOPMENT OF A HYPOTHETICAL AREA.

Corridor Concept

POSSIBLE DEVELOPMENT OF A HYPOTHETICAL AREA.

Dispersion Concept

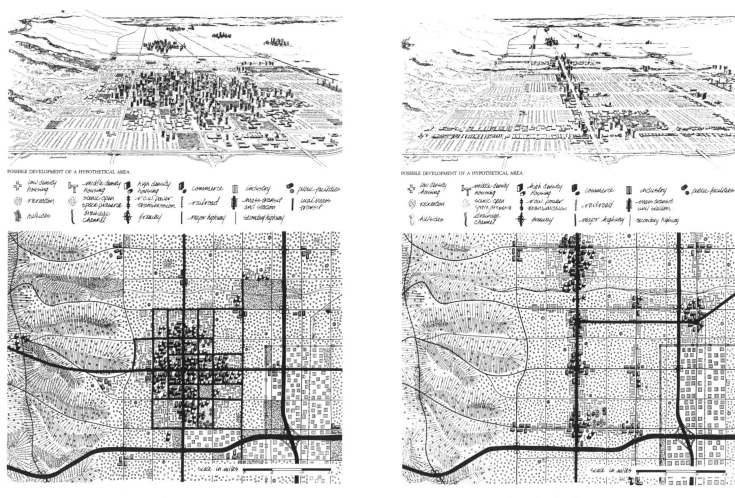

Center Concept

Low-density Concept

IMAGES OF THE CITY

Source: Figures on pages 228-229, Los Angeles City Planning Department

Predictably, the chéap subdivisions built after the close of World War II are beginning to fall apart. Hastily erected on crowded lots, this housing no longer serves the income groups for which it was originally constructed and is not substantial enough to be maintained and passed on to other people. Blight is also evident in the neighborhoods developed in the 1920's and 1930's, and obsolescence continues throughout the city. *De facto* segregation of ethnic and economic groups is reflected in residential patterns. Subdivisions and neighborhoods tend to have a single type of house on a similar type of lot for a single kind of family. In Los Angeles, buyers, and builders believe that mixtures of housing types and income groups would result in low values in residential properties and difficulties in marketing them. Since the Angeleno moves once every three to six years, this attitude is difficult to overcome. In many sections of the city it is legally supported by deed restrictions and zoning ordinances that sustain uniformity in design and community life. With the exception of luxury accommodations, apartment-house development has been exploitative—designed for maximum densities, to the lowest quality allowed by building regulations, and with little concern for amenities, open space, or other environmental qualities.

There are also other qualitative problems to be faced: visual and environmental. Despite many natural advantages, the townscape of Los Angeles is as monotonous as it is unimpressive. Local planners note that "signs and power lines clutter the streets and skyline; major streets are solidly lined with drab commercial buildings and architecturally undistinguished apartments; the hills are scarred by grading. Growth proceeds without guidelines for beauty."

Air pollution ranks as the chief environmental threat. The topography and weather are such that catastrophic pollution is an increasing danger. Both automobiles and industry contribute to the smog. Active measures are being taken to impose air-pollution standards on new industry and to force their location to areas where atmospheric conditions will dissipate the pollutants. A high-capacity mass-transit system would reduce emissions from motor vehicles and make further sense because of the extraordinarily high costs of adding new freeways and investing in traffic-control devices for moving larger numbers of vehicles than those now on the road. In addition, a mass-transit solution would ease a serious social problem for sizable numbers of poor people who are dependent on public transit for their journey to work, job opportunities, social mobility, and an improved style of life. However, mass transit has been under study in Los Angeles for forty years; and under present trends, city-development patterns, existing techniques for funding, and present-day planning assumptions, the construction of a workable system is highly uncertain.

Los Angeles has these choices in finding solutions to its environmental problems:

It can continue to cope with present trends and hope that some new technology or event will lead to a better environment.

It can "cop out" and leave the solutions to a later generation.

It can opt for changes now.

Los Angeles has decided on the latter course and has made some promising beginnings for finding what Christopher Rand has called "the ultimate city of our age." If it proves nothing else, the current quest for quality in Los Angeles may answer this question: Can the courtship of man and motor have a happy ending?

University of California Professor Ralph Knowles believes that one solution to the environmental problems of Los Angeles is a gigantic earth-moving project. Hilly areas could be reshaped to utilize the natural forces of wind and rain in order to make people more comfortable at less cost than that of indoor mechanical contrivances. Under Knowles' direction, laboratory testing of land and building shapes (in wind tunnels) has reached the point where a full-scale demonstration is feasible. Without widespread political support, however, technological innovation of this and even lesser magnitude cannot succeed. Accordingly, officials responsible for planning Los Angeles' future have taken the stand that widespread community development and expression is the best way to recreate a distinctive city.

To avoid a style of life imposed by abstraction and to obtain a plan that represented the Angeleno's aspirations, a special pathfinding committee was organized in 1967 by citizen groups and the Planning Commission to find those goals and objectives which could be documented and subjected to widespread public review. Staff papers prepared for the Environmental Goals Committee and the ensuing discussions resulted in a comprehensive statement of facts and fancies—ranging from the individual Angeleno's view of himself (it is almost illegal to be a pedestrian in Los Angeles) to an accounting of the region's economic assets and liabilities (an over-dependence on defense contracts).

Through community debate carried on at a Center for Choice (set up in different sections of the city to review the Environmental Goals Committee's work) and a polling of a cross section of public opinion, the critical issues were isolated. The Angeleno's view of life in his city in 1967—politically, socially, economically—became clear.

The general consensus indicated that most of the existing land uses and activities (business, airports, harbor, major colleges and universities, parks and reservoirs) would continue in their present location and character. Streets and freeways would remain as shown on the existing master plan. Population and employment would grow, with blue-collar jobs decreasing and white-collar jobs rising in number. Vigorous attempts to attract new capital investment would continue by both public and private groups. All these conditions would be considered the permanent context for any new plan because of present heavy investment in them or because of the importance people attributed to them in creating the city's sense of place.

From this study of community consensus, it appeared that the elements that could be manipulated to change the physical environment were: industrial and commercial land uses, land for open space and recreation, mass-transportation lines, and the density and character of new housing. These elements could be alternatively arranged into different patterns so that the combination of fixed conditions and new construction would result in distinctive forms of environmental design at the city scale. Los Angeles could select among four for its future physical form. Rejection of all four could lead to economic chaos, natural catastrophe, and civil disorder.

The Planning Commission's visual interpretation of the four options were termed: Centers, Dispersion, Corridors, or Low Density. The Centers concept envisions about thirty high-density communities in which most of the new growth would be located, cities within the city. The Dispersion concept would continue the present form of the city but use row houses and town houses to shelter the increased population. Thus, the Angeleno style of life would be perpetuated. The Corridor design would place high-density housing and commerce along new transit and highway routes. Transit-station areas would be used as the site for higher densities. The Low-Density plan would guide existing trends and patterns along the lines of present development, but population growth would be limited to approximately eighty percent of the other schemes. Supposedly with growth limits, quality would follow through renewal and redevelopment.

As the accompanying illustrations indicate, each choice implies a different physical image of the city: high density versus low density, mass transportation versus individual transportation, the spreading city versus the concentrated city, high-density corridors versus high-density centers. Los Angeles planners recognize, however, that the real choice is not the selection of one concept over the other but the selection of an emphasis, for the eventual sense of place will not be dictated by a single design but by the general direction taken.

The future of Los Angeles could not be planned without knowledge of economic and social conditions. The data could not be explained and presented to the public as available options without interpretation of the environmental images, a sense of place. It is the drawing together of these two aspects of urban planning—the facts and the environmental values—that creates an art for our time, environmental design.

Ervay Street In The Late 1920's And In 1960 Although extensive urbanization has occurred, environmental improvements in Dallas have been limited because the existing street pattern has established the design structure of the city. Any sense of place is largely restricted to what can occur within the nineteenth-century street lines. *Photos: Dallas City Planning Commission*

CORE OF CENTRAL DISTRICT – 1958
STREETS CHANGES SINCE 1875

Dallas Street System The existing central business district superimposed on an 1875 map of Dallas, Texas. Also shown are the street changes since 1875. *Drawing: Dallas City Planning Commission*

QUALITY ENGENDERED

In establishing a sense of place, perhaps the most decisive design actions today are those which affect the space set aside for the automobile and the pedestrian. Freeway and street design, traffic segregation, parking, micro-movement systems, and the pedestrian precinct are essential elements in the environment and, when neglected, result in design disasters. No larger scheme can succeed if these common elements go undesigned or are set aside as minor influences in building or community development.

These elements are as important now as they were a century ago, as seen in a comparison of two city street systems and the environmental changes related to them. Paris did and Dallas couldn't. The history of these distinctive cities reveals how the network affects the quality of the environment.

Dallas's design legacy goes back to 1841. In the turmoil and legal confusion of the new Texas Republic, conflicting land grants were issued to John Grigsby and others authorizing settlement of what is now downtown Dallas. Simultaneously, without knowledge of the latter claims, John Neely Bryan explored the site, liked the location, stayed on, and laid out a town. Starting in 1845 he surveyed and plotted the land, sold lots, and recorded deeds. About one half his original plan—blocks two hundred feet square and streets eighty-four feet wide—remains as he designed them and now serve as the core of the city.

Hesitant to press his claim against others, Bryan delayed filing his town design until 1855. In the meantime the Grigsby interests were being defended in court with as many as three hundred separate property cases being heard at one time. Additionally, several dozen other parcels adjacent to Bryan's land were granted to various parties, many of whom were nonresidents and unaware of the ownership or developments occurring on adjoining tracts.

Needless to say, this chaotic situation had serious consequences for street and site design. Accesses to buildings were located parallel to eventual property lines. Because of legal rights and law suits, little could be done to coordinate planning among the parcels. Many tracts were laid out so that streets ran at angles to each other, failed to intersect, or to have the same width in their rights-of-way. An official map (1875) shows how the uncoordinated and disjointed arrangement of streets confused and compromised the earliest development of Dallas and has affected the city's design ever since.

Though the city has grown from 2.68 square miles to almost three hundred square miles (through 606 separate annexations), the central district has remained where Bryan and Grigsby laid out their land. Dallas is now a metropolitan center with over one million people. The downtown area has changed with time, the density of buildings having risen at a geometric rate; but the physical pattern of circulation, building orientation, access, and services remains as chaotic as it was eighty years ago.

To establish a civic design tradition in the face of such an unhappy history of conflicting land claims and resulting confused street patterns would have required bold strokes on the part of an enlightened, concerned city leadership backed by an enlightened and concerned public, both of which were and are still lacking in Dallas with respect to environmental design. While several magnificient buildings have been erected in the past decade, their contribution to the designed environment has been negligible— pleasant single notes, but no symphony. Thus Dallas may continue to have no design future unless the local streets are abandoned to

Facade de l'Opéra.

La rue Turbigo.

Hausmann's Contribution To a large extent, the present-day ambience of Paris was launched when Hausmann constructed his street and boulevard system. These late nineteenth-century views show how successful he was in quickly establishing a sense of place. *Drawings: Views Of Paris, Paris, 1892*

Boulevard de Strasbourg.

slow-moving motor vehicles, and an environment solely for pedestrians is introduced at another level above—perhaps skywalks and platforms for pedestrians at the second and third stories. In this fashion the irregular disposition of buildings, which promises visual liveliness at their base when viewed at a distance (not unlike the approach to the Duomo in Florence) but disappoints at close range (again due to the presence of the automobile), might become a positive design force. Townscape could still replace grimscape in Dallas.

Like all the arts, streets-design may be categorized by opposing characteristics: the romantic versus the classical, informal versus formal, the Italianate versus the Parisian. The later style can be summed up in the prodigious work of Georges Haussmann, who took a dirty, dangerous, Medieval city and in seventeen years transformed it into a model of European town planning and a progenitor of designs whose influence is still strong and pertinent.

Haussmann's work was as bold as it was ruthless. His professed objectives were to eliminate street riots and crime in the slums, improve living conditions, prepare for increasing traffic, and construct a beautiful, monumental world capital. The degree of success is debatable, and critical evaluations of his work are no less passionate now than they were in his own day when "Osman Pascha" was mocked with vicious satire. As Brian Chapman has written: "The person, the epoch, the city are run through with contradictions and are fair game to the ironic fates of history."

Beginning in 1850, Haussmann neatly quartered Paris with east-west and north-south boulevards linking older streets with new construction. Entire neighborhoods were displaced. Slums in the heart of the old Latin Quarter were torn out. Parts of the Luxembourg Gardens were removed. The Seine was bridged in seven more places. New highways connected the railroad termini with industrial districts in all four compass directions. A circular boulevard connected the radials, and from their intersections, highways reached out into the suburbs. The streets were generously proportioned. Mobs could not barricade the routes nor traffic congest the rights-of-way. The network pleased industrialists, administrators, the army, and chiefs of police.

In addition, Haussmann's advanced planning was neither timid nor uncertain. Several of his streets now carry up to eighty thousand vehicles a day—designed at a time when twelve thousand vehicles was a high traffic count. Comparable advanced planning today in North America would result in thirty-six-lane highways.

Further, Haussmann's planning was as broad geographically as it was farsighted. Administratively, he extended boundaries of the city to include eleven neighboring communes, so that his plan would encompass the "natural" area; thus his plans had a broad context. Finding it difficult to get an unencumbered field of vision for survey measurements, he ordered timber towers to be erected higher than the tallest rooftops, from which his surveyors could accurately plot maps not only of Paris but also of its environs. His desire for efficiency and action, expressed in Draconian methods, led to the first comprehensive examination of a contemporary city.

What did this large-scale planning, demolition and removal, and building effort really give Paris? First, in a century of ugly industrial urbanization, Haussmann showed how a large metropolitan city could be beautiful, demonstrating the special amenities which come about when architecture of the street and architecture of adjacent buildings complement one another. While his rigid, opulent, imperious styles were arrogantly conceived and hardly models for emulation in democratic societies today, the buildings did reflect civic pride and were honest attempts to capture the spirit of the day. True, not all that was done can stand the scrutiny of time. The application of the formula in some Parisian quarters produced dull and montonous townscapes. In many sectors the fifty thousand trees, planted with special machinery invented to move large specimens, are the most memorable civic design qualities. But Paris remains a beautiful city.

Second, though of no less importance, Haussmann accomplished social purpose through political power, for while the *grands boulevards* were set in place, so too, were parks, central food markets, a drainage and sewer system, and a complete replacement of the antiquated and unhygienic Parisian water supply. Cholera was eliminated, and death rates declined.

What are the lessons for today? As in Dallas, do nothing and hopelessly constrain later generations? Take a chance on a design autocrat: Find our own Haussmann? The answer does not lie somewhere in between but rather involves ability to capitalize on inevitabilities and to use a known course of action with vision, imagination and a sense of social responsibility.

FREEWAYS AND EXPRESSWAYS

Parallel analogies can be drawn between Haussmann's construction of major streets and boulevards in Paris and aspects of current American urban expressway planning. Both are seemingly ruthless actions aimed at giving public benefits, without full regard to other consequences. But where Haussmann had to answer to a single man, highway planners today face the dilemma of designing for many clients. Increasingly, that clientele is forcing those who are making highway decisions to consider not just the obvious requirements for right-of-way locations, but also the environmental and community development impact the proposed highways may have on adjacent communities and the larger urban area in which they are situated.

Lines of transport into and within the city established the design skeleton of the nineteenth-century city: The freeway and expressway through and around the city have exerted a similar influence today. They are rather recent innovations, built to move large volumes of traffic at high speeds through and around congested urban areas. They differ from large local streets in that access points are limited. Specially designed ramps allow easy entrance and exit. Median strips divide the opposing traffic lanes. Overpasses and underpasses produce free-flowing movement at intersections. Pedestrians, parking, and nonmotor vehicles are barred from the rights-of-way.

The argument for freeway construction lists numerous benefits from these roads for a community: efficient distribution of traffic, immediate access to specialized centers of activity, economic stimulation to business and industry, community revitalization. The latter values accrue by the opening up of new land for development and the facilitating of relocation from the areas of the central city in decline. Thus manufacturers can acquire space for modern production lines and transport, and as noted earlier, job opportunities bring residential communities. The outward drift goes on in locked step. The migration of the middle class to job opportunities and new housing in turn creates pressure on the central city to become competitive with the outlying areas and thus develop itself to meet the demands imposed by the automobile: The freeway fosters a fresh and rapid cycle and expanding circle of urban development whenever and wherever it is constructed.

In addition, freeways are safer than the normal surface network: They have three times fewer accidents and fatalities. Initially, they also afford certain economies over the ordinary roads: shorter travel time and savings in automobile fuel and maintenance costs through less frequent shifting of gears and braking. For commercial vehicles, wage costs per mile traveled are substantially reduced.

The drawbacks of freeways are, however, equally significant. The new roads are expensive to build. The Bureau of Public Roads estimates that construction costs are $500,000 a mile in rural areas and $3.5 million a mile in urban places. These figures are averages. Land acquisition costs and construction for the Inner Belt in Boston, for example, have been estimated to be $27 million a mile.

Decision-making for freeways has two liabilities: first of all, in layers of computer printouts, difficult for the average citizen to read and the body politic to refute. The very length of the road and the complications of engineering administration force the design and location decisions to be made somewhere other than at the local level. The fine-grain distinctions in community development that are desirable for local and varied solutions are sometimes lost in freeway planning.

Secondly, the overriding considerations to date in making expressway decisions have not been safety or environmental impact but an estimate of the value of the travel time saved, and here the freeway advocates are most vulnerable. Average journeys to work on the East Coast, for example, range from 7.8 miles in the Baltimore area and 7.0 miles in the New York region to 4.4 miles in Metropolitan Philadelphia. For the average commuter, the savings in travel time from door to door on short runs using expressways may seem marginal for the money expended.

Freeways are increasingly criticized for yet other reasons. They injure the adjacent land uses because of noise and air pollution. Further, they have usually illogically divided communities and have caused callous removal of homes and industries along the rights-of-way during construction; yet in the end they are sometimes as congested as the networks they were intended to improve. The extent of the criticism may be evident in strong local protest to existing interstate highway designs in twenty-five American cities— a protestation strong enough to delay the Federal highway program and alarm Congress.

How effective is the protest? On one hand, a shrewd observer of government recently said that "official attitudes toward highway construction have profoundly but almost silently changed. The era when government viewed (them) as an unquestionable good is over." On the other hand, a century ago in his *History of England*, Macauley ranked road building after the alphabet and the printing press in importance among inventions, and Adam Smith asserted that "highway construction was the greatest of all improvements." And barring a revolution in transport, it looks as though freeways and expressways will continue to be built, including most of the twenty-five hundred urban miles remaining in the interstate system. Given the current growth in population, automobile ownership, urbanization, and the exercise of individual options in the market place to buy and operate automobiles, no substantial reduction in known freeway construction can be anticipated or is necessarily warranted. Freeways are essential to an improved environment, and though their disadvantages and negative effects to date have been substantial, they should not be consid-

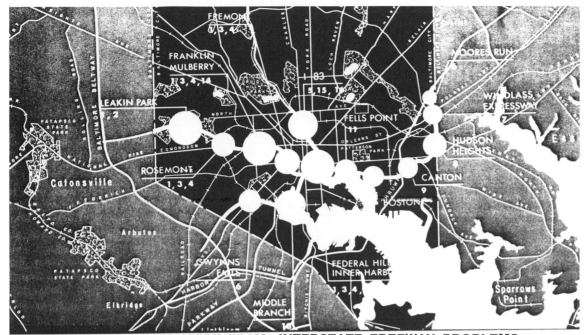

URBAN DESIGN CONCEPT TEAM: INTERSTATE FREEWAY PROBLEMS

ENVIRONMENTAL

1. Neighborhood disruption.
2. Park use disruption.
3. Dislocation of people without guaranteed relocation.
4. Bisection of neighborhood—a barrier to local movement.
5. Limitation of area developments.
6. Potential park open space destroyed.
7. Fragmentation of land development opportunities.
8. Neighborhood impact.
9. Conflict with industrial area.

10. Waterfront amenity disrupted.
11. Historical area negative impact.
12. Impact of highway on inner harbor project.
13. Potential conflict in area ripe for development.

TRANSPORTATION

14. Local street congestion.
15. Inadequate distributor.
16. Unsafe turn overloads existing highway.
17. Local traffic-expressway conflict.
18. Confusing, unsafe, overloaded.

Team Problems A graphic description of the environmental and transportation problems encountered by a multi-disciplinary team attempting to resolve location and design issues related to a proposed interstate freeway in Baltimore, Maryland. *Drawing: DeLeuw, Cather & Company, Engineers*

Interface Study Team It is now generally recognized that freeway and expressway designs cannot be produced without an interdisciplinary team. The Interface Study Team diagram suggests the basic skills that should be present and at work at all stages of the study. Since there are no shortcuts to the collaborative process, the advantages of interchange and dialogue among people who recognize a common problem but see the solution in different ways can only be realized if a core group of specialists is constantly at work. Obviously, detailed studies are carried out by specialists. The Interface Study Team thus has these special responsibilities: defining the problems, especially those which require multi-discipline staff work; identifying the kinds of solutions expected from staff-level work; insuring representational views from the several disciplines in all phases of the work; formulating the policies of the staff work; and reviewing the results and recommendations that are produced by the staff.

Symbolically, the Interface Study Team is thus seen as a cogwheel that intermeshes all efforts towards one goal: a comprehensive effort by a balanced study team, constructively involved with the formal clients and the informal communities of interest. *Reprinted from: A Proposal For The Inter-Corridor Connector Study, 1968, Sverdrup & Parcel and Associates, Inc., Engineers and Architects; Dober, Paddock, Upton and Associates, Inc., Planning Consultants*

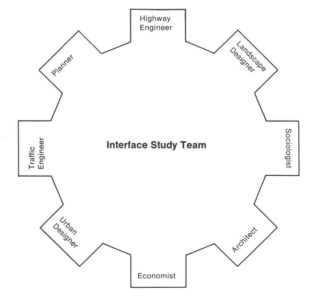

Prudential Center, Boston, Massachusetts City auditorium, high-rise office building, and hotel constructed over the New Haven Railroad and Massachusetts Turnpike rights-of-way. *Architect: Charles Luckman Associates. Photo: Prudential Center*

Proposed Treatment Of Air-Rights Over The Inner Belt In Cambridge, Massachusetts, 1965 *Designers: The Architects Collaborative. Photo: Robert D. Harvey*

AIR-RIGHTS DEVELOPMENTS

Office, Hotel and Restaurant Complex, Newton Corners, Massachusetts, 1969. *Architects: Davies and Wolf*

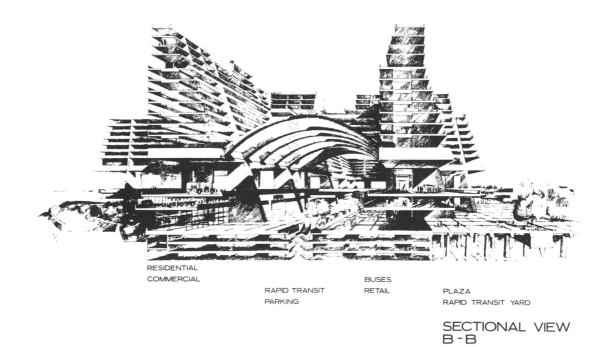

RESIDENTIAL

COMMERCIAL

RAPID TRANSIT   BUSES

PARKING   RETAIL   PLAZA

          RAPID TRANSIT YARD

SECTIONAL VIEW
B - B

Alewife Brook Park, Cambridge, Massachusetts Proposed air-rights development at the intersection of State Highway Route Two and the extension of the Dorchester-Harvard Square subway line. *Planners: Dober, Paddock, Upton and Associates, Inc., And The Architects Design Group*

ered beyond remedy. The urgent issue is not whether freeways should be part of our future but how to make them better. To find answers to these questions may require a moratorium on expressway construction in some localities, and increase in investment in planning and design, and an overhaul of administrative techniques for integrating highway, transportation, and urban planning. In any case, if freeways are to be acceptable, they must be constructed in sympathy with local planning, in the context of natural and man-made amenities and in addition to, rather than at, the price of alternative forms of transportation.

In future designs the voices of reason will be the voices of balance and comprehensiveness. The need for movement networks may be simultaneously minimized as the ability to move is maximized. A more conscious effort may be made to relate circulation improvements to styles of life in which environmental design considerations will be paramount. To reiterate, the freeways will be built; the qualities of the designs they generate can change.

There are no reasons why freeways cannot be used as positive design elements for eliminating obsolete land uses and structures and reshaping the form of the city as long as planners take concomitant action to solve the social and economic problems created by displacement for freeway construction. The rights-of-way may be dragged below surface or underground and entire communities rearranged and built with land reserved for mass transportation, parking, and in-town employment centers. Spine City, as it might be called, could turn an obtrusive engineering artifact into Motopia. Since immense expenditures would be needed to humanize the corridor, one might assume that Spine City would be limited to those segments of the freeway which ran through dense and strategically located urban areas.

Other types of freeway projects are on drawing boards all over the country. The Baltimore Harbor Freeway is now being planned by a consortium of designers, planners, and others in anticipation that a task-force approach may best fit the expressway into the urban environment. In Boston, air-rights development studies for freeways have been completed by a number of private and public groups. Three of these lie within five miles of each other. Their execution would dramatically mark a departure from the indifferent and unimaginative planning that has previously torn up sizable parts of the region.

In addition to multiple-use planning, which capitalizes on the presence of the freeway, consistent attention to the design of the right-of-way itself can give the freeway a positive quality. Proportions, scale, harmony, texture, color, and rhythm of all the elements that make up the expressway can be orchestrated as an art form giving pleasure in itself. The idea of design in motion, noted earlier, should not be foreign to the engineer and right-of-way planner. The body of architectural and design experience needs only to be tapped to transform an everyday object into an amenity.

Although the expressway is the most imposing element of city circulation, let it be reiterated that there should be no satisfaction in self or community until comparable design attention is given to all the urban streets and their design quality, for they may cleave through the townscape with as negative an effect as any badly planned expressway. They are not beyond improvement in both detail and context; especially pavement life—the pedestrian's environment. As the expressway is fitted in, the side-street panorama should be uncluttered. The perfected environment may be beyond reach, but these things can be done: Preserve and enhance what can and should be saved from insensitive highway and street design; build the new networks with vision and grandeur; plan so that the journeys along them are swift and safe; and at the points of destination, separate vehicles from the pedestrian's environment.

The Adelphi An early scheme combining air-rights development and traffic separation. The landing along the Thames served as a break in transportation. Warehousing and the movement of goods was kept to the lower level; and on the streets above, an extensive housing complex was constructed free of the commercial activities which occurred below. *Drawing: Greater London Council*

TRAFFIC SEGREGATION

The desirability of separating pedestrian traffic from vehicular traffic has been recognized since ancient times. Daily life in ancient Rome was made least hectic when Julius Caesar proscribed heavy, horse-drawn vehicles from entering the "limits of continuous habitation" from dawn to dusk. Reconstruction of Roman cities indicates that the location and design of streets took into consideration the nature of traffic and pedestrian activity. In addition the arcade afforded pedestrians all-weather protection from street traffic by extending the building over the sidewalk.

Venice is the earliest and most continuous example of successful traffic segregation. Canals handle transport, and the land largely remains a pedestrian precinct. The absence of vehicular traffic is so noticeable that Venice and other canal cities, such as Amsterdam and parts of Copenhagen, draw tourists from all over the world to enjoy the ambience and exciting and unusual environment.

Ideas for separating pedestrian and vehicular circulation have intrigued designers of both utopian and practical bent. In one of his sketchbooks Leonardo da Vinci shows a design for a multilevel road that had both functional and social purpose. He noted that the upper level was intended for gentlemen and the lower for service.

The enterprising Adams family (the Georgian architects) erected the first modern project that used traffic segregation as a device for designing a handsome and profitable real estate development (1771). Using sloping land and a length of three blocks along the Thames, they designed a central building flanked on both sides by two streets of houses, which ran to the Strand where they overlooked the embankment below. A substructure of warehouses built into the embankment with a wharf adjacent supported the whole economically. Thus commerce was assigned to the ground levels, and the residential precinct sat above, separate and removed from the bustle and heavy commercial traffic but enjoying a commanding prospect of the river and London life.

As one might expect, the introduction of steam-powered vehicles and the accompanying rise in urban densities gave incentive for a number of ideas for handling different kinds of traffic in distinct bridges and tunnels. These helped keep horses and trains and people separate from each other but did little to improve the appearance and environmental conditions of the areas through which the trains would pass. The costs of elevated and depressed rights-of-way were enormous, so sizable swaths of industrial slums were created simply because segregation of railroad traffic from the general environment and middle-to-upper-class surroundings could not be accomplished. A number of patented designs were filed for controlling and separating traffic around the railroad stations, including underground streets for carts and drays, and bridges and platforms for pedestrian movement; but few were built, and the cityscape suffered accordingly.

The design of middle-class Victorian subdivisions, using principles of traffic segregation, departed from earlier street patterns; and so pervasive did the design become that one overlooks the fact that it is of fairly recent origin. House lots were laid out with landscaped streets and parks on the front sides and narrow service roads to the rear for tradesmen and servants and service traffic. As in the street plan for Back Bay Boston, this arrangement reflected a style of social life as well as traffic control, determining the location of interior rooms as well as exterior architecture and site plan.

A high point in civic design and traffic segregation was reached in 1856, when Olmsted and Vaux designed Central Park so that the pedestrians and horses and different speeds of traffic would have separate paths and roads. The designer's idea of assigning to bridges and tunnels different kinds of traffic was not widely applied, however. The typical central street scene was a frightening mixture of people, bicycles, horse carriages, and trolley cars. The automobile replaced the latter around 1920 with appreciable change in the nature of the congestion but not in the degree and its effect on the environment. As early as 1924, prescient architects such as Le Corbusier diagrammed solutions that separated automobile traffice from other aspects of community life.

In 1928, Clarence Stein and Henry Wright pooled several simple ideas in a plan for Radburn, New Jersey, a residential community using the neighborhood concept for "life in spite of the automobile." (See page 47) Buildings faced in two directions, the front side inward to a landscaped setting, the back outwards towards the roads and service areas. Row housing was arranged in superblocks whose cores contained parks and other open spaces. The superblocks were bounded by a road and utility network which gave access to the service side of the house. Pedestrians and vehicles were separated: An internal footpath connected the houses with community-wide facilities and continued under the roads by tunnels where necessary.

Only a small portion of the Radburn scheme was built, but it has enormous worldwide influence, especially in Europe where several generations of town planners have enlarged on the idea in applying it to the design of new towns and satellite cities. In the United States the Federal Government applied the Radburn concept to several resettlement communities during the Depression, although those designs are now more memorable for their landscape architecture than their circulation solutions. Baldwin Hills (Los Angeles, 1941), on which Stein was consulted, was a more successful lineal descendant, providing sites for 627 units on an eighty-acre landscaped site. The significant aspects of the plan have been its capacity to absorb minor changes in architectural fashion, the excellent landscape which unifies the design, and the accommodations made for parking. Each housing unit now has space for three cars per family with little loss of the original amenity.

The Radburn concept has had its critics, because it has proved easy in practice for such street design to end up a nondescript mixture of cul-de-sacs, parking, clothesline, and service yards. Interior walkways, landscape, lighting, and the additional utility runs that result can be costly. Visitors may find the site layout not as attractive as approaching the house directly from the street. Unless fencing and screening are introduced, there may be some loss of privacy along the front and rear lots. However, all these problems can be overcome.

The Radburn idea of traffic separation was not commonly used in the housing boom after World War II. Paul Ritter notes that the opposition to the idea was irrational. It meant "the loss of human lives and limbs and robbed humanity of much pleasure. The tragedy of bad planning is its continuing effect." During recent years market forces have begun to play a role in reviving the Radburn idea in America and in new-town development in Europe. Cluster housing and town houses have proven to be economical uses of land, especially helpful in preserving open spaces. Safety and security now motivate home buyers. The horizontal or ground-level separation of traffic has become an important feature in subdivision design and sales.

The widespread demand for traffic segregation is not confined to contemporary subdivisions and housing development. Many large-scale environments are absolutely dependent on separating the pedestrian area from the motor vehicle. These environments include shopping centers, college and university campuses, and central-city redevelopment.

Among contemporary American environmental prototypes, the shopping center is important because of size and number. It may be the most significant innovation in retailing since the invention of money. It has changed the face of downtown and the shopping habits of most Americans.

The emergence of the shopping center parallels shifts in retailing practices and styles of urban life. The first chain-store groceries were situated in city neighborhoods, occupied about five thousand square feet of land, mostly on one floor, and offered about five hundred different items. Though they were often as big as the old grocery store, they still offered personal service, sometimes credit, and a familial rapport between the storekeeper and his customers. The keen competition in the food market today, however, requires selling space five times as large as the old neighborhood store for ten times as many items. Capital requirements and management skills alone have eliminated many local grocers. A degree of personal contact has been lost in favor of efficiency, low prices, and a wide range of goods.

The size of the typical food store today in turn requires a large economic base for support: The contemporary supermarket needs about five thousand families or a population of around twenty thousand for economical operations. These customers are drawn from a greater distance than the corner grocer's and arrive by automobile rather than by foot. Parking has to be provided. At this scale of development the supermarket does not fit easily into a residential neighborhood, so that a special district must be set aside, usually with good automobile access and circulation. This in turn makes it desirable for smaller stores, specialty shops, and services to gather nearby, taking advantage of the built-in lure of the food store (everybody has to eat) and the traffic network. The community shopping center is thus brought into being: ten to twenty acres of land, 100,000 to 200,000 square feet of building, with 2.2 square feet of parking space for every square foot of gross leasable retail space.

Unquestionably, the separation of pedestrian traffic from other forms of circulation is one of the shopping center's appealing characteristics. Victor Gruen points out that noises, fumes, odors, and the constant movement of vehicles destroy or certainly diminish "the attainment of a psychological climate conducive to shopping enjoyment." Yet dangers, confusion, and distraction come with the traffic generated by the shopper and the stores supporting services. Whether in a large center or small, the latter accounts for a good deal of motion and bustle. Garbage and trash are collected. Goods and fixtures are picked up and delivered. Furthermore, centers are not static developments but constantly changing as new merchandising ideas are introduced, renovations and alterations are carried out, and tenants come and go. All these activities are part of the everyday business of retailing.

Careful designs in shopping centers shield the customer from these intrusions. In the most profitable centers the customer is brought comfortably into a large parking area which encloses a compact group of stores. He then walks along a well-signed and well-lit route to the entrances to the center. At that point the automobile world is left behind. Sometimes the entire center is enclosed for year-round protection from the weather. Whether it is enclosed or not, there are conscious attempts to divide the overall interior space into smaller, pleasantly proportioned parts filled with sculpture, plant material, information signs, and occasionally concerts and public events. The strategically arranged interior environments expose all the stores and departments to the maximum amount of foot traffic. Carefully located sales areas, displays, rest rooms, benches, landscape, encourage movement from place to place within the center. Walking is meant to be enjoyable. Shopping becomes fun, and the customer stays, spends, comes back often. The automobile is entirely excluded; service traffic is kept underground or to the periphery. Colorful, stimulating, commercial, the shopping center is a positive case of how traffic segregation is effected and used to meet a single functional objective: selling goods and services.

The immediate environs of the typical shopping center are usually not as well-designed. The surrounding parking is as extensive as it is a stunning eyesore. Further, foot traffic from the adjacent districts is rarely encouraged. The large centers, of course, draw customers from as far as thirty miles away, so walk-in shopping is not expected. But at the neighborhood scale, "the-island-in-the-sea-of-parking" concept needs challenging. The opportunity exists for a design which has the functional advantages of the center and those of an old-fashioned street. A galleria, or mall-like design connecting two neighborhoods, with parking on either side of it, would seem an appropriate alternative, well worth exploring, especially in a new-town development.

In college and university planning, vehicular pedestrian separation is now a well-accepted design principle. About two new campuses a month open in North America to accommodate anywhere from several hundred to thirty thousand people in a micro-environment whose quality depends on fluent and safe circulation and the segregation of various forms of vehicular traffic from pedestrian traffic.

New college campuses show a strikingly provocative variety of solutions in terms of architectural style, density, capability for expansion, use of site, respect for climate, topography and landscape, as well as educational and social organization. Within this variety there are three basic forms: the traditional campus, the cluster campus, and the unistructure.

The unistructure is a continuous and consistent educational environment which reads in plan like a single building. The original design for the Massachusetts Institute of Technology is an early example. Scarborough College (Toronto), Simon Fraser University (Vancouver), Free University of Berlin, and Kingsborough Community College (Brooklyn) are recent variations on the same theme. The cluster campus is a device for allowing an institution to seem to become smaller as it grows larger. At the University of California/Santa Cruz, for example, the concept consists of twenty colleges offering largely an undergraduate curriculum each built around a discipline or academic emphasis such as fine arts, political science, or languages. The colleges are served by a set of central facilities and offices for general administration. The traditional campus design pattern consists of a central academic area within which all the major, common facilities are located and most of the undergraduate instructional buildings. Depending on the size and purpose of the institution, various quadrangles or precincts may be located nearby for student housing, graduate and professional schools, and recreation and sports facilities.

All three design concepts have similar circulation requirements. Direct and easy access from the outside community is needed for visitors and everyday traffic, both mass transportation and the individual vehicle. Traffic that reaches the campus must be efficiently distributed to parking areas as close to the prime destination point as possible. From there movement inside the campus is essentially by foot, sometimes by bicycle. The distances covered are in part set by the class schedule. To a reasonable extent, even the nonformal patterns of movement can be predicted. Accordingly, controls and designs can be imposed not only to eliminate the acoustical nuisances and visual intrusion of the motor vehicle but also to use traffic segregation as a functional objective in decisions regarding land-use locations and building sites. The easier it is to get from one part of the campus to another, the more efficient the campus design becomes.

The amount of land available for construction, density of buildings anticipated, climate, and environmental design objectives determine the degree of traffic segregation. Some campuses allow a mixture of slow-moving service and emergency vehicles to use the same routes as people. In contrast at the University of Illinois in Urbana, a separate bicycle-path system runs parallel to the pedestrian walk system. Kingsborough Community College combines vehicular access, service, and parking in a separate and distinct environmental zone. At Davis and Elkins College (Elkins, West Virginia), dramatic changes in the topography have been exploited. Buildings cross small ravines and valleys, connecting parts of the campus with protected, all-weather links. Service and through traffic is kept to the valley floors: the pedestrian environment sits above, surrounded by a generous landscape. Judith Monk's prototype design for Olmstead College (San Diego) adroitly inserts a small institution into a canyon, preserving the landscape and using the roof tops of the building for access and parking.

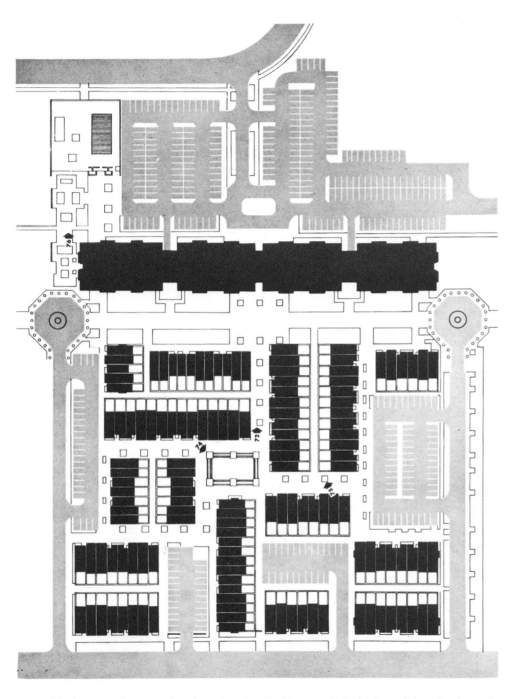

Superblock As in the Capitol Park project in Washington, D.C., high- and low-density residential buildings can be combined to form a superblock, and traffic and parking segregated from the pedestrians' environment. *Reprinted from: Design Of Housing Site, Robert D. Katz, Urbana, Illinois, 1966, C. W. Smith & Associates, Architects, For Capitol Park*

Traffic Segregation And Residential Districts A contemporary refinement of the Radburn principles as applied in one of the British new towns (Basildon). *Drawing: Basildon New Town Report, 1966*

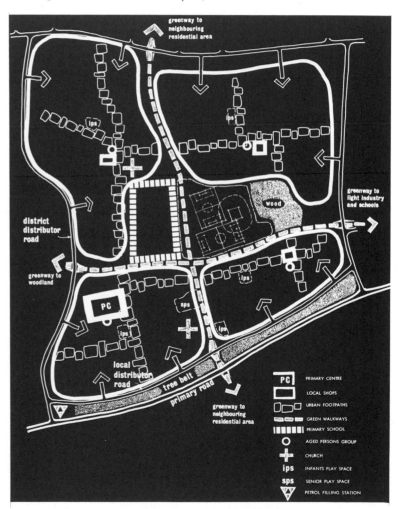

Guilford Town Center (British Columbia) The complete elimination of vehicular traffic from the pedestrians' environment encourages an architecture and landscape which creates a strong sense of place. *Architect: Francis Donaldson. Photo: Warrington*

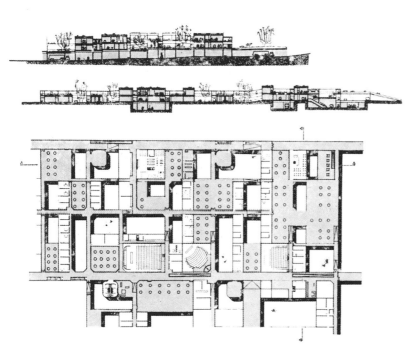

University Of Berlin A systematic approach to traffic segregation, which establishes design structure and a unique campus design. *Drawing: Candillis, Josic, Wood, Architects*

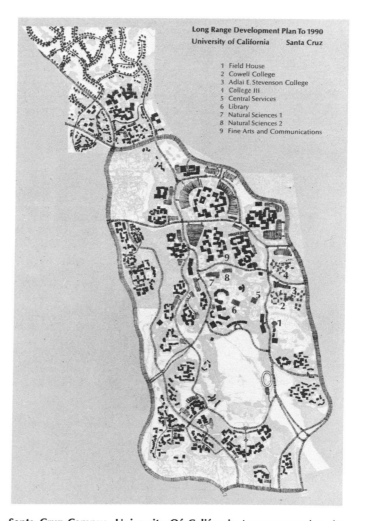

Long Range Development Plan To 1990
University of California   Santa Cruz

1 Field House
2 Cowell College
3 Adlai E. Stevenson College
4 College III
5 Central Services
6 Library
7 Natural Sciences 1
8 Natural Sciences 2
9 Fine Arts and Communications

Santa Cruz Campus, University Of California Long-range plan diagram showing how the location of circulation elements and traffic segregation help establish the boundaries of the smaller environments that comprise the total campus. *Photo: John Carl Warnecke, Architect*

College In The Mesa To preserve the character of the open space and to segregate vehicular traffic from the pedestrian's environment, Judith Munk has proposed placing an entire college in one of the canyons on the University of California, San Diego campus. Rooftops of a continuous structure would be used for grade parking. Natural features in the canyon floor would be used for landscape elements. A very large building program would thus not dispossess an unusual ecological zone. *Design concept: Judith Munk. Drawing: Richard C. Stauffer*

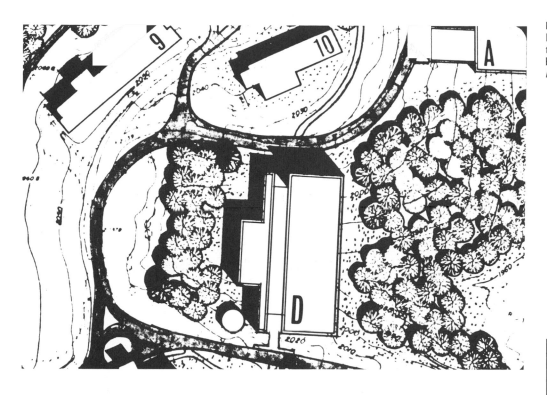

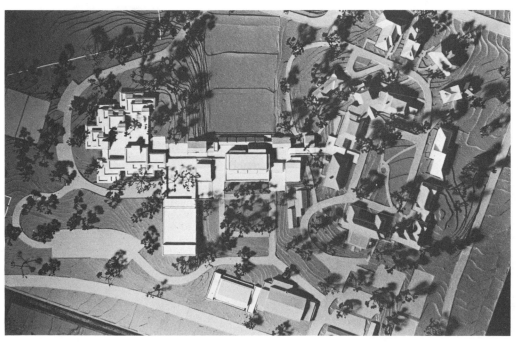

Library Bridge, Davis and Elkins College Building (D) is a proposed library which will span a steep ravine and serve as a bridge between two parts of the campus, segregating pedestrian traffic from roads used by automobiles and service trucks. *Drawing: Dober, Walquist and Harris, Inc., Planners*

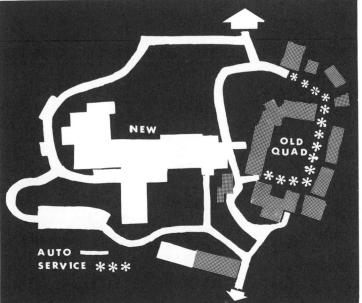

Long-Range Plan, Mt. Vernon Jr. College Automobile traffic is kept to the periphery of the campus. Because the site is small, all new construction is concentrated in one area. Pedestrians will move between the old quadrangle and the new buildings largely inside the new construction. *Photo and drawings: Dober, Walquist and Harris, Inc., Planners, Hartman-Cox, Architects*

249

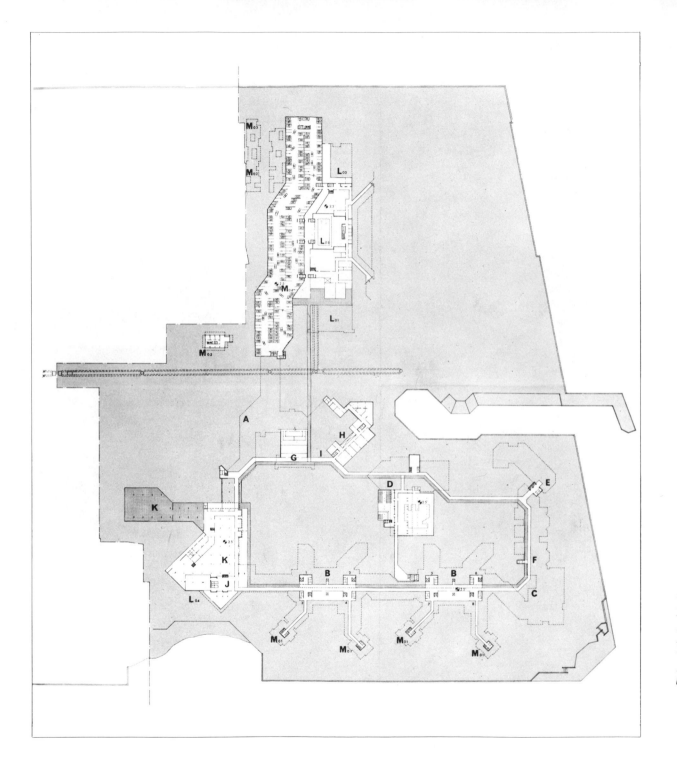

Kingsborough Community College, 1968 Current principles of traffic segregation have been well recognized in this high-density scheme for an urban campus. The service level of the megastructure contains parking and tunnels for the movement of goods and materials. The plan view shows the organization of the pedestrian precincts free of both through traffic and service traffic. (Pages 250-251.) *Drawings: Katz, Waisman, Weber, Strauss, Blumenkranz, Bernhard and Warner, Burns, Toan, Lunde, Associated Architects ; Dober, Paddock, Upton and Associates, Consultants*

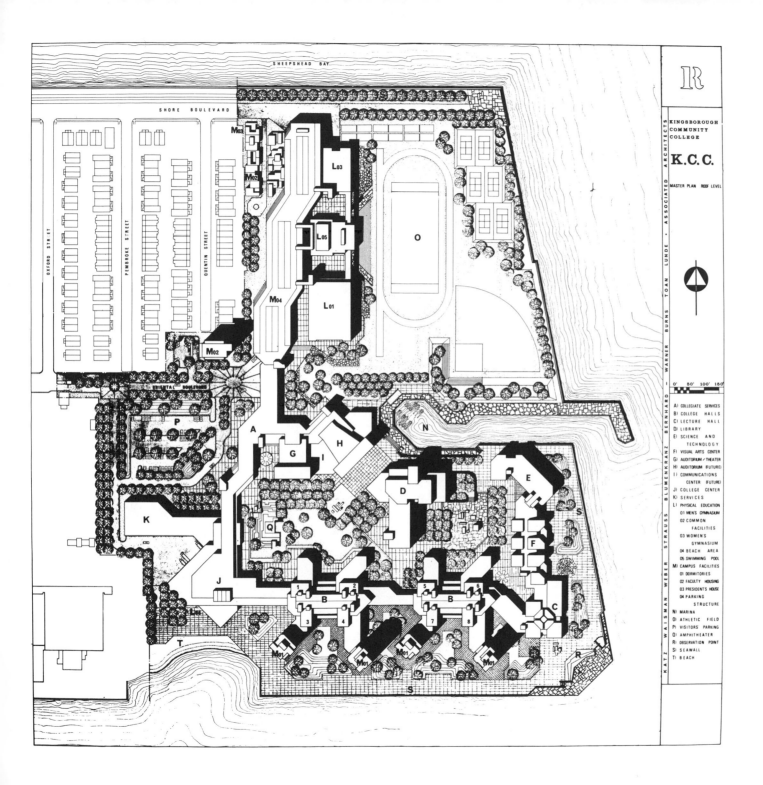

SHEEPSHEAD BAY

SHORE BOULEVARD

OXFORD STREET

PEMBROKE STREET

QUENTIN STREET

ORIENTAL BOULEVARD

M03

L03

M02

L05

M04

L01

M02

N

O

P

A

H

G

I

D

E

K

Q

F

S

J

N

1 2 5 6

B B C

3 4 7 8

M01

M01 M01

M01

L04

T

S

R

KATZ WAISMAN WEBER STRAUSS BLUMENKRANZ BERNHARD WARNER BURNS TOAN LUNDE · ASSOCIATED ARCHITECTS

KINGSBOROUGH
COMMUNITY
COLLEGE

K.C.C.

MASTER PLAN ROOF LEVEL

0'   50' 100' 150'

A) COLLEGIATE SERVICES
B) COLLEGE HALLS
C) LECTURE HALL
D) LIBRARY
E) SCIENCE AND
     TECHNOLOGY
F) VISUAL ARTS CENTER
G) AUDITORIUM / THEATER
H) AUDITORIUM (FUTURE)
I) COMMUNICATIONS
     CENTER (FUTURE)
J) COLLEGE CENTER
K) SERVICES
L) PHYSICAL EDUCATION
   01 MEN'S GYMNASIUM
   02 COMMON
        FACILITIES
   03 WOMEN'S
        GYMNASIUM
   04 BEACH AREA
   05 SWIMMING POOL
M) CAMPUS FACILITIES
   01 DORMITORIES
   02 FACULTY HOUSING
   03 PRESIDENTS HOUSE
   04 PARKING
        STRUCTURE
N) MARINA
O) ATHLETIC FIELD
P) VISITORS PARKING
Q) AMPHITHEATER
R) OBSERVATION POINT
S) SEAWALL
T) BEACH

Victor Gruen's Plan For Fort Worth *Photo: Victor Gruen & Associates, Architects*

Ancient Oxford Streets Closed To Automobile Traffic *Drawing: City Of Oxford Redevelopment Plan, 1964-1981*

Modern Stockholm *Photo: A. G. Annerfalk, Swedish National Travel Association*

Linjbahn, Rotterdam *Photo: Copyrighted Doeser Fotos, used by permission of Netherlands National Tourist Office*

Campuses with traffic separation are microdesign experiments that point out how the general urban environment may be made better. Page 13 shows a simple idea of how a community school building can weave through several residential blocks, carrying pedestrians over the through streets, making both environmental improvements and social progress in one effort.

Many visitors to the United States are visibly awed by their first views of central city. Skyscraper and street scene are powerful examples of technical skills and incredible investments in planning, designing, constructing, and maintaining public and private environments. Yet the activities contained therein are largely controlled by a ground plane circulation system whose location, width, and interconnections were designed for horse and carriage traffic. As noted earlier, this mixing of incompatible activities (of man and motor) in a limited space is inefficient and unattractive and unsafe.

At first glance the visual solidity of central city suggests a permanent arrangement with little likelihood of any chance to unscramble the mess and put it back together again. In the sense of total reconstruction, this is true, barring natural or man-made catastrophes. But some degree of change is a universal condition in central city. Fashions in finance and real-estate practices lead to construction and reconstruction of large and small parcels. Through the process of vacating areas for speculative reuse or urban renewal, decaying, unproductive land uses and buildings are removed and replaced with new uses. Difficult and nonfunctional land areas, passed over in the first waves of urbanization, can be brought into good use. Mandatory highway and mass-transportation construction opens up new possibilities for contemporary development. All these events as well as catastrophes, such as areas bombed

out in war, can be seized as opportunities for traffic segregation. The environmental results are worthwhile, and the price is not beyond our ability to pay, as some post-war European examples can demonstrate.

Because it has such a distinctive environmental character with strong cultural constraints on how much of the city pattern can be disturbed, the traffic-segregation solution of Oxford, England, is noteworthy. A major throughroad is planned to cross century-old Christ Church meadow. The care being exercised in setting the alignment and the nature of the design should strengthen the spines of American planners and give incentive to the clients they serve to encourage better designs.

Under a scheme devised by the landscape architect G. A. Jellicoe, the new road will be laid in a cutting across the meadow, thus preserving the traditional views. This involves moving the Cherwell river to the east and sensitively raising two earth mounds on either side of the gradient required to bridge the river. As the architect said, "a neat cut, insert the road, and join the parts together (so) by (optical) illusion . . . they become one." Construction costs are estimated as twice that of a surface road in the same location.

Crossing near Polly Bridge and St. Aldgates, the road will continue underground, again preserving a traditional view: Tom Tower, Christopher Wren's memorable addition to the Christ Church Quadrangle and a major landmark for travelers entering Oxford from the south. Westward, the road alignment allows variations in grade sufficient for a major pedestrian precinct to be built above and separate from the road with an intervening level of parking and access points for service. In the historic district, Magdalen Street will be closed to all vehicles and Cornmarket restricted to local-store service traffic only

and blocked completely to all motor traffic at one end. Truck traffic will be kept off High Street and buses rerouted along the streets parallel to it. These latter traffic-segregation measures illustrate an important principle: Dramatic reconstruction is only one alternative, and a combination of modest traffic-segregation solutions can also improve the general environment.

Coventry, the Detroit of Great Britain, exemplifies a straightforward solution to traffic segregation. A major ring road carries bus traffic and provides service access to the buildings. An extensive shopping precinct is set aside for pedestrians: a series of open spaces with fronting shops and intermittently an upper level of stores and recreation areas served by a balcony. Bridges from building to building and parking area to parking area take most of the pedestrian traffic off the ground plane. Bicycle and pedestrian paths join the city center to adjacent residential areas.

Park Hill, an area situated at the edge of central city in Sheffield, was developed as a series of high-rise residential slabs situated on a steep slope. The various units are tied together by a balcony-like street, which gives access to the apartments. Parking and vehicular service is kept to the periphery of the site. Local shops and recreation areas are developed at the bottom of the hill and tucked into the lower parts of the buildings. A high-deck footbridge will carry pedestrians from the project across the city's major ring road into the adjacent city center. The project as constructed suffers because of cost constraints, which have resulted in crude detailing and somber materials; but the idea of traffic segregation is worked out exceptionally well.

While all of the early post-war British new towns separated traffic horizontally, the latest designs have been based on vertical separation with vehicles on one level and the pedestrian environment on another, as in Cumbernauld and as will be seen later in this section. These measures fascinate British planners and designers, and some of the best ideas in traffic segregation continue to come from Great Britain.

Among the impressive aspects of German redevelopment has been the reservation of the Medieval street pattern for pedestrian traffic. The Hohestrasse in Cologne, totally demolished during the war, has been rebuilt with its irregular, quaint, crooked lines intact. Service to the shops is carried out during the early morning and at night. During the day, vehicular traffic is kept out by bollards inserted at either end of the street. The same principle has been applied in Essen and Bremen; in the latter case, public transport is allowed to pass through the area. The vitality and success of these experiments have led other European cities to treat some of their ancient streets in the same way; the Stroget in Copenhagen is a notable example.

In Rotterdam, whose central city was leveled by bombing in World War II, the Linjbahn has become a world-famous model for separating vehicular and pedestrian traffic. The L-shaped mall is lined on both sides with sixty-eight small shops designed, along with their outdoor paving, on a uniform grid. Pedestrian access is from the mall side and vehicular access from the rear. Arcades and covered canopies give protection from the weather and unify the store fronts. Freestanding display cases, seats, kiosks, flagpoles, sculpture, and lighting attractively fit together, scaled to the size of the shops and pedestrians. To the west of the mall, highrise apartment houses are grouped together; offices and large department stores to the east.

Equivalent measures have been taken in the United States to develop central-city malls. By closing several blocks to vehicular traffic and turning the space over to the pedestrian, downtown retail stores hope to compete with the shopping center.

Pedestrian streets, malls, and precincts such as Coventry's, all of them horizontal solutions, are successful but only partial answers to traffic separation. Essentially, they are small-scale projects with functional limitation and some important visual problems. These schemes depend on service access at odd hours. Furthermore, loading and unloading take place in the rear of the premises only, and sizable parking must be provided within reasonable walking distance. Considerable pockets of visual blight and desolation follow, and because vehicular accommodations are separated from pedestrian traffic horizontally, the pedestrian has to walk through these areas.

A practical way to overcome these defects is to segregate traffic vertically; that is, create distinct environmental zones for various activities and place them one on top of the other. The combinations are almost limitless. Space may be reserved for parking underground, on rooftops, or in the middle of buildings. Service traffic may be segregated from through traffic by assigning them different levels of rights-of-way. The pedestrian movement may be on the ground or elevated. Again, the technique is old and not without examples in recent urban development, as noted earlier concerning Cumbernauld and as the following examples of decks, platforms, and skyway will show, designs described here and illustrated later.

Utopian and visionary schemes for modern traffic segregation appear in the work of Le Corbusier and Ludwig Hilberseimer who among others recognized the environmental impact and design significance of the motor vehicle; but the first genuinely comprehensive design for vertical separation is Victor Gruen's plan for the center of Fort Worth, Texas (1955). It clearly sorted out and channeled private automobile traffic, service vehicles, and public transportation to underground peripheral destination points. Six multi-story garages handled sixty thousand automobiles. Subsurface operating space accommodate twenty-six hundred service trucks a day. A covering deck about a mile square was then free for commercial buildings of various sizes and functions. These in turn framed and formed a magnificent set of pedestrian spaces. Moving pavements, escalators, and sidewalks eliminated the need for other surface vehicles inside the pedestrian zone.

A prescient work, intelligently illustrated, Gruen's plan influenced and excited designers worldwide, even though the plan itself has never been carried out. Idiomatically translated, his seminal ideas can be seen in such completed projects as the Charles Center in Baltimore, Maryland.

The competition for replanning West Berlin (1958) elicited a number of designs for platform cities, giving the whole ground area to the motor vehicle and public transport and placing the rest of the city above it. One of the British entries showed how a series of fifty-acre superblocks could be linked together by an expressway and subway system with access into the blocks from boundary roads. Each block could hold space for ten thousand cars and could support a daytime population of twenty thousand people, a density five times that of Manhattan.

Stockholm's central-city plan (1960) indicated that Sweden's formidable planning efforts were not limited to developing satellite towns dependent on horizontal traffic separation. With the highest car ownership of any city in Europe and a vigorous and exciting central city, Stockholm had to make for an increasing amount of automobile traffic despite excellent public transportation carrying close to a half million riders a day. The long-range plan calls for a highway pattern basically a mandala form (a cross superimposed on a ring) surrounding and cutting through the central area. Near the central intersection extensive redevelopment has begun, with forty-five thousand off-street parking spaces to be provided within a mile-and-a-half radius. (As noted above, Gruen's scheme called for sixty thousand spaces approximately within one mile.)

Serelgatan, one of the first completed projects in central Stockholm, consists of five high-rise office buildings with underground parking for seven hundred automobiles. The base of the office (at ground level) is extended for shops connected by pedestrian bridges to additional commercial space across the way. The intervening space is used solely as a pedestrian concourse which leads to an improved subway station.

Even Manhattan is not beyond improvement. Proposals made by a consultant group to the City of New York show how the coordination of known redevelopment projects and scheduled highway construction could be aimed at segregating pedestrian traffic from through traffic along selected city streets. Further, the waterfront could be extended through landfill and the sites created used for new trade, commercial, educational, transportation, and residential centers. These generally would consist of a service level for parking and access covered with a pedestrian plaza open to the water.

The rendering on page 174 evokes a quality of grandeur and urbanity that seems favorably New Yorkish in these schemes. The disposition of high- and low-rise buildings around the open spaces creates a public domain and yet allows some degree of privacy. The magic of design prevails. The

drawing hints at diversity and unity. The brutal technology of motor vehicle is notably absent. Walking, sitting, standing, gathering take place on a simple stage with visual accents provided by caprice, season, style, and fashion—all changeable in the contained environment.

This design structure is not dependent on details. It emerges from two basic concepts: traffic segregation and the play of mass against open space. The variations are almost endless. A series of such units strung along the Manhattan waterfront can produce coherence in the overall form of the city, while subtle changes in each unit may engender the desired distinctiveness.

In the Whitehall area of London, ceremonial and government center of the British Empire and an extraordinary environmental setting, the planners have boldly reorganized traffic patterns so as to accommodate, over a long period of time, several million square feet of new government offices. The use of vertical and horizontal separation of traffic will enable the historic space around the Houses of Parliament and Westminster Abbey to be eventually free of all motor vehicles and at the same time allow significant increases in parking and service to new and old buildings.

The design underlines the advantages of bringing together as large a construction program as possible into a single site, rather than dispersing the normal increments of change over a wide area. As shown in the illustration, staging of new and old becomes an important part of the design solution.

Further, conceptualization of design without an appreciation of the step-by-step environmental effects can ruin the most imaginative ideas. The Whitechapel planners were highly conscious of this issue. Their design is evolutionary but has a strong sense of completion at each step. Tourists, visitors,

Londoners, and working population in the Whitechapel area will be given all the amenities they have a right to expect of working, viewing, resting, shopping, and transport. As the density of new buildings increases, accompanied by traffic separation designs, the tide of ground-level motor traffic will recede, gradually returning the area around the national monuments and office buildings to the pedestrian.

Such projects are not masterpieces in traditional design terms, i.e. conceived and executed in their entirety. Rather they aim at creating a large, unified civic-design with genuine respect for the historic environment. They set the direction for specific design solutions, so that individual designers may adjust the design aspirations to the particulars of a *terra cognita* with some hope that reasonable continuity will be maintained over time. The very act of traffic segregation is the master stroke which sets the design structure or skeleton, of contemporary central city, upon which others can build as use and reuse dictate.

Project 200 (Vancouver, British Columbia) was a bold attempt to develop a high-sensity area on a difficult site in a manner sensitive to the striking environment. The project concept illustrates how approximately eight million square feet of mixed urban activities can be accommodated on a twenty-two-acre site. The requirements for reserving a right-of-way for a possible freeway, the continuing operations of the railroad trackage, and the requisite parking and service access called for a complete separation of several kinds of movement system vertically as well as horizontally. The resulting design established a pedestrian spine along the harbor, connecting to the existing city design fabric at two main juncture points: a civic plaza and a major shopping mall. The plan restores the water's edge to

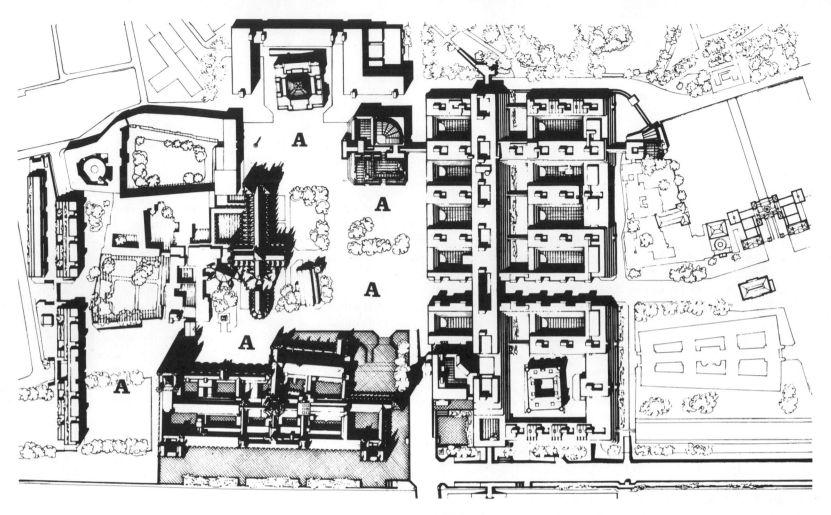

Whitehall, London Design proposals for enlarging government offices in the Whitehall area of London. The Houses of Parliament, Westminster Abbey, and other historic structures and the open space around them will be freed of all through traffic. The proposed reconstruction would include ramps and bridges that would make areas marked (A) exclusively pedestrian precincts. *Drawing: Leslie Martin, Architect*

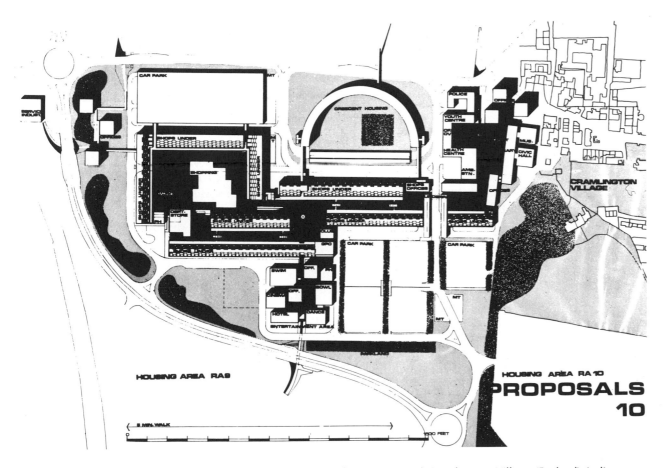

Organizing Car Parking Design proposals by Tom Hancock for the extension of Cramlington Village (England) indicate contemporary attitudes about handling large amounts of parking. Environmental zones are created for highways and roads, car parking, and the functional areas of daily life — housing, commerce, recreation, and community services. *Drawing: Tom Hancock, Designer*

city use, weaves the natural amenities (views and vistas) into the overall design and ties the project area back into the city visually and functionally. See pages 120–123.

Executed designs for residential sectors, shopping centers, college and university campuses, and central-city development in which traffic segregation is a basic environmental-design principle increase in number and size each year. The validity of the approach needs no further explanation as the accompanying illustration indicates. But there are also additional values in traffic segregation, especially for civic design, by encouraging a richer mixture of land uses and environments which previously had been separated because of traffic conditions. American cities show sharp cleavages between major land uses: residences, cultural institutions, retail and office space, and factories. Through custom and fear of environmental deterioration, zoning laws and planning policies have kept these land uses apart from one another. With the exception of rare, nonconforming areas, recent American urban development projects have been designed as a series of specialized districts exclusively given over to a single or dominant activity. Supposedly in the interests of visual harmony and health, a monotonous development pattern has been maintained. In addition, the journey from place to place has been extended, since no one district can satisfy a range of services, job opportunities, and excitements.

Building technology has, however, reached the stage where a number of functions can mix safely and pleasantly, enclosed in a single area. Traffic-segregation designs can and have already been applied to separate service and motor-vehicle areas from the function use areas. As a result, more interesting mixtures of activities can be realized and with higher densities than experience of the past several decades seemed to indicate was desirable or possible.

Full implementation of this potential awaits inexpensive forms of large-scale parking and the introduction of some form of mechanical horizontal movement systems, two technical advances not yet in sight economically. Without the former there are strong economic limitations on the kinds of uses that can be brought together into a single project and the densities that can be maintained. The latter technical solution is needed to engender a new family of building types to do for the horizontal plan what the elevator and escalator did for the vertical.

PARKING

Parking is the limbo space lying between the hell of heavy traffic and the heavenly precincts set aside for pedestrians. Like limbo, parking is a special environment, useful, but filled with uncertainties as to whether it is a punishment or a reward. First of all, there is the problem of space. Each automobile takes up about four hundred square feet of space including access. The same amount of space would be sufficient for sheltering five general office workers or two professionals or ten theater seats or a classroom for twenty-five students. The land-consuming aspects of parking exacerbate the designer still further because the average automobile arrives at its parking space well below its designed carrying capacity.

The economics of parking is a major problem. In theory there should be adequate parking at either end of the circulation system, and the amount of parking space should be consonant with the density of adjacent activities. The theory breaks down, however, because parking increases in cost by orders of magnitude at the very places where it is needed most.

In addition, on the four basic kinds of parking solutions—grade-level parking lots, parking structures above ground, underground garages, and garages combined with other space in a single structure—typical construction costs run, respectively, from two to seven, fourteen, and twenty dollars a square foot. At the upper ends of the cost scale, the amortization, insurance, maintenance, and operating costs of underground or building interior solutions for parking run about $1,000 a year for each space or about an eighth of the yearly income of the average automobile owner. For economic reasons alone, unsubsidized space for indoor parking at high-density destinations is quite limited. Some expense may be saved by using deck parking and grade parking, but they impose visual and other problems, as discussed earlier.

A third major problem in parking design is in decision-making itself. In central-city areas, the amount of parking available seldom turns out to be a function of demand but rather a balancing out of several independent and understandable aspects of social and economic behavior: the driver's ability to pay; the prestige of on-site parking; the necessity of on-site parking (hospitals and newspapers, for example, cannot operate without parking); the community's tolerance for enforcing parking regulations; the presence of mass transportation; variations in automobile usage due to differences in employment groups, community-development patterns, and journey-to-work habits; the degree of land obsolescence near destination points, which may encourage land to be used for off-street parking at reasonable rates; and zoning codes and ordinances. In the latter instance, while many communities have detailed requirements for off-street parking in new office and commercial structures, some communities have abandoned such regulations on the grounds that the downtown parking supply can never satisfy the need. Like whiskey to an alcoholic, one parking space is too many and a thousand not enough.

Yet another factor enters into the balance that determines the amount of parking space in central city: private and public subvention of parking. If they have rapid enough turnover in the spaces to amortize costs, downtown stores will provide parking at nominal rates to attract customers. City parking authorities will subsidize a certain number of all-day parkers to hold "downtown" economically together long enough to come up with a more substantial solution to street and land-use obsolescence. But here again decision making does not typically follow a planning rationale. Most such subsidizing measures come about through seat-of-the-pants judgements and not incisive analysis and action. Traffic control, parking, transportation, street and highway engineering, redevelopment and long-range planning authorities are jealous of their traditional prerogatives, in competition with one another for limited public funds and under political pressure from different groups. A consensus on the nature of the parking problem and the solutions available is a rare instance of cooperation among public servants. Yet the problem is not so overwhelming that nothing general or specific can be suggested or illustrated that will encourage concrete solutions.

Like every other facet of the environment influenced by the motor vehicle, parking cannot be designed out of context, and at the same time it is clear that continued central-city revitalization must encompass large-scale parking solutions. Past development of highspeed highways into the central areas has not been accompanied by equivalent measures to provide for the auto-

Lower Manhattan Plan Traffic segregation proposals to be undertaken in the renewal and reconstruction of lower Manhattan. Also see p. 174. *Drawing: Lower Manhattan Plan, 1966. Whittlesey, Conklin, and Rossant, Architects; McHarg, Wallace, Todd and Associates, Planners*

mobile at its destination, and as new circulation networks are introduced, comparable parking measures must take place. Thus a $15-million-a-square-mile expressway should include funds for parking in the area it serves.

There are other positive measures of a fairly specific nature that can be taken to cope with the parked automobile in central city. As much as possible, large groups of people commuting singly via automobile on a regular basis from the fringes to downtown can be intercepted at some distant point and encouraged to change to mass transportation. Efficient use of downtown parking space can be stimulated by charging reduced rates for small cars, automobiles used for car pools, and short-term parkers. Consideration can be given to keeping to the periphery of downtown unique land uses which generate large amounts of traffic and parking and yet do not make substantial contribution to the life of downtown, thus reserving the core of the city for activities which engender a high-density, low-parking, pedestrian-oriented environment. Thereafter the presence of parking downtown should not of itself be a frightening prospect. Parking decks and parking garages need not demean the cityscape. The design of the parking structure can be approached as an architectural problem and sensitively integrated into the environs.

Underground garages in Union Square, San Francisco; the Piazza Diaz, Milan; and Mellon Square, Pittsburgh, have provided parking for automobiles below surface with a landscaped city square above. Parts of Hyde Park, London, and the Common in Boston have been similarly used. In the latter instance, the garage was neatly tucked away, although the restored landscape is inadequate, and pleasant pedestrian routes to and from the garage have unfortunately been omitted from original design and present construction.

Technical considerations dominate architectural solutions: stability and structural requirements, ventilation, fireproofing, critical dimensions for turning and moving up and down ramps, and optimum space utilization. Accordingly, a distinctive architecture of parking has emerged, especially above ground. It is also possible, however, to select materials and manipulate massing to complement the environs. Sometimes the base and roof deck can be landscaped, and depending on site and situation, the structures can be placed so as to serve as a backdrop for more important pieces of civic architecture. In tomorrow's cities the top surfaces of parking structures may serve as the base plane around which new buildings can be clustered, the garage roof thus producing a pedestrian environment separate from automobile storage and movement, as in Constitution Plaza, Hartford, Connecticut. Obviously, the larger the project, the more substantial the design opportunities.

Parking provisions in middle city—the land between the core and the suburbs—are especially vexing. Demands generated by residences and visitors rise as density increases and mass transit declines, yet the latitude for improvement in parking is limited because of land occupancy and old street patterns which are not susceptible to change. Further, the introduction of off-street parking regulations in these areas may endanger the designed environment to the point that obsolescence may accelerate. Several cities have had the chilling experience of losing many of their front lawns as bans on off-street parking were enforced and homeowners thereupon paved their front yards. Finally, an American Society of Planning Officials advisory bulletin (September 1966) has noted that "there is little reason to believe that in the future people will be willing to forego the convenience of owning a car for the dubious privilege of living in the older parts of the metropolitan area." Thus some kind of middle-city parking solution without the blighting effects of large parking lots and congested streets is called for.

Since the individual property owner can do very little to alleviate the parking situation, one general approach to its solution is to make parking a city function—a utility like water and sewage disposal or a service like police and fire protection. Then well-designed off-street parking facilities may be developed on vacant land or on city-purchased property acquired either through eminent domain or as property comes up for sale on the open market. A well-landscaped quarter-acre lot in a five-acre city block can yield about thirty spaces, enough to take some demand off the street and other properties. The location of the lot and its design require careful study of the environmental character of each residential district, but this is a task well within the skills of present planning technology. Neighborhood participation in the decision-making may well help allay fears about decline in property values because of an interior parking lot and perhaps even stimulate additional support for the aesthetic treatment of the parking area. Granted that there are more urgent issues deserving public debate than the local parking problem, nonetheless it is an important issue—like the common cold, a minor but universal nuisance deserving care.

Such small ground-level parking lots and also simple decks (minimum lot size: ten thousand square feet) may be financed through a user's fee. Parking charges ranging from $20 to $40 a month are not unusual in an apartment-house district, and a comparable charge might be levied by municipal parking facilities. Other funding techniques include special assessments and direct municipal subsidy of development costs. In Milwaukee the city charges $4 a month for overnight street parking. As the monies are collected, an equivalent sum is then applied to constructing off-street facilities, although these are in downtown locations.

Even in new residential areas the problem of parking is not absent; parking standards are probably already below actual need. The typical current subdivision regulation and zoning ordinance calls for one off-street space per dwelling unit, but in its *Land-use Intensity* guidelines, the Federal Housing Administration now advocates two off-street spaces per single-family detached house. In the Federal ratios, however, the number of parking spaces to housing units decreases as the number of units per acre rises. Yet pace-setting, medium-to-high-price subdivisions now build three off-street spaces per housing unit plus parking for visitors, and there is concern lest this ratio be used inappropriately. The application of such standards in high-density subdivisions would reduce the feeling of open space, and automobiles might dominate the street scene. On the other hand large parking lots would seriously interfere with separation of pedestrians and motor vehicles. Hence, the parking dilemma is the same in the fringes as it is in central city: The automobile is an exorbitant user of land while at rest, and the solutions to the visual problems it creates are not cheap.

Because parking tangibly affects the designed environment, some local ordinances now regulate the size of the parking space (both minimum and maximum), its location on the house lot, the access to the space, the surface materials used, and whether the space shall be enclosed by a carport or a free-standing garage. All these measures suggest a growing concern about the motor vehicle and its effect on community appearance and community design.

EXAMPLE OF SKETCH DESIGN

Endless streets, with their overhead wires and unrelated collection of signs, poles, cans and fixtures, look confused and uninspiring.

Cleaning up the street would help its appearance, while the creation of a strong terminal feature would limit and define the space.

Should traffic eventually go into tunnels the streets could become handsome promenades. People could enjoy them rather than be crowded on the sidewalks. At peak hours they would fill the major streets.

Cleaning Up The Clutter A lack of a consistent parking program results in street designs that are cluttered and visually unsatisfactory. Design proposals made by Toronto's planners show how environmental design techniques can be used to coordinate everyday urban development in such a way that the parking problems can be resolved gradually. *Drawing: Toronto Metropolitan Planning Board*

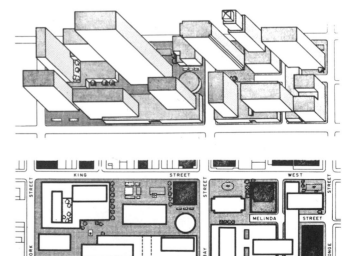

Each new development should be reviewed to see how far it can help implement the plan, by creating walkways and plazas and enhancing the streets and its surroundings. A sketch design showing what might be achieved could be a guide in this review. It should be flexible, without binding legal status, and be adapted as downtown develops.

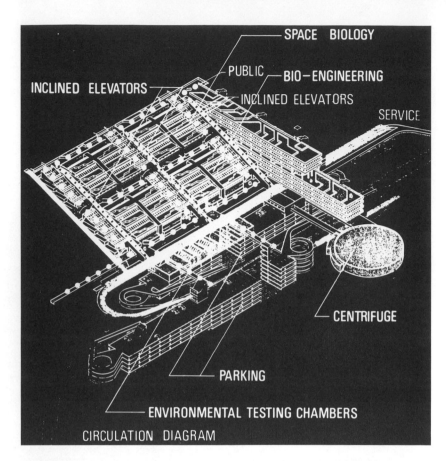

SPACE BIOLOGY

PUBLIC — BIO-ENGINEERING

INCLINED ELEVATORS — INCLINED ELEVATORS

SERVICE

CENTRIFUGE

PARKING

ENVIRONMENTAL TESTING CHAMBERS

CIRCULATION DIAGRAM

Connecting Link Designers of the Environmental Physiology Laboratory (1968) at the University of California, Los Angeles, used inclined elevators to connect parking structures on the floor of the valley with buildings and test chambers sited on the valley slope. *Drawing: Henry C. K. Liu, Designer, George A. Dudley, Consulting Architect*

Form Generator The use of a micro-movement system could encourage a whole new family of building types. Planning studies for the Acorn Park facilities (Arthur D. Little, Inc.) suggested that an economical four-story structure over 1,500 feet in length could work if a moving corridor were used to carry people as efficiently horizontally as elevators do vertically. *Drawing: Dober, Paddock, Upton and Associates, Inc.*

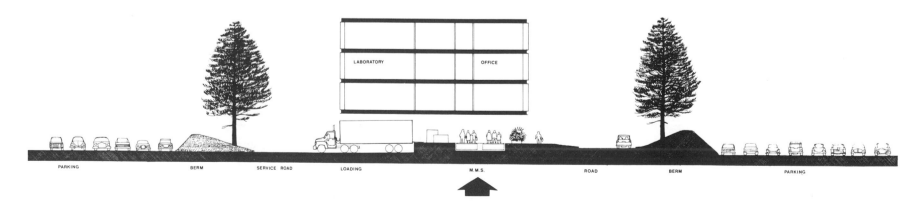

LABORATORY   OFFICE

PARKING   BERM   SERVICE ROAD   LOADING   M.M.S.   ROAD   BERM   PARKING

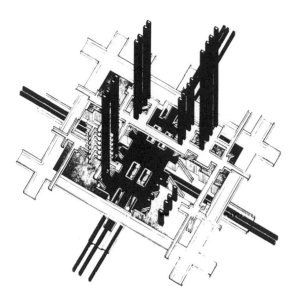

The Access Trees Design studies by the New York Regional Plan Group have suggested that the internal circulation problems of mid-Manhattan may be resolved by integrating high-speed transit stations with high-capacity vertical lift systems. Large numbers of people can be moved easily and congestion reduced. *Drawing: Ray Y. Okamoto*

MICRO-MOVEMENT SYSTEMS

The environmental objectives of separating man from motor may lead to a new form of transport for short intracity trips at speeds faster than foot travel. Indeed, any significant removal of automobiles from the central business district, the development of high-density superblocks, or city designs that depend on large garages at the thresholds of urban places (with the anticipation that the automobile won't be used for internal travel) will all require a Micro-Movement System (MMS). Further, as the transportation architect Brian Richards points out, the introduction of micro-movement systems is imminent because of increasing traffic and congestion, public demand for better transport, environmental improvements that require MMS solutions, a scale of urban development which makes such systems practical, and the rising cost of operations in the traditional forms of movement, such as buses, subways, and taxis, which may in turn encourage transportation solutions that require fewer people to run and maintain the systems.

Although MMS is the most neglected form of transportation technology, there have been notable attempts in the past to introduce such systems into urban areas. Prior to the introduction of the motor bus (which still remains the cheapest, most flexible form of mass transportation), a number of interesting experiments were undertaken to develop mechanical solutions for moving people short distances. These included elevated tracks which carried small cars and moving platforms. A prototype of the latter was built for the Columbian Exposition in Chicago (1893). It took the form of a mile-long ellipse, with two parallel platforms. The first ran at three miles an hour and carried a passenger from a stationary position to the second parallel platform which was moving at a faster speed and also had benches for sitting.

Expositions and fairs have been popular places for MMS experiments. The Paris Exhibition (1900) included an elevated conveyor system two and half miles long with ten stations for getting on and off. It was operated twelve hours a day for eight months and carried almost seven million people with no serious accidents or mechanical failures.

The Wembley Exhibition in London (1924/25) had an MMS system installed whose operating cost was so small that it could be run free of charge in the second year. Both New York World's Fairs (1939/40 and 1964/65) operated conveyor systems through exhibitions, with high volumes of traffic and genuine comfort. The Swiss Exposition in Lausanne, Seattle's World's Fair, and Montreal's Expo '67 contained examples of highly publicized and well-used monorail systems constructed in the last several years. Charging twenty cents a ride, the Swiss model paid for all design, construction, and operating costs six months after opening. The Seattle model still makes money (though a very short run) and is a popular tourist attraction in winter and summer.

One promising horizontal movement system for New York City, never carried out due to engineering and labor-union considerations, was the proposed continuous platform for linking Times Square and Grand Central Station along an underground route that paralleled 42nd Street. The system could carry thirty-two thousand people an hour in each direction. The design was rejected ostensibly because the depth of the right-of-way might interfere with future subway construction. In 1953 a second attempt was made with an improved version designed by Stephens and Adamson for the Goodyear

Total Design The integration of all forms of circulation, vertical and horizontal, individual carrier and mass transit, is technically feasible today. Kenzo Tange's design proposals for the reconstruction of Skopje, Yugoslavia, suggest the scale and character of human habitation under such conditions. *Model: Kenzo Tange, Architect*

Tire and Rubber Company. A nonstop system fully automated, it included a series of cabins on a continuous track, entered and exited at normal walking speeds and then accelerated to fifteen miles per hour. Difficulties with the transport unions prevented installation and operation of a full-scale model.

Another attempt is certainly warranted. The Times Square District shows signs of being the next major new office zone in the city. Just to the southwest lies the Port Authority's Bus Terminal, a major city threshold with a present ability to handle up to forty thousand passengers an hour. The linking together of these two areas with Grand Central Station by an MMS solution could have a profound effect on the environment of central Manhattan.

Micro-movement systems are obviously subskeletal elements in any design structure, and can be divided technically into two categories: the continuous and the intermittent. The former includes the escalator and moving sidewalk and the latter such devices as the elevator and monorail. Vertical systems are of course common enough occurrences, but the number of horizontal systems in operation is far below proven need, as any traveler using the major airports in Chicago, Boston, or Philadelphia and carrying his baggage from plane to taxi would attest.

To succeed, MMS designs must meet these design, financial, and operational objectives: MMS should be readily available at the points of contact between intersecting and connecting forms of mass transport, and they should tie together and/or travel through heavy-density areas. MMS should furthermore operate on a continuing basis at low operating cost. Initial construction and maintenance should be funded by a public agency or authority, and use by the public should be as "free" as police or fire services.

The systems should be capable of carrying high volumes of traffic at speeds up to fifteen miles per hour. Finally, the technical solutions should be safe, quiet, and attractive.

What kinds of MMS devices will be used to meet these objectives will depend, of course, on the character of the environment into which they will be fitted. Inside the core areas of cities, average trips are typically short distances; and as noted, intermittent as well as lengthy and continuous systems are available. Ideally, a completely complementary and integrated horizontal and vertical system would be desirable.

The above comments hint at mechanical systems that follow designated routes, such as conveyor belts, linked cars, moving sidewalks, elevators, escalators, and similar devices. An alternative, microscale but not an MMS solution, is a set of minicars for public use in an environmental zone which has been cleared of private automobiles. Such a car would presumably be a battery-type vehicle, so the air would be free of pollution and noise. Slow speeds would reduce accidents. Theoretically, the small size of the vehicle would enlarge the capacity of streets and parking areas, and the present environment could be used without the introduction of new technology. Since no new kinds of machine or areas are therefore needed, the major problem with the minicar has been one of operations and users' fee.

Traffic engineers in Paris (1958) considered special licensing for minicar drivers. Ticket-holders would be able to use the vehicle anywhere inside the minicar district. Other research groups have suggested coin-operated vehicles or a special ignition key which also turns on an automatic recording device for keeping track of mileage and user. Data-processing devices are advanced enough to allow automatic billing.

The scramble system of minicars has certain merits, but in comparison Micro-Movement Systems have these same and additional advantages: operation in a protected, all-weather environment, higher capacity per structural unit, and susceptibility to incorporation into some larger design context.

Today's trends are leading towards a three-level environment for circulation and movement systems: an underground level for services and through transport and traffic, the street level for local traffic and emergency vehicles, and a plaza or deck level for the pedestrian zone. Some form of MMS technology could operate at any of the three levels. But just as the elevator stimulated a whole new concept in building form, the MMS may also stimulate yet another family of building types and environments, with its ability to move large numbers of people horizontally through a comfortable, attractive, and safe environment. Although the analogy is not precise, the words "horizontal skyscraper" come to mind as one image of the future.

Constitution Plaza, Hartford *Design: Charles DuBois, Coordinating Architect. Sasaki, Walker and Associates, Landscape Architects. Photo: Travelers Insurance Company*

Valingby Town Center A well-praised example of a mid-twentieth century urban space. Consistent attention to design details in paving, street furniture, lights, and landscape have attracted designers from all over the world. *Photo: Richard P. Dober*

Pocket Park Constructed on the site of the old Stork Club, this well-used pedestrian precinct has encouraged a number of other cities to develop pocket parks. *Designers: Zion and Breen, Landscape Architects. Photo: Richard P. Dober*

THE PEDESTRIAN PRECINCT

For every site there is an ideal use, and for every use an ideal site. Sites and activities have natural affinities for each other. With this in mind, the planner's search for clues and cues through site analysis and a cognizance of human behavior is the first step in designing an appropriate environment for the pedestrian.

What of the human parameters? Modern man will walk for pleasure and necessity, but the distance covered will depend on topography, climate, and inclination; or the range may be set by custom more than anything else. It is said a woman carrying packages will walk no more than six hundred feet when going from a store in a shopping center to her car. A ten-minute walk is a conventional distance for college and university students traveling between classes. A well-conditioned athlete can run a mile in approximately four minutes and walk the same distance in six. Typical pedestrians, however, move along at a much slower pace: from six to ten feet per second. In addition, if direct routes are not designed in urban areas, pedestrians are generally likely to make them, thus pedestrian paths have to be as direct as possible in urban areas. Finally, even in this age of the automobile, people spend more time walking than they do in an automobile. Extensive travel by foot is here to stay, and the design of the pedestrian precinct is a relevant task for all environmental designers.

Site analysis involves appreciation of two sets of facts and possibilities: natural conditions and artifacts. The first are already existing and embrace the general character of the environment: earth and ground forms, landscape materials, topography, existing views and vistas, and direction of wind and angle of sun. Appreciation of artifacts describes and evaluates what man has done and might do—land uses, circulation, structures, utilities, legal restrictions—both on the site and in the environs. All the physical features may be summed up in plans, models, or drawings.

The spirit of site analysis is to let no scene or object of the environment go unobserved, to delve into the implicit and the explicit, to reveal the personality of the place—its ambience, its townscape. (The latter is a coined word which Gordon Cullen uses to invoke the poetry of environmental design.) The idea is to describe and define so as to prepare to adjust plans to conditions inimical or encouraging to future development. With all this information absorbed, the program for new uses and construction can be applied with discrimination and sensitivity; the pedestrian's environment can be rooted in a rich and fruitful design context; site and activity can blend.

What elements of design can help insure that site and activity will be well mixed? Surely, the design of the pedestrian path in the city should take into account the probable amount of traffic, the direction in which it wants to move, and the nature of the activity on either side of it. Beyond considerations of function and safety, however, the designer has a whole palette of design elements for giving the pedestrian environment a richness and detail not available in design for driving. First of all, the variations and combinations of spaces and buildings are as many as there are designers and viewers. Through the careful design of pedestrian movement, the accidental, the fortuitous, the contrived, and the deliberate can be brought together into a harmonious design relationship neither conforming, confining, nor inhibiting.

One of the unfortunate results of the automobile's intrusion into the pedestrian world has been the removal of the subtle and the accidental from much of the central city. Loss of these elements has accompanied discrediting of the American street as an environmental design and its replacement by rather sterile substitutes. Some such design elements—plazas and malls—typically shut out the city around them. The visual connections to the outside are inadequate and boring. The designs are totally dependent on the new architecture that surrounds them and do not draw on the existing cityscape. Several well-praised new city spaces are like cemeteries for the living—collections of marble and granite monoliths, some are horizontal for sitting, others vertical for interest, all are surrounded by manicured plant material, reflecting good taste and impeccable design skills but loss of life.

The sequence of movement through space is a critical element in designing for the pedestrian. There is the threshold (transition from automobile to foot traffic), where safety is paramount. After that, the progression improves or deteriorates as vistas are opened and closed, the contents of the spaces play their role in enriching the visual field, and the architectonic elements fit together comfortably or provocatively as the function of the place may demand. Words hardly serve the cause of townscape well; but procession, recession, undulation, and anticipation hint at dynamic designs which emerge from slight alterations in building alignments, variations in setbacks and projections, and temporal qualities that come about as signs, seasons, and light change the tone and appearance of the visual world. Different kinds of enclosures, large and small spaces, interior and exterior can be threaded together. Foreground and background architecture can be synchronized for orientation, grandeur, unity. Sun and shadow, floor and wall, surface and textures may invite movement, repel, or be neutral. In some respects these design effects can come about only with age, requiring as they occasionally do, succeeding generations to add their own mark and distinctive qualities to the whole. In other instances controls must be set on behavior to insure the designer's intention. Some critics feel the ability to perceive, appreciate, and complement movement sequence has been lost in recent environmental design, in the sections describing a sense of place, there are examples that suggest otherwise.

Changes in elevation—from slow, safe, shallow ramps to steep steps precariously inserted into man-made or natural embankments—are typical problems challenging the designer creating a pedestrian precinct. But the ways of doing so are many, as a brief description of some of the traditional methods shows so well.

There are four ways to move pedestrians vertically through the environment. The first are mechanical contrivances: chair lifts, elevators, and escalators. The second are architectural elements which are part of a building complex, such as the steps that extend the palaces of Versailles into the space around them or the stairway that serves as a plinth to the Capitol Building in Washington, D.C. Third are the cantilevered, detached or almost detached, free-flying flights a la Piranesi which connect one level with another —staircase or ladder. Fourth, are the carved out, sculpted, scooped, or cut steps passing through rock, earth, or wall. Each has its special reason for being, reflecting both the direction of movement, pace, degree of formality or informality, and the public or private character of the buildings and areas the steps serve. Outdoor steps test the designer's ingenuity. The bag of tricks—tried and true formulas applied for design effects —seems inexhaustible in open space. On the classical sites of Europe, from the Acropolis to St. Paul's Cathedral, risers range from four to seven inches, the going from twelve to nineteen inches. Aesthetic delight and memorable design qualities come from a careful interplay of what is correct and what is audacious.

Emotional impact can be a major element in design for changes in elevation. Motion upwards can be exhilarating, impressive, challenging, a rewarding expansion of views and vistas. The sky, overhead plane, and expanding horizon become the visual interests. In the downward direction, vision is oriented to the base plane and the horizon diminishes. Here, the very forces of gravity can be used to induce emotional reactions. In a design for the Franklin Delano Roosevelt Memorial, Hideo Sasaki and his associates brought the pedestrians slowly up to the edge of the central area and then suddenly down to the cenotaph. To achieve a heroic or monumental feeling for pedestrians approaching a public building, the foreground spaces are sometimes tilted slightly upward or downward—a simple device used once again around Boston's new City Hall.

The quality of paving material in the pedestrian precincts can establish design structure. Hard surfaces occur where people gather, walk, and carry on activities in concentrated numbers. Soft-surface areas are complementary spaces, more seen than used. The reason is simple: maintenance.

Hard surfaces need not be dreary, colorless, and dull. Designers have been continually intrigued with the plastic qualities of various paving materials and continue to experiment with them. Artificial stone, pressurized iron slag, and coal tar have been tried in Germany. The French have used devitrified glass in Lyons. Ceramic paving blocks once decorated Budapest, oyster shells New Orleans, compressed grass has been laid in Richmond, iron blocks in New York, wood in many places in many forms. In 1890, a patent was given for a noiseless stone block made of paving stones encrusted in waste-fiber and bound by elastic bituminous. The block was used but not successfully and had the same fate as India rubber, slag blocks, coal tar, and leather as paving materials. The accepted "palette" for hard surfaces today includes concrete, asphalt, and various kinds of stone. Simple or intricate patterns can be laid depending on how the materials are mixed, scaled, and set. Changes in textures, color, and width can subtly direct pedestrian movement in predetermined patterns, identifying byways from main paths, arrest movement, or speed it up.

Thus a pedestrian's environment is structured by the elements which define his space, the lines of circulation that carry him through it, and the plastic qualities of the surface on which he moves or rests. Of equal importance are the design elements that fill the space and modulate it: lighting, landscaping, signs, and signals, and street furniture. The latter comprises utilitarian objects such as mailboxes, trash bins, bollards, benches, and other paraphernalia.

Design, selection, and placement of street furniture will add or detract from the overall unity of outdoor spaces depending on material, scale, color, and family resemblance between the various elements. How well they fit together is a test of the quality of the design. An analogy may be made to nautical design: Whether custom-made or stock model, simple as an oar or complex as a radar set, all the parts of a well-designed boat blend together. An accommodation of function without disparity can make memorable designs on shore as well as at sea.

In ancient times life in urban places generally came to a halt when the sun went down. Streets and public places were not

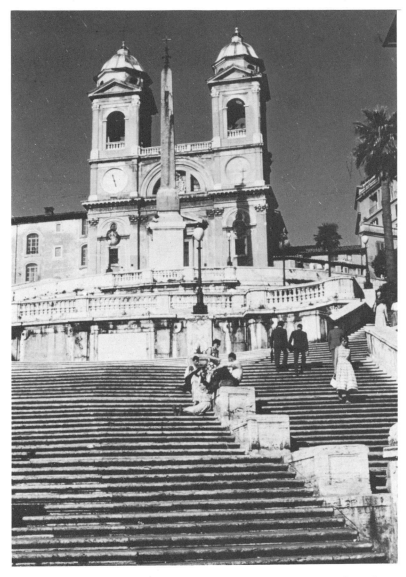

Floor Space As a Design In the main, contemporary designers have failed to reach the levels of excellence achieved in the Renaissance approach to floor surfaces of pedestrian precincts. The amenities of contemporary public space are diminished accordingly. *Photo: Richard P. Dober*

Stairs A historic device for creating pedestrian precincts, stairs have been neglected in recent urban development. Changes in elevation have been treated as a functional problem in circulation and the opportunity to create a sense of place neglected. *Photo: Richard P. Dober*

Simple Signs Signs are necessary to inform, direct, and control the movement of people. Good graphics measurably add to a sense of place. *Photo: Richard P. Dober*

Use Each lot is for the exclusive use of the owner, guests or lessees for residential purposes only. Artists, artisans and craftsmen may pursue their trade within the property boundaries so long as actual business transactions are not conducted thereon. All properties shall be maintained in clean condition and good repair.

Rentals You may rent or lease your property subject to the Mililani Town covenants ("protections").

Carports and Driveways At least two parking spaces shall be provided for each residential lot, in a carport enclosed on at least two sides. Mililani Town, Inc. homes have driveways and two car carports which may be expanded as designed to three car capacity. You may keep personal automobiles, boats or a truck of up to one ton capacity in the carport. Mililani Town's residential streets are designed for beauty, privacy and quiet. Driveways shall be placed upon minor streets only in order to protect residents from traffic hazards.

Vehicle Maintenance Repairs on personal vehicles may be performed within the boundaries of your property so long as such repair is not visible from the street or neighboring property. Minor maintenance on personal vehicles may be performed in the carport.

Storage Open storage shall not be visible from the street or neighboring property. Enclosed carport storage is included with Mililani Town, Inc. homes.

Trash Enclosures Trash or refuse, or such containers, shall not be placed in areas visible from the street or neighboring property. Mililani Town, Inc. homes include a "pau" (a specially built trash enclosure).

Clothes Drying Enclosures Clothes drying shall not be visible from the street. A "huna muumuu" (a specially built covered clothes drying yard) is included with all Mililani Town, Inc. homes.

Fences Fences, walls and hedges constructed or planted by Mililani Town, Inc. shall be maintained, except those between private lots which shall be jointly maintained by adjacent owners or may be removed upon the agreement of adjacent property owners.

Landscaping Trees planted by Mililani Town, Inc. within ten feet of the property line on any lot shall not be removed or cut down without the prior approval of the Design Committee of the Mililani Town Association.

Signs You may post one sign of specified size on your property for the purpose of sale or rental, for legal purposes or a temporary construction sign. A residential identification sign is permitted with a maximum face area of one square foot.

Noise You may install an exterior sound device for security purposes only, but no other exterior sound devices are permitted.

Antennas You may install an outdoor antenna on your property so long as it is placed on normal grade in the rear yard and does not exceed ten feet in height. Rooftop antennas are not permitted.

Pets You may keep a reasonable number of generally recognized house pets unless they are deemed a nuisance to the neighbors.

Fires You may use an exterior barbeque or imu fire for the purpose of cooking but other fires such as trash or plant refuse fires are not permitted.

Exterior Lighting You may light your property so long as the source of light (bulb or lamp) is not visible from the street or neighboring property.

Alterations You may make alterations to your property. Major alterations or additions exceeding $500 in value or visible from the street or neighboring property must be approved by the Design Committee. Existing improvements may be refinished to the original colors, but new colors must receive prior approval of the Design Committee.

Building Materials There shall be no use of second hand lumber, garish or reflective materials other than glass or reasonable hardware when visible from the street or neighboring property.

Summary The foregoing is only a summary without independent legal significance and reference is made to the full Declaration for a complete statement of the "protections" applicable, it being understood that in the event of any discrepancy or omission in this summary as compared with the full Declaration, the provisions of the full Declaration shall control. Any representations to the contrary by any Sales Representative or other person is without authority and shall not be binding in any way on Mililani Town, Inc. A complete copy of the full Declaration is available at the office of Mililani Town, Inc. for inspection, and will be given to each purchaser. He must read and accept the Declaration.

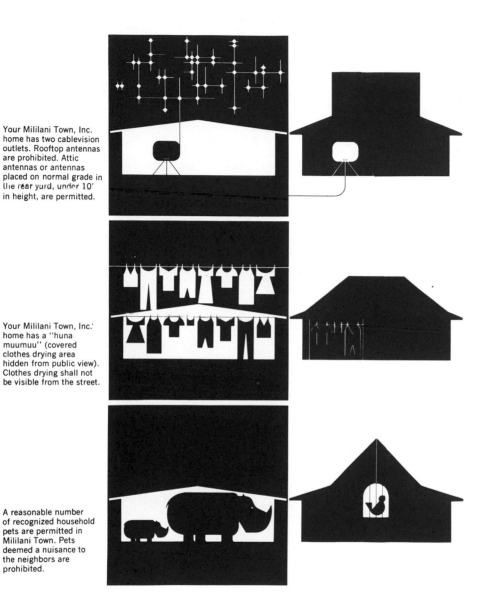

Your Mililani Town, Inc. home has two cablevision outlets. Rooftop antennas are prohibited. Attic antennas or antennas placed on normal grade in the rear yard, under 10' in height, are permitted.

Your Mililani Town, Inc. home has a "huna muumuu" (covered clothes drying area hidden from public view). Clothes drying shall not be visible from the street.

A reasonable number of recognized household pets are permitted in Mililani Town. Pets deemed a nuisance to the neighbors are prohibited.

Your Mililani Town, Inc. home has a "pau" (specially built trash enclosure). Trash and garbage cans shall not be visible from the street or neighboring property.

You may install an exterior sound device for security purposes only. Speakers or other exterior sound devices are prohibited.

Small cooking fires are permitted on your Mililani Town property. Trash or plant refuse fires are dangerous and are prohibited.

Design Control A sense of place depends in part on the small things that make up the everyday scene. To maintain the strength of the original design concept, developers of Mililani Town, Inc., have established deed restrictions that control the location and size of outdoor elements. *Drawings: Vernon DeMars, Architect*

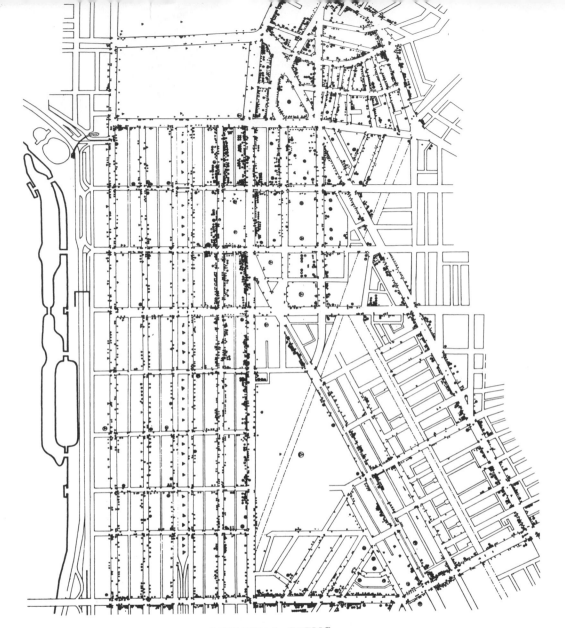

MAP 3: EXISTING INTENTIONAL SIGNS

verbal
● explicit
○ implicit

non-verbal
▶ explicit
▷ implicit

⊘ ⓑ above first story
□ non-visual

Communication Overload Signs are necessary communication devices in an urbanized society. Used selectively, they can give design structure and enliven a sense of place. The key word is *selectivity*. As a MIT study of signs in cities has shown, the major problem is one of excess and overload, as seen in this study of Back Bay Boston. *Drawing: Department of City and Regional Planning, Massachusetts Institute of Technology*

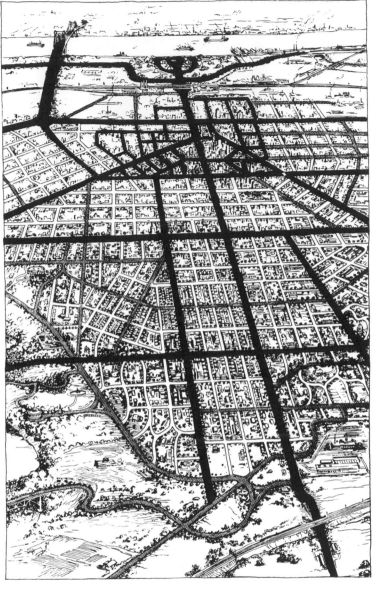

Plan Showing Classification of City Streets
According to Lighting Requirements.

Business District   Primary Thoroughfares   Secondary Thoroughfares   Residence Streets

Lighting Lighting determines the night-time mood of the urban place. At the largest scale of design, it can also help reinforce the overall design of the city — another old idea still worth using. *Source: Catalogue, Lightolier Company, 1928*

illuminated. Travel after dark was filled with risk. The wary pedestrian carried his own torch and protection. Oil lamps, illuminating gas, and street lighting are of relatively recent origin, changing the style of city life.

Lighting can do many things, however, besides facilitate after-dark travel and provide safety and security. A well-designed scheme for illumination will reveal the organization and structure of the pedestrian world, identify significant places and buildings, delineate appropriate routes for movement, and create a special aesthetic effect appropriate to the mood, character, and quality of the area.

The concept of using lighting effects to give design structure to urban spaces has been advanced by engineers, architects, and utility companies. Early promotional work of lighting equipment manufacturers advocated a hierarchy of street lamps for picking out and reinforcing a city's circulation network. Custom-designed stanchions and lamp hoods were used to mark special streets in the City Beautiful Movement, so as to give an urban place a smart, modern look. Similar designs were made for individual homes and housing developments.

Technological improvements in lighting efficiency have increased the illumination power of street lamps and at the same time reduced operating costs—through the use of mercury vapor, sodium, and fluorescent tubes and bulbs. Unfortunately, however, the application of these devices in the pedestrian world has been at the price of grotesque visual effects. The yellowish and green hues these lights cast may be fine for street safety, but they distort and destroy all other colors. Accompanying these "improvements," the tendency to provide equal illumination at every point along a street, sidewalk, or building elevation eliminates shade and shadows and the texture of ma-

terials. The subtleties of natural lighting, changing as they do with time and season, are replaced with a rather mechanical, humdrum, and commonplace experience.

The same technology, however, has reached the point where expressive lighting schemes can be introduced. Filters and color-corrected lenses, combinations of vapor and fluorescent lighting, the use of down-lighting and up-lighting concealed light sources, headlights and spotlights, all can be arranged for adequate but subtle and effective illumination. Though the art of environmental illumination uses the same principles of design as architecture and landscape architecture, it is a specialist's field. Illumination is more than canned sunshine and as such can give the modern pedestrian a visual world uniquely his own at night.

A pedestrian environment without trees and plants is a poor place, for landscaping civilizes, moderates, and rewards. Most landscape designs are architectonic. The plant material is moved from a natural ecological situation to one cultivated by man. The selection of materials is done for design effect but conditioned by the requisites of survival and maintenance. Part of the design is the art of anticipating what happens when the seasons change and when plant material grows into full form. Within these limits there are many good alternatives.

When properly prepared for transport, protected en route, and well planted on arrival, trees twenty-five feet and higher can be relocated. Post-war urbanization in Russia was notable for its tree-planting schemes. Special months were set aside to encourage universal participation, somewhat like Arbor Day in the United States. Russian technicians moved trees up to thirty years old and at all times of the year. In residential districts, tree densities up to sixty an acre were not uncommon. Americans have been timid in this

regard, despite numerous reasons to fill the environment with plant material.

Generally, it is safer and more effective to plant trees and shrubs in groups than individually. Formal planting of trees in lines and straight rows creates a compelling design. There is the danger, however, that the failure of a single plant will ruin the symmetry unless replacement in kind and size is made. Plants with equivalent design characteristics may have unequal resistance to disease, insects, and rough usage. Street trees, especially, need to be well selected. Automobile fumes, abrasions from passing vehicles, and generally low maintenance are typical hazards they have to overcome.

The choice of natural ground covers for public places is a distinct design problem until one discovers that there are green covers other than grass that grow well. Depending on climate and location, there are variations of pacysandra, vinca minor, ajuga, and hedera helix that will give a rich, dense, green color all year round and with less maintenance than required by grass.

The function and scale of planting need careful consideration. The placement of plant materials can be done arbitrarily to create an incident in space, but the better designs are purposeful and carefully determined. Plant material can enclose space, serve as a screen, a fence, a backdrop, or visual picture frame, as do buildings and structures. Plant material can be used for dramatic effects, to indicate a special juncture or turning point in space, to mark boundairies and edges. Plants can be chosen for color or shadow, be used grandly or intimately, selected for people to touch and smell as well as contemplate at a distance.

Occasionally, the landscaping can be started in advance of new construction by planting in accordance with a general, long-range plan. Often the best landscape is that which is preserved and woven into the design, again underlying the necessity of site analysis as a prelude to design. But whether the plant material is found on the site or imported, there are few places where the pedestrian's world will be diminished by trees, flowers and shrubbery.

Finally, a word about signs in the pedestrian environment. On a per-capita basis the expenditures for outdoor advertising, ubiquitous and sometimes visually overwhelming, are fairly modest, about $2 per person per year; but this represents only a fraction of the investment made in outdoor communications. The omnipresent signs and signals make the pedestrian world a communications network. Many of the messages, sometimes important ones, never get through, their visual impact being too far reduced through redundancy, irrelevancy, and incongruency or lost from the outset in the visual clamor. Nonetheless, public and private signs and signals continue to fill the urban scene, and improvements in this hectic and often contradictory visual world lie not in eliminating the devices but rather in using them effectively.

Kevin Lynch believes that we must organize the visible city as an information carrier if people are to operate within it with efficiency or pleasure. The common reasons for signs and signals are sufficient in themselves to create a need for a better communications network: location of public property and institutions, location of goods and services, control of movement systems, handling of public and personal emergencies—all the things that mark the presence and activity of people. But the opportunities for communication are even greater, including perhaps the expression of social attitudes on community issues such as the sensing and valuing of differences in economic, social, cultural, and interest groups. Finally, signs and signals can further a sense of orientation in time and space, supporting and sustaining other design efforts to make the image of the city discernible and thus they reinforce the sense of place.

POSTSCRIPT

I am encouraged to think that man's instinct for survival and continuity is matched by his instinct to create communities—observe the play of children or the action of semi-literate peasantries flooding urban places in the wake of population growth and urbanization. It is this instinct that challenges environmental designers everywhere. While need and aspiration has by far outraced society's ability to respond with the traditional methods of planning, architecture, and engineering, there are aspects of historic and current professional practice that are still valid to the times and the problems. Accordingly, from the many ways I could view environmental design, I have chosen three themes: how an improved habitation might arise from innovation and invention in handling everyday environmental opportunities, the desirability of ordering the designed environment by creating design structure, and the delights that arise from a strong sense of place. While experimental methods of design will certainly come in the future, each of these themes is likely to remain universally useful at all scales, for all men, for all time.

ACKNOWLEDGMENTS

Dictionaries suggest that the verb ACKNOWLEDGE has, like many words, several meanings. Acknowledge is to admit the truth, recognize authority or claims, express appreciation, reward as a service. In that order I do the following.

This book is not a work of scholarship but a collection of comments, ideas, conjectures, and statements whose gathering together into one volume has been stimulated by travel, work, leisure, and to an extent that they probably never realized, questions from clients, friends, colleagues, students, and children. Accordingly, while a long and proper bibliography might sustain some of the points made in this book, there is no reason to include one. The reader wishing to explore some of my themes further will have little trouble in finding extensive bibliographies. I have however in text and captions noted the sources of a good quotation or a paraphrase that made a point better than I could. To the extent possible I used those statements in what I considered to be the spirit of the original rendering. Writers thus cited can be considered authoritative, but I do admit that I am open to claims of having misinterpreted their relevancy and meaning. Mea culpa.

My appreciation for the help given me is by no means limited to the following, but Michael Spear, Radoslav Sutnar, William Ryan, Robert Beal, Lawrence H. Fauber, Percy Seitlin, and Charles Hutchinson were most responsive at critical moments. My professional partners of the last four years, Charles W. Harris, Lawrence W. Walquist, Jr., James A. Paddock, and W. Robin Upton, have made substantial contributions. Some of the materials on large schools/small sites are from an unpublished study prepared by George F. Connolly, Earl R. Flansburgh, and myself, supported in part by the Educational Facilities Laboratories, Inc. I am grateful to Jonathan King for permission to use them. A Milton Fund Small Grant (Harvard University) allowed me to briefly examine the constructive reuse of wasteland, a subject to which I hope to return. Miss Caroline Shillaber of the Harvard University Graduate School of Design Library and her assistants were not just helpful, but encouraging. It is personally gratifying to know that they soon will move to a new building commensurate with the quality of service they perform and the materials they have gathered. Without Johanna Madden this book would not have been completed; and if it reads well, the credit is due to Lorraine Lyman.

Finally, if there are rewards to be gained or given for writing this book, they should go to my children and wife, who patiently waited and also served in ways known only to them and me.

RPD

INDEX